Resilience: A New Paradigm of Nuclear Safety

Joonhong Ahn · Franck Guarnieri
Kazuo Furuta
Editors

Resilience: A New Paradigm of Nuclear Safety

From Accident Mitigation to Resilient Society
Facing Extreme Situations

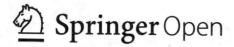 Springer Open

Editors
Joonhong Ahn (deceased)
Formerly at Department of Nuclear
  Engineering
University of California
Berkeley, CA
USA

Kazuo Furuta
Resilience Engineering Research Center
The University of Tokyo
Bunkyō-ku, Tokyo
Japan

Franck Guarnieri
Centre for Research on Risks and Crises
  (CRC)
MINES ParisTech - PSL Research
  University
Sophia Antipolis
France

ISBN 978-3-319-86471-6   ISBN 978-3-319-58768-4   (eBook)
DOI 10.1007/978-3-319-58768-4

Printed on acid-free paper

This Springer imprint is published by Springer Nature
The registered company is Springer International Publishing AG
The registered company address is: Gewerbestrasse 11, 6330 Cham, Switzerland

Prof. Joonhong Ahn (1958–2016)
We dedicate this book Resilience: A New
Paradigm of Nuclear Safety to the memory of
Prof. Joonhong Ahn of the University of
California Berkeley, who served as
co-organizer of the underlying workshop and
co-editor of this book.
Joonhong Ahn was an extraordinary
scientist, educator and human being, deeply
committed to the peaceful and safe use of
nuclear power for the development of
emerging regions of the world, particularly
Asia. After the disaster at the Fukushima
Daiichi Reactor complex in 2011, he traveled
to Japan and neighboring countries more
than fifty times to provide counsel and
assistance on the remediation efforts,
advising governments, industry and academia
on the path forward to help reestablish
confidence in the paradigm of nuclear power

for the region.

A hallmark of Ahn's scholarship in the nuclear fuel cycle, waste management and nuclear chemistry was his visionary engagement with the social sciences; enduring societal goods could not be attained solely by technical optimization, but technical progress informed and guided by a deep humanity. Not surprisingly, he was at the very headwaters of the development of the engineering ethics program at Berkeley, and cofounder of the Institute for Resilient Communities at Lawrence Berkeley National Laboratory.

Poignantly, it was the very evening of the final day of this workshop, when all the participants returned home sharing a great sense of satisfaction and optimism, that Joonhong felt the first abdominal pain that signaled the illness which soon was to claim his life. It was so characteristic of him that after his diagnosis he visited me to inquire, completely dispassionately, what he should do with his remaining time to best serve his profession, the university and his students. Without any hesitation I advised him that, above all the achievements of his remarkable career, this workshop embodied his true legacy, and that he should see to the completion of the proceedings and their wide dissemination. I thank his co-organizer, Prof. Franck Guarnieri of MINES ParisTech, and all the colleagues for carrying the baton for the final stretch, and over the finish line. To Prof. Joonhong Ahn, we dedicate this book, and recommit ourselves to carry on his vision and work.

Karl van Bibber[1]

[1]Chairman, Department of Nuclear Engineering, University of California Berkeley
karl.van.bibber@nuc.berkeley.edu

Foreword

The consequences of the Fukushima Daiichi nuclear accident in March 2011 sparked a debate about the nuclear safety. While releases of large amounts of radioactive materials resulted in no casualties due to radiation, the impact particularly on local communities is substantial and manifold. Although local communities want to be ensured that effective actions are being taken to allow them to go back to their normal life as early as possible, the lack of understanding for the transport of radioisotopes in the environment and eventually the uptake in humans as well as in the biological effects of low dose radiation has made it difficult for various stakeholders to develop concerted efforts to accelerate recovery. These challenges are compounded by the eroded public trust for government and operators.

To address this need, a multidisciplinary initiative is carried out by scientists at UC Berkeley and Lawrence Berkeley National Laboratory (LBNL), MINES ParisTech, and Tokyo University to provide the necessary guidance for effective assessment and remediation efforts and to provide trusted, unbiased, and nuclear-industry-independent perspective to build trust with local and global communities.

Based on such on-going initiatives, there are two emerging questions: (1) how integration between understanding for natural scientific processes and understanding for a society at different scales and regions can be achieved for the objective of accurate monitoring, and then ultimately (2) how such accurate monitoring and public participation can and should be integrated in decision-making processes for achieving a resilient society.

To address such questions and to develop a research plan, we hosted, in March 2015,[2] a two-day international workshop in UC Berkeley. In this workshop, first, we share various observations about "damages" in a severe nuclear accident and

---

[2]International Workshop on Nuclear Safety: From accident mitigation to resilient society facing extreme situations. March 2015 (22–24). Supported by: France/Berkeley Fund—Department of Nuclear Engineering (NE) University of California, Berkeley (UCB), and Lawrence Berkeley National Laboratory (LBNL)—Centre for Research into Risks and Crises, MINES Paris Tech—Center for Japanese Studies, UC Berkeley—Resilience Engineering Research Center, The University of Tokyo.

then address the central questions: How can we utilize knowledge of natural science and engineering in monitoring system's exogenous and endogenous conditions with a suite of performance measures that reflect different needs of resilience by different stakeholders after an accident, and in developing recipes that enable a resilient society? The discussion will focus on (1) state of the art for measurement methodologies and (2) challenges that must be overcome.

This book includes most of the lectures given during the 2015 workshop in UC Berkeley.

It begins with an introduction which gives an overview of the concept of resilience.

Then, the book is divided into five parts. An "Epilogue" synthesizes and concludes the book.

Part I is about "What are damages in nuclear accidents?" This part provides information for the reader to understand that corrective actions following a nuclear accident must be based on the definition of damage to be prevented. The difficulty lies in how we define the "damage." Therefore, we propose to elaborate an original framework for the identification and characterization of severe nuclear accident damages.

Part II is focused on the "Measurement of Damages," because we need to have accurate and effective methods to measure those, as situations evolve rapidly in the aftermath. In this section, authors introduce currently ongoing efforts for applying qualitative/quantitative methods and develop methodologies and tools to explore effective integration and application in decision-making process.

Part III considers the "Barriers against Transition into Resilience" because recovery from nuclear accident refers to social representation of risks and cannot be limited only to technical issues or crisis management guidelines. In this view, the authors consider the various impediments and their interactions to the transition into resilience of the many actors of the civil society.

Part IV includes contributions made by Ph.D. candidate students from UC Berkeley, MINES ParisTech, and Tokyo University.

Contents

Contributors

Hiroyasu Abe Department of Nuclear Engineering and Management, The University of Tokyo, Tokyo, Japan

Ivana Abramovic Department of Nuclear Engineering, UC Berkeley, Berkeley, USA

Aissame Afrouss Centre for Research on Risks and Crises (CRC), MINES ParisTech/PSL-Research University, Sophia Antipolis, France

Sophie Agulhon MINES ParisTech/CRC, Paris 8 Vincennes-Saint-Denis University/LED, PSL Research University, Sophia Antipolis Cedex, France

Joonhong Ahn University of California, Berkeley, CA, USA

Alexandra (Sasha) Asghari University of California, Berkeley, USA

Romain Bizet CERNA-Centre for Industrial Economics, MINES ParisTech/PSL-Research University, Paris, France

Hortense Blazsin Centre for Research on Risks and Crises, MINES ParisTech/PSL Research University, Sophia Antipolis, France

Massimiliano Fratoni University of California, Berkeley, CA, USA

Yasumasa Fujii Department of Nuclear Engineering and Management, The University of Tokyo, Tokyo, Japan

Kazuo Furuta School of Engineering, Resilience Engineering Research Center, The University of Tokyo, Tokyo, Japan

Ken Goldberg University of California, Berkeley, CA, USA

Franck Guarnieri Centre for research on Risks and Crises (CRC), MINES ParisTech - PSL Research University, Sophia Antipolis, France

Kathryn A. Higley Oregon State University, Corvallis, USA

Tatsuya Itoi The University of Tokyo, Tokyo, Japan

Kohta Juraku Tokyo Denki University, Tokyo, Japan

Taro Kanno Department of Systems Innovation, School of Engineering, University of Tokyo, Tokyo, Japan

Ryoichi Komiyama Resilience Engineering Research Center, The University of Tokyo, Tokyo, Japan

Justin Larouzée Centre for Research on Risks and Crises (CRC), MINES Paristech/PSL Research University, Sophia Antipolis, France

François Lévêque CERNA-Centre for Industrial Economics, MINES ParisTech/PSL-Research University, Paris, France

Xudong Liu Department of Nuclear Engineering, University of California, Berkeley, CA, USA

Giorgio Locatelli University of Lincoln, Lincoln, UK

Dipta Mahardhika Department of Systems Innovation, University of Tokyo, Tokyo, Japan

Christophe Martin Centre for Research on Risk and Crises (CRC), MINES ParisTech/PSL Research University, Paris, France

Hiromu Matsuzawa Department of Nuclear Engineering and Management, The University of Tokyo, Tokyo, Japan

Charlotte Mazel-Cabasse Berkeley Institute for Data Science, Berkeley, UC, USA

Naoto Mitsume Department of Systems Innovation, University of Tokyo, Bunkyō, Japan

Kohei Murotani Department of Systems Innovation, University of Tokyo, Bunkyō, Japan

Delvan R. Neville Oregon State University, Corvallis, USA

Brandie Nonnecke University of California, Berkeley, CA, USA

Masahiko Okumura Japan Atomic Energy Agency, Chiyoda, Tokyo, Japan

Ryan Pavlovsky Etcheverry Hall, University of California, Berkeley, USA

Dominique Pecaud Université de Nantes, Nantes, France; Centre for Research on Risks and Crises (CRC), MINES ParisTech/PSL Research University, Paris, France

Aurélien Portelli Centre for Research on Risks and Crises (CRC), MINES ParisTech/PSL University Research, Paris, France

Kimiaki Saito Lawrence Berkeley National Laboratory, Berkeley, CA, USA

Kyoko Sato Program in Science, Technology, and Society, Stanford University, Stanford, CA, USA

Naoto Sekimura The University of Tokyo, Tokyo, Japan

Ryuma Shineha Seijo University, Tokyo, Japan

Mikihito Tanaka Waseda University, Tokyo, Japan

Sébastien Travadel Centre for Research on Risks and Crises (CRC), MINES ParisTech, PSL – Research University, Sophia Antipolis Cedex, France

Adrián Agulló Valls Industrial Management Engineering, Universitat Politècnica de Valencia, Valencia, Spain

Kai Vetter Applied Nuclear Physics, Lawrence Berkeley National Laboratory, and Department of Nuclear Engineering, University of California, Berkeley, CA, USA

Haruko Murakami Wainwright Lawrence Berkeley National Laboratory, Berkeley, CA, USA

Rin Watanabe University of Tokyo, Tokyo, Japan

Tomonori Yamada Department of Systems Innovation, University of Tokyo, Bunkyō, Japan

Shinobu Yoshimura Department of Systems Innovation, University of Tokyo, Bunkyō, Japan

The Fukushima Daiichi Nuclear Accident: Entering into Resilience Faced with an Extreme Situation

Franck Guarnieri

Abstract A transdisciplinary concept, resilience has emerged from monodisciplinary approaches and finds its foundations in various domains such as materials science, ecology, psychology, sociology, ethology, medicine, etc. Although the concept has been a work in progress in the scientific community for several decades, it was only adopted by the safety studies community in the 2000s. The Fukushima Daiichi accident has accelerated its popularity and led to an abundance of theoretical and methodological references, ideas and concepts, processes and approaches that are more-or-less operational.

Keywords Fukushima Daiichi · Nuclear accident · Resilience · Entry into resilience · Extreme situation

1 Introduction

Unfortunately, dramatic nuclear accidents are a source of information, knowledge and learning for researchers in the safety studies. As their career progresses, members of the community learn more: safety is driven by accidents, catastrophes and disasters of all kinds. While this is obviously a tragedy, it is clear that we do not really know what else to do.

The accident at Three Mile Island on March 28, 1979 taught us that inappropriate actions can lead to core fusion, that serious accidents necessarily involve all stakeholders in civil society, that the defenses of a nuclear installation must be 'deep', and that each accident contains in its genesis 'precursor' scenarios that, if they can be identified, may help to avoid the situation becoming worse [1].

The Chernobyl accident on April 26, 1986, taught us that 'global' public opinion about nuclear safety cannot be ignored [2]. In particular, it showed us that opera-

F. Guarnieri (✉)
Centre for Research on Risks and Crises (CRC), MINES ParisTech - PSL Research
University, Sophia Antipolis, France
e-mail: franck.guarnieri@mines-paristech.fr

© The Author(s) 2017
J. Ahn et al. (eds.), *Resilience: A New Paradigm of Nuclear Safety*,
DOI 10.1007/978-3-319-58768-4_1

1

tor's cannot act as the sorcerer's apprentice, blithely dispensing with basic safety requirements. It also taught us, a posteriori, that operators must learn from both their own experience and that of others. The concept of 'safety culture' emerged from the Chernobyl accident. It has since been widely popularized and reused to demonstrate that the safety of a facility is the top priority for all operators [3–5].

What have we learned from the Fukushima Daiichi accident?

Two Commissions of Inquiry were established: one at the initiative of Prime Minister Naoto Kan, the other by the Japanese Diet. Both Commissions recognized that the nuclear accident at Fukushima Daiichi was "a man-made disaster" and not simply caused by the earthquake and the giant tsunami that occurred on March 11, 2011 [6, 7]. The Commissions' reports were voluminous, and supplemented by international analyzes [8, 9]. Everything came down to the facts, causes and consequences of the accident. Finally, and as usual, everyone agreed (without actually explicitly saying so) that the accident could be seen as useful, whether in terms of how to regain 'control' of a system that was out of control, or as a way to learn from each other's mistakes, or as example of the intertwined decisions that were taken at multiple levels (local, hierarchical, organizational, inter-organizational, political, international, etc.). In other words, as a way to say that all of this could have been avoided if only…

All of these Commissions produced a long list of recommendations that are both sensible and helpful in improving safety. However, fundamentally they offer nothing new—all are consistent with a 'normative' vision of nuclear safety. Although it cannot be disputed that they are an invaluable source of knowledge and progress, their very nature creates a form of myopia. This myopia is so pronounced that it eventually produces expert and techno-centered analyses that only take into consideration standards, laws, regulations and procedures. It excludes all forms of humanity from a human activity, while accidents are clearly a departure from the logical course of events.

Does this mean that the accident at Fukushima Daiichi will have taught us nothing? Obviously not! Rather, it places the concept of resilience, which has become particularly fashionable in many disciplines, at center stage in nuclear safety. The concept was integrated into safety sciences in the early 2000s. It has taken pride of place in the context of the Fukushima Daiichi accident. Entering into resilience assumes the system has survived, if not it has perished!

This chapter is organized into five sections. It identifies and describes the determinants of the entry into resilience in a socio-technical system that is the victim of an unprecedented accident which, without an adequate response, will obliterate the system itself. The first section revisits the notion of the accident, looking at it in terms of the 'extreme situation'. The second introduces the concepts of resilience and entry into resilience. The third and the fourth sections respectively discuss the link between the entry into resilience and notions of time and space. The last section discusses the human and organizational determinants of an organization's entry into resilience. The Fukushima Daiichi accident, and in particular the decisions taken by the engineering teams on the site between 11 and 15 March 2011 serve as a case study.

2 From Nuclear Accident to Extreme Situation

Fukushima Daiichi is a Japanese nuclear power plant, which on March 11, 2011 suffered, like the rest of Eastern Japan, the effects of a terrible earthquake followed by a devastating tsunami. Before becoming a nuclear power plant, the site was a training camp for the Japanese kamikaze during the Second World War [10]. At the time of the accident, the Director of the plant was Masao Yoshida. The Daiichi plant is a neighbor to the Dai Ini plant, where Naohiro Masuda was the Director, and whose handling of the crisis was presented as a model of good management [11]. Like Masuda, Masao Yoshida knew perfectly well how to handle the situation (which we will later term 'extreme') he and his men were faced with, despite the fact that the damage and losses were far more extensive at Daiichi than Daini. Nevertheless his actions came under severe criticism. But that is the subject of another story [12].

The management failures that occurred in the handling of the crisis at Fukushima Daiichi are not the first of their kind: the accident at Three Mile Island, not to mention Chernobyl had already highlighted the inability of crisis management procedures to cope with 'unthinkable' situations. From this, it seems almost reasonable to conclude that few real lessons have been learned [13].

There appears to be a kind of illusion of safety, a fact underlined by the pertinent observation of a TEPCO manager who explained the accident in terms of overconfidence, lack of imagination and various biases [14]. Despite all efforts to overcome them, these three challenges lead to accidents that Perrow [15] describes as "normal" or "systemic", due to complex interactions and tight coupling within the system. However, the phenomenon described by Perrow is not new. The historian Fressoz [16] describes its origins in the nineteenth century with the development of the railway system; catastrophes have proven to be inherent in systems where "nobody is able to anticipate and ward off the effects" [17].

While it is recognized that failures are inevitable, contemporary approaches to 'beyond-design-basis' accidents in nuclear safety [18] must address ongoing challenges. In the aftermath of extensive damage, actors must adopt innovative, improvised solutions to return to a safe situation. Similarly, they must draw upon resources that are not part of the usual frameworks and patterns for problem solving [8]. It is clear that this postulate is far from being accepted. The numerous investigations into the accident repeatedly reaffirmed the benefits of the concept of defense in depth [19], despite its inability to effectively evaluate events that lead to the emergence of new sources of vulnerability [9].

The situation at Fukushima Daiichi goes beyond coping with an unthinkable disaster, and concerns "a state of emergency which seems to have no end", or even "a slow, diluted catastrophe, an ongoing catastrophe" [20]. This has consequences for how we handle the phenomenon. The Fukushima accident is not limited to the period from 11 to 15 March 2011; the site is subject to ongoing natural threats (from another earthquake or tsunami) and the facilities remain severely damaged.

When question of when the accident will end raises the issue of uncertainty in a post-accident context where many parameters cannot optimized. One of the lessons learned from the Fukushima Daiichi crisis is that many countries now consider the post-catastrophe phase of preparations for a "return to normal" to be equally as strategic as crisis management preparations [9]. Confirmation comes from the Director of the Fukushima Daiichi plant, Masao Yoshida. Yoshida testified that the operator, TEPCO, had not provided any emergency measures in the case of a power failure, which meant that there was no way to cool the reactors [21]. Even today, certain events remain unexplained and there are clear cases of ambiguity.[1]

Therefore, rather than the concept of the accident, a more useful concept appears to be that of the 'extreme situation'. Gilbert [23] argues that "some situations have become real black boxes" due to a blatant lack of investigation and analysis. Safety approaches are limited to compliance with risk analysis principles and standards, and therefore tend to underestimate extreme phenomenon that cannot be planned for.

A situation is called 'extreme' when conditions are radically different from those of so-called 'normal' life and are unusually intense, becoming excessive, or even unbearable [24]. Dealing with the extreme situation pushes people to their limits; to the edge of the abyss [10]. The individual, group, organization, company, or more simply, the system is faced with extreme violence, a radical shake-up of life as they know it.

The extreme situation leads to the destruction of identity, the loss of benchmarks and frames of reference. The explanation is simple; identity is shaped or manufactured by external relationships (specifically, compliance) with current social norms, adherence to common and therefore shared values, responses to social expectations, and dependency or even subordination between actors in the system. From the moment the (existing) value system is shattered, a change occurs—and a new system appears.

The concept of the extreme situation therefore places the individual and the organization that must face the unthinkable at the heart of the analysis [24]. The unthinkable takes the form of three 'entities' that become, through the forces of nature and human weakness, uncontrollable and 'unleashed'. Following a period of 'devastation' and predictions of 'certain death' [25], the actors involved in the Fukushima Daiichi accident began a phase of 'coping', which enabled them to mobilize multiple resources in order to survive in the short term [26]. They then began a return to an acceptable situation, despite extensive damage, widespread pollution and the hazards that endangered, and continue to endanger the site.

---

[1]For example, the causes of the malfunction of the core of the reactor cooling system (Reactor Core Isolation Cooling) of Unit 2 remain unclear [22].

3 Entry into Resilience: A Way to Cope with the Extreme Situation

The concept of resilience is only relevant following damage, loss, an accident, trauma, etc. Pre-event is the domain of prevention, prudence, or even precaution. To be able to discuss resilience, you first have to survive. Even this is not enough, as, while it can be a sustainable situation, survival is fragile. The concept has positive connotations. Resilience is an asset, it represents progress; being or becoming resilient does not mean returning to the nominal, pre-shock state. This is anyway impossible because the system, whatever form it takes, remembers the event (albeit for a limited period of time).

Although the concept is in fashion, there is no universal definition that can be applied to all domains. That said, the English term 'resilience', itself derived from the Latin verb *resilire* (to bounce), is made up of *re* (again) and *salire* (rise), which implies a retroactive effect [27]. While in the 1970s the term was associated with the ability to absorb and overcome the effects of significant, unexpected and brutal disruption to ecological systems [28], hybrid definitions have since emerged in many disciplines including geography [29], psychology [25], sociology [30], organizational sciences [31], ergo-psychology [32], etc. Within this smorgasbord of definitions, two fundamental ideas prevail: community,[2] and the process.[3]

In the absence of a consensus, resilience can be defined as the capacity of a system to absorb disturbances and reorganize itself during ongoing changes [33]. It is probably more relevant, especially in the case of an accident as serious as that at Fukushima Daiichi, to place less emphasis on states of equilibrium as "frontiers as a function of the domain of attraction" because paradoxically, highly fluctuating instability can also foster entry into resilience [28]. In practice, a system can be very resilient, yet fluctuate significantly and therefore be fairly unstable. This approach seems more relevant in the case of Fukushima Daiichi where, given the enormity of the shock, it was more important to preserve relationships in the socio-technical system than to return to the previous equilibrium as quickly as possible, which in fact proved to be an unrealistic expectation [34].

Contemporary views of nuclear safety see the concept of resilience as a post-crisis process, part of a community dynamic that stresses organizational adaptability [35]. It has a predictive dimension that helps the organization to overcome adversity and get back on track [27]. However, this predictive dimension, and *a fortiori* entry into resilience, must not neglect the role of probability, uncertainty [13] or even a "surprise" dimension in the success (or failure) of its implementation [36]. Here again, there is a reference to a conscious capacity to "navigate" and "negotiate" [37] in order to cope with an extreme situation.

---

[2]In ecology, a community is a group of organisms belonging to populations of different species making up a network of relationships.

[3]A process is a system of activities that uses resources to transform inputs into outputs.

The term 'entry' expresses "movement from one place to another" [38]: there-fore there is a 'transition' that must be taken into account. The concept is also linked to the question of "the place through which we enter" [38], and thus the direction of the transition. Time also plays a part, as the 'entry into' begins with the exercise of "a practice" [38] that implies a "change of state" [39].

The change of state is explained by unforeseen sequences of complex interac-tions that could not be predicted [15]. Entry into resilience must therefore be perceived as an 'exploratory zone' in which it is necessary to have a better understanding of the interactions in order to improve how they are managed. From this point of view, a learning phase is necessary. It is difficult to rationalize this exploration phase either upstream or during the process as it resembles "cognitive DIY that does not belong to the scientific world (which does not mean that it cannot be effective)" [40]. In this case, rationality could be described as procedural, with "simple but strict rules, which certainly does not completely eliminate risk, but reduces it to a level below that resulting from substantive rationality" [41].

Entry into resilience therefore translates into the creation of a new system [33] when ecological, economic or social conditions make the initial system untenable. In this case, the variables and scope that define the system must be modified. Nevertheless, the potential for the loss of a certain degree of resilience is inevitable, in the context of the dynamic interactions found in adaptive complex systems. Such multiple interactions were thrown into sharp relief during the Fukushima Daiichi crisis [8] and complicated the implementation of appropriate responses.

It is important to note that older definitions include an element of 'privilege', linked to the capacity to 'enter into'. The chambers of the king of France could only be visited at specific times [42]. By extension it could be argued that this 'entry into' also embodies a situation where the parties involved must demonstrate in advance their capacity to access the privilege: in other words, education, training, experience, professionalism, etc. The *curriculum vitae* of Masao Yoshida [12] is illuminating in this sense as it demonstrates his competence and expertise, making him the right person in the right place.

Finally, the notion of 'entry into' finds support from the biological metaphor of the membrane, "a generic organ that links the interior with the exterior, the past and the future, using the dual mode of qualification/interpretation of the future through the past, and the integration of the future using the encoding of the past" [43]. This metaphor enables the introduction of the relationship between time and space.

Prigogine [44] places the question of time "at the crossroads of problems of existence and knowledge". In this sense, his argument is similar to that of Heidegger [45] for whom the question "of the meaning of being" cannot be examined without an interpretation of time as the possible horizon for any under-standing. If it is accepted that resilience is a process, and that the entry into resi-lience is a moment (an instant or a short duration…), then it cannot be argued any other way than that time is one of the determinants that shapes its nature. If it is accepted that time is "at the crossroads", the question of space necessarily arises. Time and space are therefore the most important, fundamental and essential determinants to be examined.

4 Entry into Resilience and Time

Time is "huge and complex" [46]. Classically, it is approached from three angles: chronology (a sequence of events that follow one another), simultaneity (events that occur at the same time) and duration (a measurable period). Typically, a linear and cumulative concept is contrasted with one that takes into account multiple, discontinuous temporalities. For example, [47] summarizes it as, "rather than simulate a linear story that is 'in progress', priority should be given to these flashbacks, these evil blows, these lightweight catastrophes that perturb an empire much better than major upheavals. Give priority to non-linearity, reversibility, everything that relates, not to an unfolding or an evolution, but a winding game, a reversion in time. Anastrophe versus catastrophe. (…). Everything happens in loops, in tropes, in reversals of direction". The challenge is to know how to manage these various feedback loops, "the analyzed object's victorious ruse" [48] especially within a system in crisis.

The concept of crisis can be defined as a runaway phenomenon, the translation of an acceleration, or a loss of control. Entry into resilience aims to halt this phenomenon using a more powerful natural or artificial mechanism, which will initially slow it down and then stop it [49]. More specifically, in the case of an extreme situation the focus is less on a search for the causes of this runaway phenomenon, than to understand the "delirium of forms and appearances", this "endless cycle of metamorphoses" where everything "explodes into connections". It could be said that entry into resilience helps to "slow down, stop at certain points this total correlation of events" [48]. It is therefore essential to have feedback loops [50] that act retroactively on the source. From this perspective, entry into resiliency can be seen as an attempt to create a return to the origin of the crisis so that the system can be brought back under control and become more fluid.

The challenge is therefore to find the right tempo[4] that allows the technical objects[5] [51] in the system sufficient time to independently enter into resilience. This tempo is punctuated by 'phases' that have a role to play, but "by phase, we do not mean a moment in time that is replaced by another". It is more accurate to say that "in a system there are phases when reciprocal tensions are in equilibrium; the current system composed of all phases taken together is that the full reality, not each phase taken by itself, a phase is only a phase when compared to another. (…). Finally, the existence of a plurality of phases defines a neutral, balanced center, which provides the conditions for the existence of the diphase" [52]. An overall harmony with respect to time management must therefore emerge. In the time management context, it is in the relaxation phases where minor or major factors are

---

[4]Here, the concept of tempo is not limited to the speed of execution but the pace at which a set of actions unfolds.

[5]"The configuration of technical objects specifies a certain division between the physical and social world, assigns roles to certain types of actors—human and nonhuman—and thereby excludes others, permits certain modes of relationship between these actors, etc." (p. 49).

synchronized. In practice, there is period of relaxation that is proper to the technical object, which should be respected.

This issue of tempo is generally adapted in improvisation phases. Weick [53], drawing on the work of Berliner [54] and his study *Thinking in Jazz: The Infinite Art of Improvisation*, argues that at the organizational level improvisation involves reworking pre-composed material and considering it in relation to unanticipated concepts that are designed, formed and transformed under specific performance conditions. Jazz is not spontaneous, intuitive music but flows from the experience of 'musician/actors' and the disciplined application of a vast musical knowledge. This analysis can be expressed in the context of the improvisation of actors involved an extreme situation, who must rely on a repertoire of training resources, experience and a shared vision [55].

A dual temporality must be also managed [56]. The first requires a short-term response that should not limit the adaptability of the system in the long term. The second involves establishing a balance between the need for resources to be mobilized in the very short term and other resources that need to be saved for the long term [32].

Finally, time cannot be discussed without the notion of duration [57]. Bergson opposes the idea of the duration of consciousness and scientific time (defined in terms of measurable periods). Instead, time is the measurement of repetition in space. It is thus a way to reduce an evolutionary phenomenon to spatial coordinates. "Pure duration" is unconstrained thought in the timeline of the same evolutionary phenomenon. It implies something that ensures continuity between successive states. According to Bergson this is the consciousness of the observer, which he calls "intuition", because it is not perceived through a projection of the evolution of the system described in spatial terms, but by a thought that is inseparable from its object.

5   Entry into Resilience and Space

The idea of space, inseparable from that of time, is classically considered in the broad and Cartesian sense that defines it as a "scope, a medium in which the observed phenomena occur or abstractions are the object of study" [58].

The viability of a technical object is intimately linked to its "associated milieu" [59] or even a "mixed milieu, both technical and geographical" [52]. The severity of this local challenge was severely underestimated in the spatial management of the Fukushima Daiichi accident. It has even been argued that in general, the remote interaction between the various organizations that should have been responsible made it "pitiful" [60]. An example is the relationship between Yoshida (who considered senior managers to be disconnected from reality) and TEPCO head-quarters—when his request for 4,000 tons of water was understood by management as 4000 tons of drinking water [61]. Frustrated, the plant's director gave greater weight to the local situation, to the extent that he decided to ignore certain orders from headquarters. The notable example is when he decided to continue (after checking with his subordinate) operations to inject seawater to cool Reactor 2 [62]. However, most of the local initiatives taken by Masao Yoshida only proved relevant in a posteriori debates and controversies. An example is his decision to move skilled workers from Reactor 1 to a more secure building further away for their own protection. Paradoxically, some of these workers criticized the decision, which they perceived as demeaning [61].

It can be difficult to identify the perimeter of the local zone, and *a fortiori* that of the zone for entry into resilience. Nevertheless, it seems very likely that it should be as close as possible to where the accident took place. Thus, for example, the support of the United States (which is also indicative of an ambiguous state of dependency) was not always very useful. An example is that American authorities initiated the assembly of four water pumping systems for each of the plant's ponds. However, on March 18, 2011, when the parts had already been assembled in Australia and were waiting to be transferred by plane to Japan, the Americans learned that the Tokyo Fire Department had been able to build a pump that was similar to the equipment already used at the site [63]. This failure can be explained by the argument that the nuclear industry is "essentially a centralized energy" [59]. A crisis of the magnitude of Fukushima Daiichi requires a level of decentralization that is not found in its culture. The extent of this centralization is most clearly seen at the time when the crisis began (between Friday, March 11 and March 12 at 10 am), when TEPCO's most senior managers, Tsunehisa Katsumata (Chairman) and Masataka Shimizu (President), were respectively on business trips to China and Nara (an historical Japanese city) [64]. Their absence meant that TEPCO was unable to make quick decisions. This trend towards centralization often results from an exaggerated fear of panic, which must be avoided at all costs, but that actually occurs very rarely [65].

The importance of the local management of the crisis at Fukushima Daiichi has been highlighted in numerous analyzes and reports [9]. Following the catastrophe, some countries re-evaluated the size and nature of their emergency evacuation zone, according to the country and the geographical context of its nuclear plants. Developing cooperation with neighboring countries in the case of emergencies has become a new priority, in particular joint maneuvers and exercises.

Local solutions must therefore be established to facilitate and accelerate entry into resilience. These solutions can, for example, be based on the pertinent criterion of "concentration" [65] (of energy, populations and decision-making powers).

In this context, the concept of Community Building Recovery Corporations (CBRC) [66] seems to be relevant. It is based on three levers: change, community and leadership. Economic reconstruction (i.e. post-accident) is established on a local level with the support of citizens in a combined public-private approach. The resources that are mobilized are primarily local, guided by the principle of a high level of creativity in their implementation. The Fukushima accident demonstrates the usefulness of the CBRC concept, as the main challenges concerned not only electricity and information, but also the more basic problems of a lack of food, access to toilets and additional emergency rescue teams [8], which could have been better solved locally.

However, in a nuclear crisis decentralized management is not sufficient for success. The Chernobyl catastrophe was primarily managed in a decentralized way and clearly ended in failure. "The accident analyzes give the impression that the workers at the bottom of the ladder ran the system in their own way, trying to cope with the facility, to gain time and to fix things, incredible acrobatics" [40]. Moreover, the arrival of engineers from Moscow proved ineffective as they were highly specialized in electro-mechanics and not the operation of a nuclear reactor (and knew even less about nuclear safety). Consequently, this failure of decentralized management at Chernobyl led operators and regulatory authorities to favor a technical approach that limited the room for maneuver of operational staff. However, the Fukushima Daiichi accident poses a challenge to this approach.

Beyond the issue of decentralization, entry into resilience also raises new questions on the theoretical level, notably regarding the path of this "entry" and the definition of a space dedicated to it. This requires a change in the formalization of objectives, and consequently the trajectory of the socio-technical system [32]. The approach differs from ecological definitions of the concept of resilience that tend to focus on the capacity (or not) of a community or group to confront and overcome adversity, and therefore to enter into resilience. A perspective framed in terms of trajectory places the emphasis on essential and vital functions that must continue after a crisis, something that seems more relevant in the framework of successful entry into resilience. Moreover, this trajectory must move in the direction of a "basin of attraction" in which systems tend to reach equilibrium [33]. This equilibrium must play the role of attractor in guiding the trajectory of entry into resilience and result in "stability landscapes" that are defined by clear boundaries. Stabilization can be disrupted [67] by both external elements (for example, conflicting orders from headquarters or the government) and internal elements (for example, lack of data, or undue, and ultimately mistaken reliance on information provided by a sensor).

In practice, the path is more like a rotary (roundabout) than it is linear [68], as each actor can enter or leave at any point to enter into resilience, while a kind of 'insider/outsider' dynamic can be implemented in neighboring communities. The trajectory can also be oriented in such a way as to fill the structural voids at the borders of organizational units [69], which become strategic hubs. A risk related to the orientation of this trajectory is 'destinationism' [41], which consists in having an excessive focus on the target to be met. However, in the case of an extreme situation that is highly unstable, it is sometimes necessary to adjust the target. Finally, establishing a clear and limited objective presupposes that complex, uncontrolled interactions can be understood.

The capacity to implement a "pivot strategy" is essential. This strategy can increase "resilience capital" [70] and as a result facilitate entry into resilience. To extend and improve resilience, organizations must also encourage heretical questions at operational levels [71].

We now address the question of the formalization of a space dedicated to resilience and entry into it.

"Scoping" [72] is designed to clarify the boundaries of the organization in relation to a given area. In the case of a nuclear accident, although the management of the organization is limited to certain target areas, it is important to maintain a holistic view of the crisis. The Cynefin model [73], for example, distinguishes four categories of context that can help in understanding an extreme situation:

- Chaotic context: there does not seem to be any link between causes and their effects. The available concepts are useless. An open-minded approach, focused on investigation and exploration is essential before taking decisive action and analyzing its effects. Under time pressure, action must be both rapid and carefully considered.
- Complex context: there are multiple interactions between causes and their consequences but it is impossible to precisely model or predict the impact of an initiative on the system. In this context, an insightful and flexible exploratory analysis should be carried out.
- Knowable context: the relation between cause and effect can be assessed but there is not enough data to be able to evaluate the situation.
- Known context: the relation between cause and effect is predetermined.

From the perspective of trajectory, entry into resilience consists in moving from a chaotic context to a complex context. The situation at Fukushima Daiichi will only be fully brought under control when a known context is achieved. Current conditions at the site are still complex and fragile: the return to a chaotic context cannot be excluded if, for example, another major earthquake occurs. This shows that entry into resilience does not guarantee a sustainable trajectory in the long term.

It is important to note that the transition from one context to another or, more generally, entry into resilience implies a period of adaptive research. During this period, actors establish trade-offs as a function of boundaries between acceptable and non-acceptable performance [74]. If it proves impossible to avoid crossing a boundary, an error or accident may occur. There is therefore a 'dead point', beyond which entry into resilience reaches a point of no return and will fail. This point represents a threshold where any system that "crosses this fine line of reversibility, contradiction, questioning" will "enter, living, into its own frenzied contradiction..." [48] and ultimately perish.

Therefore, entry into resilience implies that the actors know how to delimit the intervention space and can establish boundaries beyond which the situation is irreversible. In an emergency, it is likely that in most cases the concept of entry into resilience and its relation to boundaries is metaphoric, and cannot be quantified in terms of risk [32]. Nevertheless, the understanding of risk must be communicated internally to actors at all levels of managerial responsibility [75].

We should not overestimate aspects of local culture in the management of resilience and thus the establishment of a resilient space. It is more important to observe the influence that certain forms of culture have on an organization.

For example, entry into resilience involves a shift from a pathological and bureaucratic culture (notably embodied in the relationship between TEPCO and Japanese political powers), to a generative culture that enables the organization in crisis to activate resilience in the best possible conditions [76]. It should be noted that this argument refers more to general boundaries than a definition sensu stricto of measurable organizational boundaries.

In summary, the questions of the path for entry into resilience and the definition of its space are acute. Going beyond the models of Snowden [73] and Westrum [76], a more accurate analysis of the Fukushima crisis will help to formalize and theorize trajectories and spaces necessary for entry into resilience by paying greater attention to the details of how operations unfolded.

6 Human and Organizational Factors of Entry into Resilience in an Organization

Many models [15, 18, 32, 65, 69, 74, 75] describe the response of organizations to disruption at all orders of magnitude. Most rely on a call for strategic and operational, or even financial resources. Some also examine the safety management system and attempt to determine if a system is resilient or not. Although these contributions are clearly important, in our view they overlook the dynamic character of the resilience process. In fact, there are very few models that describe the mechanisms and determinants through which resilience emerges in response to a traumatic event.

Two key processes in the mobilization of resilience can be cited: the reconfiguration of resources and their mobilization [77].

Resilience may emerge when "resources are sufficiently robust (…) or rapid" [68] to be able to slow down and counteract the negative effects of a severely degraded environment. A key concept that quickly emerges is the "conservation of resources". According to the theory of the same name (Conservation of Resources, COR) [78], stress emerges when resources are limited, lost or when individuals cannot replenish them despite significant effort [68]. COR theory argues that actors must invest in resources to guard against the loss of other resources. In some ways, entry into resilience makes sense in this context: it makes it possible to anticipate the loss of valuable resources and thus curb fears of a runaway situation that could lead to the loss of the system itself.

In terms of resource mobilization, the work of Powley [79] on the concept of "resilience activation" is particularly interesting. Entry into resilience is established as a result of three mechanisms that follow each other in chronological order. Resilience is activated during a critical period. Its smooth operation implies first and foremost "liminal suspension", which in turn activates "compassionate witnessing" and then "relational redundancy". It is in these conditions that a resilient organization is created.

Liminal suspension activates the resilience of the organization in two ways. First, it provides a temporary holding space in which its members have time to readjust on a psychological, emotional and relational level. Members of the organization can help and support each other without work constraints in a "holding space for pain". Second, the crisis can shift the positions of actors, leading to the creation of new relational structures and the strengthening of existing relationships. The social positions of some members of the organization may shift as new relations that create solidarity are formed. Liminal suspension activates resilience by undoing social and organizational structures for a period of time in which actors lose their 'status' and new, deep bonds are formed that challenge previous relationships. It also creates time and space for relational structures to shift, thereby "loosening control" to reduce "defensive perceptions" between members of the organization [80]. However, liminal suspension does not make the distinction between roles obsolete.

"Compassionate witnessing" implies empathy for others. This phase follows on from the new relational structures that emerged in the previous phase. Compassionate witnessing can be activated in two ways within an organization. The first is to demonstrate mindfulness in relationships with others. This dimension is not only limited to cognitive aspects, but also implies identifying the emotions of others, leading to an appropriate response to their emotional, physical and social needs. The second is sharing and connecting. In particular, sharing experiences helps the organization's members to restore order and bounce back. Compassionate witnessing activates resilience by adapting the organization to the response of its members. In this case it does not coordinate behavior, but rather emphasizes the importance of emotions and thoughts in the interpersonal relations that enable actors to heal from the trauma.

"Relational redundancy" refers to interpersonal connections that overlap and intersect, and extend beyond the boundaries of the social reference group. There are two aspects to relational redundancy. First, there is an informational connection that relates to the ability of members of the organization to share critical information about safety. Secondly, overlapping social links provide a holistic and panoramic view of the entire set of interactions within the organization. Relational redundancy therefore activates resilience through intersecting interactions that ensure the persistence of relationships within organizational systems. From this perspective, the informational field widens when actors share critical information within the system or with their immediate social reference group. Counter-intuitively, resilience is activated not through principles of organizational efficiencies, but through principles of redundancy and excess relational capacity. From this perspective, each actor in the system plays a strategic role in finding and transmitting critical information to other members of the organization. This abundance of connections favors the emergence of a new kind of network that creates multiple opportunities for members to share information, express emotions and therefore enable organizational resilience.

7 Conclusion

Characterizing the determinants of entry into resilience is a major challenge. It is clear that very little is known about it. Some disciplines (for example, psychology and ecology) have made more progress than others. These disciplines have the advantage that research is based on a clinical approach, and they are able to examine extremely well-documented cases. However, the challenge that remains is to generalize the examination of a specific and particular situation. In the fields of management science and human resources, the work of Powley [79] is particularly interesting. The author proposes a data analysis methodology and a model for determinants that are very relevant to our understanding of the mechanism for entry into resilience. The field of safety sciences has made its own contribution. The work of Hollnagel [32] provides a definition of the concept of resilience and its application to the measurement of the performance of a safety management system on the one hand, and, on the other hand, four meta-categories (respond, monitor, anticipate, and learn) that offer other ways to investigate the issues in more depth. While the number of conceptual models that are aimed at standardization is legion, the fact remains that little is known about the actual teams, groups, and organizations that must face the unthinkable. This lack of knowledge favors an approach in the field that tries to immerse itself as deeply as possible into the mechanisms and processes that come into play when facing an extreme situation. The Fukushima Daiichi accident offers an extraordinary field for researchers by virtue of the profusion of data sources and the information that has already been made public and which will eventually be made public in the future. One example is the hearing attended by the plant's Director [12]. His very personal testimony helps us to understand what motivated the responses of engineering teams who were faced with an unprecedented situation and offers an initial starting point for thinking about the concept of engineering in extreme situations [81].

References

1. United States, *President's Commission on the Accident at Three Mile Island, Kemeny, J.G., The Need for Change: The Legacy of TMI* (United States Government Printing Office, 1979)
2. P. Tanguy, Les leçons de Tchernobyl. Conclusions of the SFEN-SNS colloquia held at Paris 15–17 April 1991. Revue Générale Nucléaire, (2), 157–157 (1991)
3. IAEA, International Atomic Energy Agency, Summary Report on the Post-Accident Review Meeting on the Chernobyl Accident, International Nuclear Safety Advisory Group (Safety Series 75-INSAG-1), Vienna (1986)
4. IAEA, International Atomic Energy Agency, Safety Culture, International Nuclear Safety Advisory Group (Safety Series 75-INSAG-4), Vienna (1991)
5. IAEA, International Atomic Energy Agency, Examples of Safety Culture Practices (Safety report series No 1), Vienna (1997)

6. The National Diet of Japan Fukushima Nuclear Accident Independent Investigation Commission, *The Fukushima Daiichi Nuclear Power Situation Disaster: Investigating the Myth and Reality* (Ed. Routeldge/Earthscan, New York, 2014)
7. ICANPS, Investigation Committee on the Accident at Fukushima Nuclear Power Stations of Tokyo Electric Power Company, Executive Summary of the Final Report (2012)
8. The National Academy of Science, *Lessons Learned from the Fukushima Nuclear Accident for Improving Safety of U.S. Nuclear Plants* (The National Academies Press, Washington, 2014)
9. The Fukushima Daiichi Nuclear Power Plant Accident: OECD/NEA Nuclear Safety Response and Lessons Learnt. OECD 2013. NEA No. 7161
10. R. *Kadota, On the Brink: The Inside Story of Fukushima Daiichi* (Kurodahan Press, 2014)
11. R. Gulati, C. Casto, C Krontiris, How the other Fukushima plant survived. Harward Bus. Rev.
12. F. Guarnieri, S. Travadel, C. Martin, A. Portelli, A. Afrouss, *L'accident de Fukushima Daiichi; Le récit du directeur de la centrale*, vol. 1 (Presses des Mines, Paris, L'anéantissement, 2015)
13. E.M. Geist, What Three Mile Island, Chernobyl, and Fukushima can teach about the next one. Bull. Atom. Sci. (2014)
14. K. Benedict, The myth of absolute safety. Bull. Atom. Sci. (2014)
15. C. Perrow, *Normal Accidents, with a New 'Afterword'* (Princeton University Press, Princeton, 1999)
16. J.-B. Fressoz, L'apocalypse joyeuse, une histoire du risque technologique. Seuil (2012)
17. F. Tourneux, *Encyclopédie des chemins de fer et des machines à vapeur* (Renouard, Paris, 1844)
18. M.A.B. Alvarenga, P.F. e Melo, Including severe accidents in the design basis of nuclear power plants: An organizational factors perspective after the Fukushima accident. Ann. Nucl. Energy **79**, 68–77 (2015)
19. International Nuclear Safety Advisory Group, *Defence in Depth in Nuclear Safety* (IAEA, Vienna, INSAG-10, 1996)
20. M. Ferrier, Fukushima, récit d'un désastre. Folio Gallimard (2012)
21. The Japan Times, Heavy control consol falls back into Fukushima fuel pool: Tepco, August 29 2014
22. S. Mizokami, Y. Kumagi, Event sequence of the Fukushima Daiichi accident, in Reflections on the Fukushima Daiichi nuclear accident (Springer, 2014), pp. 1–17
23. C. Gilbert, Quels risques pour la recherche en sciences humaines et sociales. Dans Bourg D., Joly P-B., Kaufmann, A. (Dir.), Colloque de Cerisy, Du risque à la menace. Penser la catastrophe. PUF (2013)
24. G.N. Fischer, Le ressort invisible. Paris, Seuil, republished by Dunod (1994)
25. G.E. Richardson, The metatheory of resilience and resiliency. J. Clin. Psychol. (2002)
26. L. Pearlin, C. Schooler, The structure of coping. J. Health Soc. Behav. (1978)
27. A CARRI Report, Definitions of community resilience: an analysis (2013)
28. C.S. Holling, Resilience and stability of ecological systems. Annu. Rev. Ecol. Syst. **4**, 1–23 (1973)
29. N. Adger, Social and ecological resilience: are they related? Prog. Hum. Geogr. **24**(3), 347–364 (2000)
30. S. Saint-Arnaud, P. Bernard, Convergence or resilience? A hierarchical cluster analysis of the welfare regimes in advanced countries. Curr. Sociol. **51**(5), 499–527 (2003)
31. P. Reinmoeller, N. van Baardwijk, The link between diversity and resilience. MIT Sloan Manage. Rev. (2005)
32. E. Hollnagel, D. Woods, N. Leveson (eds.), *Resilience Engineering—Concepts and Precepts.* Ashgate, (2006)
33. B. Walker, C.S. Holling, S.R. Carpenter, A. Kinzig, Resilience. Adaptability and Transformability in Social-Ecological Systems, Ecology and Society **9**(2), 5 (2004)
34. M. Bunn, O. Heinonen, Preventing the Next Fukushima. Science, **333** (2011)

35. A. Boin, A. McConnell, Preparing for critical infrastructure breakdowns: the limits of crisis management and the need for resilience. J. Conting. Crisis Manage. **15**(1) (2007)
36. T. Aven, On some recent definitions and analysis frameworks for risks, vulnerability, and resilience. Risk Anal. **31**(4) (2011)
37. M. Ungar, Resilience across cultures. British Journal of Social Work **38**(2), 218–235 (2008)
38. A. Rey, Le Petit Robert (2012)
39. C.S. Holling, Engineering resilience versus ecological resilience, in *Engineering within Ecological Constraints* (1996), pp. 31–44
40. C. Morel, Les décisions absurdes. Gallimard Folio Essais (2002)
41. C. Morel, *Les décisions absurdes II* (Comment les éviter, NRF Gallimard, 2012)
42. P. Dibie, Ethnologie de la chambre à coucher (Vol. 4). Editions Métailié (1987)
43. J.H. Barthélémy, Cahiers Simondon, N°4, L'Harmattan, 15 April 2012, (2012), pp. 23–24
44. I. Prigogine, Laws of nature, probability and time symmetry breaking. Physica A **263**(1), 528–539 (1999)
45. M. Heidegger, *Being and Time: A Translation of Sein und Zeit* (SUNY Press, 1996)
46. M. Paty, Sur l'histoire du problème du temps. Le temps physique et les phénomènes. In Le temps et sa flèche. Editions Frontières (1994)
47. J. Baudrillard, L'illusion de la fin ou la grève des événements. Galilée (1992)
48. J. Baudrillard, Les Stratégies fatales. Grasset Livre de Poche (1983)
49. M. Foucault, *Sécurité, Territoire, Population, Cours au Collège de France, 1977–1978* (Gallimard, Seuil, 2004)
50. J. Ellul, Le bluff technologique. Pluriel (2012)
51. M. Akrich, How can technical objects be described, in *International Workshop on the Integration of Social and Historical Studies of Technology, Enschede*, vol. 3, no. 5 (1987)
52. J.-Y. Chateau, Le vocabulaire de Simondon. Ellipses (2008)
53. K. Weick, Improvisation as a mindset for organizational analysis. Org. Sci. **9**(5) (1998)
54. P. Berliner, *Thinking in Jazz: The infinite art of improvisation* (University of Chicago Press, Chicago, 1994)
55. T. Wachtendorf, K. Kendra, *Improvising Disaster in the City of Jazz: Organizational Response to Hurricane Katrina* (Social Sciences Research Council, 2006)
56. K. Tierney, *The Social Roots of Risk, Producing Disasters, Promoting Resilience* (Stanford University Press, 2014)
57. H. Bergson H (Robin Durie, ed.), *Duration and Simultaneity* (Clinamen Press, Manchester, 1999). Originally published in French: Paris: Presses Universitaires de France, 1968
58. R. Descartes, D. Weissman, *Discourse on the Method: And, Meditations on First Philosophy* (Yale University Press, 1996)
59. G. Simondon, Du mode d'existence des objets techniques. Aubier (2012)
60. P. Virilio, La vitesse de la libération. Galilée (1995)
61. M. Onoda, H. Takahashi, Fukushima No. 2. Scrambled to avoid same fate as sister site Fukushima No 1, in Japan Times, September 10 2014
62. H. Takahashi, Y. Yukiko Maeda, Y. Shinohara, Yoshida's call on seawater kept reactor cool as Tokyo dithered, in Japan Times, September 14 2014
63. D. Lochbaum, E. Lyman, S.Q. Stranahan, *Fukushima: The Story of a Nuclear Disaster* (The New Press, 2014)
64. Y. Funabashi, K. Kitazawa, Fukushima in review: A complex disaster, a disastrous response. Bull. Atom. Sci. **68**(2) (2012)
65. C. Perrow, *The Next Catastrophe: Reducing Our Vulnerabilities to Natural, Industrial, and Terrorist Disasters* (Princeton University Press, 2007)
66. R.J. Samuels, *3.11 Disaster and Change in Japan* (Cornell University Press, 2013)
67. M. Edelstein, A. Wandersman, *Contaminated Communities: The Social and Psychological Impacts of Residential Toxic Exposure* (Westview Press, Boulder, CO, 1988)
68. F.H. Norris, S.P. Stevens, B. Pfefferbaum, K.F. Wyche, R.L. Pfefferbaum, Community Resilience as a Metaphor, Theory, Set of Capacities, and Strategy for Disaster Readiness. Am. J. Community Psychol. **41**(1–2), 127–150 (2008)

69. D. Woods, Creating Foresight: How Resilience Engineering Can Transform NASA's Approach to Risky Decision Making. Testimony on the Future of NASA for Committee on Commerce, Sciences and Transportation, John McCain Chair (2003)
70. A. Winston, Resilience in a hotter world. Harv. Bus. Rev. (2014)
71. H. Smet, P. Lagadec, J. Leysen, Disasters out of the box: a new ballgame? J. Conting. Crisis Manage. **20**(3), 138–148 (2012)
72. ASIS, *Organizational Resilience: Security, Preparedness, and Continuity Management Systems-Requirements with Guidance for Use* (American Nationals Standards Institute, 2009)
73. D. Snowden, Complex acts of knowing—paradox, and descriptive self-awareness. Journal of Knowledge Management **6**(2), 100–111 (2002)
74. J. Rasmussen, Risk management in a dynamic society: A modeling problem. Saf. Sci. **27**, 183–213 (1997)
75. S. Dekker, E. Hollnagel, D. Woods, R. Cook, Resilience Engineering: New directions for measuring and maintaining safety in complex systems. Final Report, November 2008. Lund University School of Aviation (2008)
76. R. Westrum, Cultures with requisite imagination, in *Verification and Validation of Complex Systems: Human Factors Issues*, NATO ASI Series, ed. by J.A. Wise, V.D. Hopkin, P Stager, vol. 110 (Springer, Berlin, 1993), pp. 401–446
77. M.H. Schafer, T.P. Shippee, K.F. Ferraro, When does disadvantage not accumulate? Toward a sociological conceptualization of resilience. Schweizerische Zeitschrift für Soziologie **35**(2), 231–251 (2009)
78. S.E. Hobfoll, *Stress, Culture, and Community: The Psychology and Philosophy of Stress* (Plenum, New York, 1998)
79. E.H. Powley, Reclaiming resilience and safety: Resilience activation in the critical period of crisis. Human Rel. **62**(9) (2009)
80. K.M. Suttcliffe, T.J. Vogus, Organizing for resilience, in *Positive Organizational Scholarship: Foundations of a New Discipline*, ed. by K.S. Cameron, J.E. Dutton, R.E. Quinn (Berrett-Koehler, San Francisco, CA, 2003), pp. 94–110
81. F. Guarnieri, S. Travadel, Engineering thinking in emergency situations: a new nuclear safety concept. The Bulletin of the Atomic Scientists **70**(6), 79 (2014)

Part I
What are Damages in Nuclear Accidents?

Does the Concept of Loss Orient Risk Prevention Policy?

Dominique Pecaud

Abstract This chapter examines the concept of loss and damage and how they are used in a political and moral context. It takes as a starting point the nuclear accident of Fukushima Daiichi and the short, medium and long-term consequences for the human and non-human environment. It also identifies some potential elements for the transformation of the disaster into a catastrophe, and how we can develop different forms and scales of resilience.

Keywords Loss · Damage · Market · Gift · Catastrophe · Resilience

1 Introduction

The French word *dommage* (damage) has several translations into English. Two are of particular interest: first, it is the partial or total physical destruction of a living being (ranging from physical injury to death) or an object, due to an accident; second, it represents loss, as in the expression "it's a pity" or "what a pity".

This polysemy makes it difficult to create an empirical concept in the sense of a class of objects that can be characterized by the elements that compose it. But why even try? Because a deeper understanding of the anthropological meaning of damage can help to identify social dynamics that receive little attention in classical risk prevention studies [1]. In practice, the word appears to designate both the consequences of a disaster, and the losses suffered by humans and non-humans which are serious enough for it to be asserted that the disaster must not happen again. In this case, taking damage and its dynamics into account can help to go beyond the simple issue of loss.

D. Pecaud
Université de Nantes, Nantes, France

D. Pecaud (✉)
Centre for Research on Risks and Crises (CRC), MINES ParisTech/PSL Research University, Paris, France
e-mail: dominique.pecaud@univ-nantes.fr

© The Author(s) 2017
J. Ahn et al. (eds.), *Resilience: A New Paradigm of Nuclear Safety*,
DOI 10.1007/978-3-319-58768-4_2

21

The problem is this: the definition of disasters and resilience to their effects cannot be addressed if the issue remains limited to the usual collective representations and ways of thinking about risk prevention that—as disasters demonstrate—fail. A change is in order. This study advances a concept of damage and its technical, economic, political and moral usage. It addresses the Fukushima Daiichi nuclear accident and its consequences in the short, medium and long term; not just as a simple accident, but as a way of life that is both individual and universal, for a human and non-human environment that has passed into history.

Human and non-human: the expression comes from the *sociology of objects*, which [2] remind us is *an extension of the sociology of innovation and the pragmatic sociology of action*. This sociology seeks to overcome epistemological divisions such as those between the individual and the collective, or methodological individualism and determinism. To this we can add the accident and the essence. It *aims to repopulate the sociological universe* with the set of objects that, according to [3] participate in *the construction of society*, unlike the roles found in classical sociological theory: *faithful tools, critical infrastructure or, finally, projection screen*.

Loss can be defined as a system of representations and actions following an event. It is contemporary with the construction and designation of victims. This conceptual definition is part of a political philosophy. It is based on the role of organizations and various forms of authority in shaping collective action and therefore, the creation of the common good.

Rather than take a traditional approach, this chapter presents a series of viewpoints that address both theoretical and practical aspects of the concept of loss. It is argued that loss should be understood both as a moral or legal heuristic[1] and as an *actant* [5] that is able to change form depending on the circumstances in which it is used.

2 Industrial Degradation, Causal Links and Preventive Maintenance

The degradation of an environment can be seen from *outside* the subject, i.e. as an object. In this case, the issue of loss has no emotional connotations for the person who observes it. If we consider the consequences of the nuclear accident at Fukushima Daiichi, there are many reasons for this indifference: lack of information, mistrust regarding its content or source, geographical remoteness, few signs of

---

[1]In law, two elements define the concept of loss: (a) although there may be no damage as such, there is a causal relationship—it can be connected to something. "Loss is simply any harm caused by something. It is this cause that does harm and that gives rise to the legal concept of loss; in other words, loss does not exist naturally" [4]; (b) When the cause is natural, the law will only consider common-sense explanations that throw light on the causal link. Loss differs from harm, harm being defined as the result of loss.

dramatic degradation,[2] the idea that everyone is part of planet earth or humanity, little fear concerning the maintenance of one's own physical integrity, etc. The assessment of the degradation depends largely on its nature and how often it occurs (which is often underplayed), and also on a sense of belonging, solidarity or compassion that everyone feels towards the human and non-human victims of a disaster.

People do not consider the hazards of the environment in which they live to be responsible for the deterioration of their health so long as they believe that this environment is clean and healthy. On the other hand, they will try to blame their environment for their own physical deterioration. They highlight the phenomena they consider to be true, although this may be difficult to justify scientifically.

Nobody would hold the environment directly responsible when a native species is pushed out of its habitat by another species. Other reasons are highlighted, for example, the imbalance of the environment due to detrimental human activities.

These examples give rise to several questions. The first concerns the cultural emphasis given to the distinction between man and nature. The second concerns the idea that we have of a healthy environment and how to recover if we judge it to be degraded. The third concerns the anthropology and epistemology of science. What degree of scientific validity do we attribute to causal relationships between environment and health?

Machinery can also be subject to degradation that does not necessarily create human victims. While it is designed to withstand normal wear and tear, engineers may replace the entire unit, or its parts, in order to avoid a breakdown or other incident. Risk prevention policy terms this 'preventive maintenance' and the aim is to identify the potential degradation of a piece of equipment and anticipate any unwanted consequences. Preventive maintenance aims to maintain technical equipment in a satisfactory state.

Preventive maintenance is defined as "maintenance that is carried out at predetermined intervals or according to prescribed criteria and intended to reduce the probability of failure or degradation of the operation of a good" (extract from the French standard NF EN 13306 X 60-319). It can also take the form of *scheduled maintenance*, i.e. "preventive maintenance that is performed at predetermined time intervals or according to a specified number of units of use but without any preliminary check of the condition of the equipment" (extract from the French standard NF EN 13306 X 60-319), or *conditional maintenance*, which is defined as "preventive maintenance that is based on monitoring of the operation of the equipment and/or important parameters for its operation including the actions that result from it" (extract from the French standard NF EN 13306 X 60-319). The term 'important parameters' refers to key indicators of the equipment's condition. Finally, preventive maintenance can also take the form of *predictive maintenance*,

---

[2]Here we refer to a disaster rather than a catastrophe in order to respect the fact that the catastrophe is primarily the result of an intellectual construction [6]. At the time of writing, no-one knows whether the disaster will become a catastrophe.

i.e. "conditional maintenance that is carried out following extrapolated forecasts and an assessment of important parameters indicating the degradation of the good" (extract from the French standard NF EN 13306 X 60-319).

Preventive maintenance relies on data that serves as objective causes of possible damage. Even at the design stage of the nuclear power plant at Fukushima Daiichi, both the idea of an earthquake occurring near the coast, and a tsunami were taken into account. To reduce the risk of the degradation of facilities, construction standards were applied. To reduce the risks created by a tsunami, dykes, able to stop waves 5.70 m high were erected around the plant.[3] The height was based on data related to the region's history of tsunamis, seismological expertise and wave dynamics.

> The construction of the Fukushima Daiichi plant that began in 1967 was based on the seismological knowledge at that time. As research continued over the years, researchers repeatedly pointed out the high possibility of tsunami levels reaching beyond the assumptions made at the time of construction, as well as the possibility of core damage in the case of such a tsunami. TEPCO[4] overlooked these warnings, and the small margins of safety that existed were far from adequate for such an emergency situation [7].

In the case of the Fukushima Daiichi disaster, it is not certain that the tsunami (the cause) is the only source of the breakdowns (the effect) that led to the nuclear accident. Other causes can be highlighted, such as the earthquake, poor maintenance, or the organization of work in general.

> TEPCO's report says the first wave of the tsunami reached the site at 15:27 and the second at 15:35. However, these are the times when the wave gauge set 1.5 km offshore detected the waves, not the times of when the tsunami hit the plant. This suggests that at least the loss of emergency power supply A at Unit 1 might not have been caused by flooding [7].

> Since 2006, the regulatory authorities and TEPCO have shared information on the possibility of a total outage of electricity occurring at Fukushima Daiichi should tsunami levels reach the site. They also shared an awareness of the risk of potential reactor core damage from a breakdown of seawater pumps if the magnitude of a tsunami striking the plant turned out to be greater than the assessment made by the Japan Society of Civil Engineers. There were at least three background issues concerning the lack of improvements. First, NISA [the Japanese regulatory authority] did not disclose any information to the public on their evaluations or their instructions to reconsider the assumptions used in designing the plant's tsunami defences (…). The second issue concerned the methodology used by the Japan Society of Civil Engineers to evaluate the height of the tsunami. Even though the method was decided through an unclear process, and with the improper involvement of the electric power companies, NISA accepted it as a standard without examining its validity. A third issue was the arbitrary interpretation and selection of a probability theory. TEPCO tried to justify the belief that there was a low probability of tsunami, and used the results of a biased

---

[3]The tidal wave that followed the magnitude 9.0 earthquake occurred on 11 March, 2011 at 2:46 p. m. local time reached its maximum height of 23.6 m at Ofunato, in the Iwate Prefecture, north of Fukushima Daiichi (Executive Summary of Urgent Field Survey of Earthquake and Tsunami Disasters, 25 March, 2011, Port and Import and Research Institute). The height of a tsunami varies according to many criteria. It was estimated to be about 14 m at Fukushima Daiichi, while the plant itself lay at 7 m above sea level.

[4]Tokyo Electric Power Company.

calculation process as grounds to ignore the need for countermeasures. TEPCO also argued that basing any safety assessment against tsunami on a probabilistic approach would be using a methodology of technical uncertainties, and used that argument to postpone considering countermeasures for tsunami [7].

3 Loss, Damage and Victims

The term 'loss' is used to designate the consequences of degradation. In some areas, these consequences can be anticipated, while in others they cannot. For example, in an armed conflict the intentional bombing of a military building can lead to its destruction. While this may have been the intention of those who undertook the action, it can also destroy buildings or injure people who were not the target. This is referred to as 'collateral damage'.[5] In this case 'damage' is most meaningful when injury has been caused to a person or group, their property, or an environment (if there is an interest in claiming compensation). Losses can be immediate or longer term.

When it concerns a person, group, non-human living species, or an environment the term 'victim' must take on a broader meaning. Not only can it be used to designate people, but also living or non-living entities that have representatives who are able to speak on their behalf. For example, the quality of a coastline is said to be degraded following an oil spill and the media do not hesitate to describe the shoreline as a 'victim'. Bees are another example. This living species is the victim of agricultural chemicals, or, as described in a French newspaper, the "victim of the lack of biodiversity".[6] Finally, before work-related accident legislation was implemented, there were no official victims of occupational accidents or illnesses. They could not be acknowledged until dangerous working conditions or hazardous machinery was recognised. Similarly, an analysis of breakdowns due to human factors can be seen as the clumsy (or even malicious) use of machines by operators.

In general terms, there are no non-human victims. Instead, they are represented by individuals or groups who speak on their behalf and defend their interests before an authority that is responsible for estimating the loss they have suffered.

---

[5]"Collateral damage and proportionality are two inseparable concepts. The concept of proportionality in jus ad bellum reflects the balance that must be maintained between, on the one hand, military requirements and, on the other hand humanitarian interests, such as the cost in human lives. It aims to limit damage to civilians during attacks against legitimate military objectives, by weighing the military advantage that would result from the attack against the losses that it would cause to the civilian population (i.e. collateral damage). It was not until 1977 that this proportionality rule would be included in a Treaty. It is found in Articles 51, 5.b and 57, 2.b of the First Additional Protocol of 1977 to the Geneva Conventions of 1949." (emphasis added). The International Law Centre of the Free University of Brussels, https://dommagescollateraux.wordpress.com/pratique/ accessed 15 February 2015.

[6]Le Figaro, 25-11-2014.

4 Assessment of Compensation: An Anthropological Approach

Once is has been established that there has been a loss, it can be estimated in economic terms. This estimate is not an end in itself. Its purpose is to provide compensation to the victim. Objective losses must be recognized by the party responsible and the victim, while its assessment is usually carried out by scientific experts and judicial authorities. Once this is done, the victim can receive compensation. Here, the purpose is to re-establish a situation that the party responsible for the damage, the victim and any third parties (scientific and/or judicial) assess as having been compromised. The new situation should be as similar as possible to the previous situation, although full recovery may be impossible. For example, the judicial process may end with the victim receiving financial and/or non-material compensation, such as a symbolic award, which takes into account any 'damages and interest' and corresponds to the non-material harm suffered by the victim. Here, the intention is to compensate for the victim's suffering.

The issue of loss should be seen in an anthropological context, that of the links that create and sustain a society. Two paradigms can be used to characterize these links.

4.1 The Market Paradigm

The first is the market. From a political liberalism perspective, the market represents a utility. It regulates trade, guarantees individual freedom and collective effectiveness. For example, debt repayments end an unequal relationship, in which the borrower is beholden to the lender. Once the debt is repaid, both parties are free to act and a new exchange can begin. Another example is the prison sentence, which settles a debt that the convict has to society.

In 1839 the lawyer Thimus [8] used the example of the duel to illustrate the dynamic of compensation. Here, we look at it from the perspective of natural law. In a duel, the injured party seeks redress and other party risks damage to their reputation if they refuse the challenge. Doing so would not only break the symbolic link between them, it would also negate their shared values. When community membership depends on shared values, the refusal to honour a claim for compensation ends the relationship. From the point of view of the injured party, should the other party refuse to offer compensation, they are excluded from the community.

We can transpose this relational dynamic to the Fukushima disaster. In this case, who may seek redress, and from whom? Answers to these questions require explanations that relate to the society that has suffered the injury. There are various scales.

Any accident expresses a conflict that manifests as an imbalance between human and/or non-human entities. In the case of Fukushima Daiichi, it concerns two 'energy' entities: the plant and its environment. These are social and cultural constructions and embody antagonistic forces.

The nuclear power plant can be described in very simple or very complex terms. It can be seen as: a simple technical device that generates electricity; as a socio-technical system that regulates functional relationships between humans and machines, and between it and its environment; or in terms of a combination of human and non-human elements linked by a struggle for physical or political power. At the same time, it is the result of human activity, which is called into question from the moment it is unable to recover from the nuclear accident. Its energy takes multiple forms.[7]

The plant's environment is overwhelmed and transformed by natural forces: a magnitude 9.0 earthquake, followed by a tsunami.

Each of the three protagonists—the earthquake, the tsunami, and the nuclear power plant—has destructive potential, leading to an examination of the relationship between nature and the nuclear plant. In this context, the disaster represents an increase in entropy. Can it be seen as the last stage in a conflict? Using Marc Bloch's war metaphor, does the disaster that follows the accident and the disaster put an end to the conflict? "A last resort for the resolution of political disputes, war must be used to end a conflict and allow a return to equilibrium, even if fragile. However, modernity sterilises this organized violence, the capacity for destruction, the sacrifices are so great that they make it impossible to go back".[8]

4.2 The Gift Paradigm

The second paradigm is that of the gift [9]. Unlike the market, reciprocal gift-giving corresponds to an exchange that has no beginning or end. Although the giving of a gift should not correspond to a state of indebtedness, it is nevertheless the position the receiver finds themselves in. In turn, reciprocating the gift makes the new receiver indebted. The event that led to the catastrophe can be considered as part of this relational system.

Three consequences arise. First, the system creates a long-term context, assuming that gifts continue to be exchanged. Should the receiver not reciprocate, the donor has a hold over them. Second, the gift that is given in return is never the formal equivalent of the initial gift, and the new giver must demonstrate the superiority of their gift to the initial giver. Third, the power that is expressed by the gift giver and the recipient who reciprocates makes a lasting impact on their

---

[7]This observation is not based on technical arguments; it simply refers to the ability of the nuclear reactor to destroy its own cooling mechanisms.

[8]Quoted by Emilio Gentile in L'apocalypse de la modernité, la grande guerre et l'homme nouveau [The Apocalypse of Modernity, the Great War and the New Man], Flammarion, 2011, p. 171.

shared world view. The exchange system that unites giver and recipient requires everyone to benefit. The relational system of gift and counter-gift is ternary. Each of the elements of the exchange results in three obligations: giving, receiving and reciprocating. The exchange is based on the capacity of each party to receive, therefore to assess what is received and what should be given at a later date. Their shared understanding of what has been received defines the gift and the counter-gift and maintains the momentum.

The market and gift paradigms highlight different views of exchanges. In the case of the market, the relationship is unsustainable as it is ended by a payment. The relationship and conflict must be renewed, beginning again from zero. In the case of gift giving there is no end-point, although an inability to reciprocate unbalances the exchange and changes its modalities. The unbalanced relationship becomes fixed, and the party that cannot reciprocate becomes beholden to the gift giver. They are left at the mercy of the donor and the subject of their political decisions.

The reconciliation of these paradigms with the concept of loss appears to be a very useful heuristic. The question is: does loss correspond better to the market paradigm or the gift giving dynamic? This leads to another question: what, in both cases, drives the dynamic?

4.3 Consequences for the Definition of the Concept of Loss

Before considering some answers to these questions, it is first necessary to consider two important elements related to the concept of loss.

First, does the definition of loss form part of a temporally closed exchange system (e.g. a breakdown, accident and recovery) or, an open exchange system (e.g. between a disaster, an ongoing catastrophe and resilience)?[9] In both cases, we need to clearly define environmental, political or industrial compensation on the one hand, and the scope of resilience, on the other.

Second, we must not forget that loss can only be defined or assessed if victims can be identified. This usually requires a third party, a mediator who can identify both the loss and its victims. It assumes that the victims and the party responsible for the loss share an understanding of the situation that led to the loss and that there is no need for mediation. Identifying losses and victims becomes difficult or impossible when the victims are silent, either because they are incapable of speech or because they are not asked to speak. However, acknowledging the views of the parties involved is essential in understanding not only the material consequences of the disaster, but also its symbolic construction, considered to be one of the key phases of psychological and social resilience.

---

[9]Here, ecological resilience refers to the ability of the overall system to rebalance itself in the long term; it does not necessarily mean that it returns to the state that it was in prior to the disaster, which may be impossible to achieve [10].

5 The Fukushima Daiichi Catastrophe as a Total Social Fact

The relationship between people, whether they be settlements in the neighbourhood of the Fukushima Daiichi plant or humanity in its entirety (defined either as anything that belongs to the human species on the one hand, or other material and natural elements on the other hand) must be taken into account if the definition of the nuclear catastrophe is to go beyond technical aspects. As time passes, the catastrophe must be seen as a 'total social fact' and the dynamic is that of reciprocal gift giving. On the one hand, the elements that make up this relational system can be considered as an endless sequence of actions and reactions (sometimes referred to as a chain reaction), which describe an inevitable succession of events (rather than a causal chain). On the other hand, it must make it possible to clearly identify the victims of the accident, then the disaster, how victims emerge, the balance between them and the losses they have suffered, and therefore to determine the extent of their losses in the context of an exchange defined by the social (and potentially natural) limits of 'resilience'. Ecological resilience corresponds to changes in an environment that is affected by a catastrophe, which leads to a more stable situation where the consequences of the catastrophe are physically or symbolically mitigated.

6 A Choice of Paradigm?

The market or the exchange of gifts? The market is governed by rules applied by statutory law, while the dynamics of gift giving are rooted in the philosophical presuppositions of natural law. Statutory law reflects the actual legal position of a society at a given time. It is made up of regulations governing, for example, trade. On the other hand, natural law allows and protects the expression of human nature; it addresses universal principles. For example, [11] considered that it was based on two principles that preceded reason. Nowadays, this would be what is not covered by instrumental or logical rationality: self-love and altruistic pity are two practices that form the foundations for human relationships. These two principles suggest a universal human nature that falls short of the rational rules that govern collective action. However, despite the universality that philosophers and law-makers of the Enlightenment sought to give it (notably through the Declaration of the Rights of Man and the Citizen) natural law has been discussed in terms of its potential relativity in space-time. The inclusion of local and traditional cultures changes the idea that men are the product of their relationships with each other or their environment. Ways of living together become defined for both humans and non-humans, which show that the idea of human nature is historically and culturally relative. This is manifested, for example, through the definition that each individual has of human health or an acceptable level of pollution. The ensuing discussion focuses on a universal definition of health (e.g. the World Health Organisation) and negotiations to establish indicators, then standards for environmental pollution. Other debates focus on securing financial compensation for victims, the social acceptability of a hazardous technology, or the pristine condition of a shoreline prior to environmental clean-up or decontamination, etc.

The search for the foundations of natural law corresponds to the desire to define, then respect 'the' or 'a' human nature. The latter cannot be distinguished from nature itself, which is here defined as what surrounds human beings. Both the history of ideas about human nature and growing awareness of environmental degradation show the need to reduce the philosophical, political, moral and economic divide between Man and Nature. This is not without consequences. It means that:

(a) In the context of Fukushima Daiichi, Nature (made up of human and non-human elements) should be considered as an integral victim of the accident and the ongoing catastrophe, and should be able to claim compensation. This would force people to redefine their moral responsibility vis-à-vis the impact of their activities on the environment.

(b) Distinguishing the three elements that made up the events of 11 March, 2011 suggests that only the nuclear accident was the source of the damage to the environment. The accident cannot be likened to a simple natural disaster, just because there was an earthquake and a tsunami. Nature cannot be considered as a victim of itself, nor can a natural event be distinguished from a manifestation of Nature. The argument simply lends weight to an anthropocentric perspective that maintains the divide between Nature and human nature. Nature is constantly evolving, and the distinction between how this happens (e.g. low-level changes or sudden and dramatic events) may reinforce the anthropocentric tendency.

The distinction between a natural event and a nuclear accident makes it possible to think about two types of resilience: the first relates to the self-regulation of Nature, which is constantly changing and can never be considered as the return to a previous state; while the second relates to human resilience and focuses on life during and after the disaster.

An illustration of this comes from Fukushima Daiichi (or Chernobyl) where a decision was taken to establish no-go zones around the facility. This type of decision implicitly reaffirmed the divide between human and non-human beings. It designates humans as the most important victims of the ongoing disaster. However, the decision to ban fishing or hunting in areas adjacent to the no-go zone shows than there were other victims.

(c) Although the earthquake and tsunami led to the destruction of nature, humans did not see it as a victim. However, this depends on the anthropological status that is assigned to it. Moving a random stone may not lead to a claim for damages, while moving another that is part of a dyke or historic monument can lead to funds being allocated for its reconstruction. In the first case, the stone does not have victim status, it the second, it does. The same argument can be made about living species. Whether they are designated as victims will depend on the degree of domestication or their value in terms of heritage (e.g. protected and other endangered species). In all cases, the designation as a victim of a natural event depends on the degree of humanization attributed to living or inanimate natural elements. This attribution can be implied or voluntary, traditional or political.

It must be noted that the Fukushima Daiichi accident could only have been caused by humans, even if other elements contributed to material, imaginary and symbolic construction of the catastrophe.[10] First and foremost, the plant was created by humans. In simple terms, if the plant had not existed, there would have been no nuclear accident![11]

The nuclear accident is therefore unlike the earthquake or tsunami. The identification of those responsible and victims is indicative of instrumental thinking and statutory law. The accident finds its origin in the fact that humans have used nuclear fission to generate electricity, and they designed the necessary technology; while external natural forces (the earthquake followed by the tsunami) damaged the plant, they did not create an imminent danger to the environment. Radioactivity is a natural (re)action that has been exploited by humans and become an organised, industrial activity[12] designed to produce electricity. The nuclear accident and its consequences were an assault on the technology's environment and made it dangerous or unliveable for its inhabitants.

Not only humans, but also their pets were evacuated from around the Fukushima-Daiichi plant.[13]

The investigations that are being carried out into the accident and the ensuing devastation seek to determine the responsibilities of human beings. They focus on why the plant was built, why it was located in an area subject to earthquakes and tsunamis, construction techniques, the use of data to assess the risk of earthquakes and tsunamis,[14] etc. Their findings will determine who is responsible and the human and non-human victims. In time, their understanding of the disaster will become part of local and international statutory law. Investigations that seek to establish who is responsible will address questions such as: were emergency services sufficiently responsive and appropriate to the situation? Were anti-seismic building codes met? Did land-use plans take account of the risk of a tsunami? Was a full risk

---

[10]See [12]. The construction of the catastrophe can be viewed in terms of its material, imaginary dimensions (fantasies, individual and collective representations) and symbolic (collective meaning attributed to material phenomena and individual and collective representations) dimensions.

[11]This observation forms the basis for radical anti-nuclear protests. See [13].

[12]Sometimes called the domestication of nature.

[13]http://www.actu-environnement.com/ae/news/japon-evacuation-animaux-centrales-fukushima-ifaw-12578.php4.

[14]See the article "Tsunami: les ancêtres savaient [Tsumani: the ancestors knew]", published in the French newspaper Le Monde on 6 May 2011. Available online at http://www.lemonde.fr/japon/article/2011/05/06/tsunami-les-ancetres-savaient_1517972_1492975.html.

assessment carried out? Was the plant in a satisfactory condition? What controls and maintenance procedures were in place? Were they sufficiently rigorous and respected? Was the data related to the plant's design and technical operations satisfactory? In the event that these investigations and the courts identify particular weaknesses or shortcomings, the victims will receive financial and/or symbolic compensation. These 'gifts' will contribute to resilience that will act as a reciprocal gift, thereby creating new human requirements, such as gifts to the environment.

7 Conclusion

Our work looks the question of the losses resulting from the nuclear accident at Fukushima Daiichi from multiple perspectives: natural and statutory law; compensation based on the market paradigm and that of gift-giving; and the designation of victims, losses and compensation. On the latter point, it seems interesting to think in terms of a kind of non-radical environmentalism.[15]

The situation is already a reality. Tawada [16] points out that "there is no Japanese word that is an exact translation for the German work 'catastrophe'. In German, the word is used in relation to nature and politics. In the event of a natural catastrophe, politics comes readily to the mind of people." While [12] considers that "it has become impossible to clearly separate the movements of political change from those that lead to an environmental threat. It is seen in the symbolic dimension of the catastrophe, the overturning of the world of meanings leading to the revelation of a gap where legislation was supposed to take over." From this flows the importance of understanding how compensation can be used to fill the gap.

The catastrophe began on 11 March, 2011 and it does not yet appear to have reached its end. Many observers believe that it is just beginning. Current discussions focus on two areas: an objective explanation of the disaster that initiated the catastrophe; and the designation of victims and the assessment of losses. But who will be invited to the discussions and for how long? It must not be forgotten that in French the word 'loss' is equivalent to the concept of "harm or damage caused to someone or something" (*Dictionnaire Littré*). This presupposes a dramatic event that takes place over some time (with the idea that the timeframe widens—or not— its scope). It also presupposes that it is possible to identify the persons or things that

---

[15]Afeissa [14] [15] defines environmental ethics as an ethic "which produces a new object, the non-human natural world, judged worthy of moral consideration on its own merits, in other words regardless of any coefficient of utility for the existence of man and considered as a place of intrinsic value or as a holder of rights whose existence as such, command a number of moral and legal obligations." It raises several questions that this article seeks to clarify: To whom should natural rights (of Man or Nature) be given? Can it be done, and can we assign rights to non-humans without strengthening an anthropocentric perspective leading to radical monism? Who can claim these rights (humans, non-humans) and, in the case of non-humans, how does this manifest? Who attributes such rights? In particular, who speaks on behalf of whom?

have or will suffer harmful effects, either immediately or at a later stage. The definition of losses requires the identification of all persons and things marked by the event and an attempt to establish causal relationships between them. Epistemological caution should inform these discussions, their development and the decisions that remain to be taken.

This chapter is part of broader thinking on resilience engineering applied to the Fukushima Daiichi catastrophe. But is it the time to talk about resilience, when we still do not really know if the catastrophe is over. What is resilience? In the case of Fukushima Daiichi, resilience refers to the capacity of a technical system to continue operating, or to resist further damage in an unfavourable environment. It also refers to the ability of an ecosystem to recover its operations or development as they were before the system was disturbed.

Losses require the evaluation and implementation of compensatory measures that are designed to cancel out any injury to humans or non-humans who are designated as victims of the disaster. The aim of these practices is to restore harmony to the relationship between a victim (human or non-human) and the party responsible (human or non-human). In all cases, the assessment of losses and the implementation of measures designed to compensate for the harm suffered (reciprocal gift-giving) help to create shared meaning. We argue that this construction relates to resilience to catastrophe. The meaning that this evokes must form part of the preamble to such practices. A shared understanding appears to be necessary to think about resilience. A shared vision and sensitivity to, *a minima*, what needs to be considered, foreseen, done, and said is necessary for the construction of new meanings and practices linked to resilience. The definition of loss therefore presupposes many other shared intentions: to carry out repairs, and manage resilience in the best way possible. Practical and technical issues must be addressed, which must not overlap, merge or lead to the same answers. The redefinition of loss is perhaps an indicator of a consensus, if it can be accepted by the various stakeholders.

References

1. H.-P. Jeudy, *Le désir de catastrophe* (Circe-Poche, Paris, 1990)
2. R. Barbier, J.-Y. Trepos, Humains et non-humains: un bilan d'étape de la sociologie des collectifs. *Revue d'anthropologie des connaissances*, **1**, 35–58 (2007)
3. B. Latour, Une sociologie sans objet? Note théorique sur l'interobjectivité. *Sociologie du Travail* **36** (4), 587–607 (1994)
4. D. Anzilotti, La responsabilité internationale des États à raison des dommages subis par les étrangers, *Revue générale de droit international,* **13**, 13–29 (1906)
5. B. Latour, *Enquête sur les modes d'existence.Une anthropologie des Modernes* (La découverte, Paris, 2012)
6. Ailloud-Nicolas, Scènes de théâtre: Le tremblement de terre de Lisbonne (1755) et Le Jugement dernier des rois (1793), in *L'invention de la catastrophe au XVIIIe siècle. Du châtiment divin au désastre naturel*, ed. by A.M. Mercier-Faivre, C. Thomas (Droz, Geneva, 2008)

7. *The Fukushima Nuclear Accident Independent Investigation, Commission Official Report*, The National Diet of Japan (2012)
8. F.-G.-J. Thimus, *Manuel de Droit Naturel ou de Philosophie du Droit* [1839] (Nabu Press, 2012)
9. M. Mauss, Essai sur le don. Forme et raison de l'échange dans les sociétés primitives, [1923–1924], in *Sociologie et anthropologie* (Presses universitaires de France, Paris, 1968)
10. D. Pécaud, Méthodologie de retours d'expérience: concepts, finalités et conditions, in *Les conséquences du naufrage de l'Erika, Risques, environnement, société, rehabilitation*, ed. by J.-P. Beurier, Y.-F. Pouchus (Presses Universitaires de Rennes, 2005)
11. J.-J. Rousseau, *Discours sur l'origine et les fondements de l'négalité parmi les hommes* [1753] (Gallimard, Paris, 1965)
12. S. Le Poulichet, *Environnement et Catastrophe* (Mentha, Paris, 1991)
13. S. Topçu, *La France nucléaire, l'art de gouverner une technologie contestée* (Seuil, Paris, 2013)
14. H.S. Afeissa, *La communauté des êtres de nature* (MF, Paris, 2007)
15. *Ethique de l'environnement, Nature, valeur, respect*, textes réunis par Afeissa H. S. (Vrin, Paris, 2007)
16. Y. Tawada, *Journal des jours tremblants* (Verdier, Paris, 2012)

How the Fukushima Daiichi Accident Changed (or not) the Nuclear Safety Fundamentals?

Kazuo Furuta and Taro Kanno

Abstract In this chapter, the fundamentals of nuclear safety that the Fukushima Daiichi accident did and did not change will be discussed. While the most basic strategy of defense-in-depth principle is still valid, some problems have emerged after Fukushima, preparedness for all-hazards and multiple disasters, and importance of the administration of emergency response. From this observation, enhancing the resilience of nuclear systems is a critical issue after Fukushima. The safety enhancement measures considered in nuclear facilities will be reviewed referring to the elementary characteristics of systems resilience, and a new framework will be proposed for dealing with unsafe events, where unsafe events are classified into three categories.

Keywords Defense-in-depth · Residual risk · Beyond design-basis · Resilience · Category 2 events

1 Introduction

After the Great East Japan Earthquake (*Tohoku Earthquake*) and the Fukushima Daiichi accident (*Fukushima*), people used a word "unanticipated" for describing the disaster. It is true that the up-to-date seismology at the time of disaster could not foresee that such a huge earthquake and tsunami can ever occur in the area, and the main cause of the accident was insufficient preparedness of the plants against tsunamis [1]. It seems an improper remedial action, however, just reevaluating the risk of tsunamis more precisely and increasing the height of seawalls. It seems wrong also to think that the fundamentals of nuclear safety has broken down and it should be replaced with another one. Having reviewed the experiences of the

K. Furuta (✉) · T. Kanno
School of Engineering, The University of Tokyo, Tokyo, Japan
e-mail: furuta@rerc.t.u-tokyo.ac.jp

© The Author(s) 2017
J. Ahn et al. (eds.), *Resilience: A New Paradigm of Nuclear Safety*,
DOI 10.1007/978-3-319-58768-4_3

disaster, what we have to do is rather renovating the basic strategy of nuclear safety, defense-in-depth principle, from a viewpoint of systems resilience.

2   What Did not Change After Fukushima

After Fukushima, many people including the press condemned that the myth of nuclear safety was over and the thoughts of experts were totally wrong. The accident, however, has shown clearly that the most basic strategy of defense-in-depth principle is still valid, because the accident was caused exactly from the lack of defense-in-depth. The single safety barrier that had protected the Fukushima Daiichi plants against tsunamis was the seawalls. Since the largest scale of tsunamis that may possibly occur in the area is uncertain, multiple barriers should have been implemented for protecting the plants against tsunamis. In this situation, the tsunami caused by Tohoku Earthquake that exceeded the design basis was fatal.

In addition to the seawalls of an insufficient height, the areas of safety-relevant equipment in the plants were not watertight. The emergency power supply such as metal clad switchboards as well as diesel generators were located under the ground level. All these equipment were therefore submerged and lost functions just after the tsunami hit the plants. The backup systems against station blackout were insufficient, either, both in the emergency power supply and the means of water injection.

Safety barriers were not well prepared for mitigation of the consequence of an accident. Before the JCO criticality accident (JCO) [2], which occurred in 1999, emergency response that requires evacuation of nearby residents had been a taboo in Japanese nuclear development. As an aftermath of JCO, Act on Special Measures Concerning Nuclear Emergency Preparedness was enacted, and emergency response drills were enforced in each prefecture of major facility sites. It was revealed, however, in Fukushima that these efforts were totally ineffective, because the scale of accident was far beyond the prescribed scenarios of emergency response plans.

As described above, the disaster of Fukushima occurred, not because the very basis of nuclear safety was wrong, but because it was not implemented and maintained properly. Defense-in-depth is the most basic strategy of nuclear safety that had been established at an early stage of nuclear development, before sophisticated methods of risk-informed safety management were introduced. After Fukushima, some people claim that we should rely more on risk-informed methods for safety management and we should evaluate more precisely the risks of external hazards. It is necessary to do so, but only introducing more sophisticated risk-informed methods is not the final answer.

Figure 1 shows an overview of the safety management based on a probabilistic concept of risk, which is a combination of the scale and probability of damage. A certain risk limit can be chosen as the curve shown on the scale-probability plane in the figure. The region above this curve is called an unacceptable region, and that

Fig. 1 Safety management based on a probabilistic concept of risk

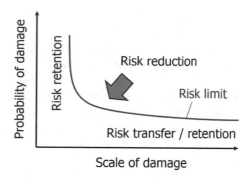

below the curve an acceptable region. If the system status is located within the unacceptable region, one must make all efforts to reduce the risk. Safety management on a probabilistic concept of risk is actually more complicated. A tolerable region will be introduced often between the unacceptable and acceptable regions with an upper and lower risk limits. As low as reasonably practicable (ALARP) principle is applied in this region. In addition, the risk profile of a complex system cannot be treated with a risk limit presumed under the assumption of normal distribution, because it shows a long tail and rare events may occur more frequently than expected. Anyways, Fig. 1 shows the first-order approximation of the risk-based approach of safety management.

Even if the system status is located in the acceptable region, however, it does not mean that the risk has vanished. The risk that still remains after having satisfied the risk limit is called the residual risk. We also have to deal with the residual risk after having satisfied the risk limit. Occurrence of an unanticipated event often leads to further reduction of the risk limit and then to elevation of safety regulation, but this process is an endless loop. Management of the residual risk is performed by risk retention and risk transfer, which are often out of the scope of ordinary safety regulation, and the strategy of their application is to be established. In a social setting, risk transfer is performed usually by insurance, and risk retention by disaster preparedness, compensation of damage, and so on. The method of risk retention, however, for damage that exceeds the scale of commercial insurance is disputable.

Renovating deterministic approaches following the defense-in-depth principle will be a key. After Fukushima, the Nuclear Regulation Authority (NRA) of Japan enforced new regulatory standards for commercial power reactors in July 2013. The new standards request enhancement of design basis, protection against earthquakes and tsunamis, and new requirements for severe accidents. In order to fulfil the standards, Japanese utility companies are now taking remedial actions to their plants and installing various safety measures such as follows [3, 4], and these measures are in line with enhancement of defense-in-depth rather than introducing new principles.

Measures to Prevent Damages from Natural Disasters: The maximum seismic motion and height of tsunamis are reevaluated based on the up-to-date knowledge of seismology, and the hazards of active faults close to the plant site are reexamined. The seawalls are reconstructed and reinforcement structures are added to the plant components, if necessary. Countermeasures are also taken against other natural disasters such as tornados, volcanic eruptions, and external fires.

Installation of Watertight Structures and Countermeasures Against Internal Flooding: Watertight doors are installed to the reactor building. The structures of ventilation openings are redesigned to prevent water invasion. Countermeasures against internal flooding are also taken by installing pipe and cable penatration seals, water protection covers, weirs, and so on.

Reinforcement of Emergency Power Supply and Water Injection: Not only permanent but also mobile equipment are installed for emergency power supply and water injection considerig beyond-design-basis situations. The capacity of these equipment are designed with enough margins to compensate for maintenance outage and equipment failres. The storage locations and connection points for the mobile equipment are diversified. Water injection to the spent fuel pits is also reinforced.

Prevention of Reactor Containment Damage: Auxiliary containment spray systems are installed, which can be supplied by permanent and mobile water injection pumps. Water injection lines to the bottom of containment vessel are considered for cooling core debris. Filtered containment venting systems are installed to prevent damage of the containment vessel from overpressure. Measures to prevent hydrogen explosion are taken by monitoring, evacuation, and recombination of hydrogen gas.

Preventing Dispersion of Radioactivity: Measures to prevent hydrogen explosion in the reactor building are taken as well. Water cannon trucks are equipped for mitigating dispersion of radioactivity in case the containment vessel or the spent fuel pits are damaged. Pollution control screens are installed at the drainage canal exits.

3 What Changed After Fukushima

3.1 All-Hazards and Multiple Disasters

Though the basis of nuclear safety did not change even after Fukushima, some problems have emerged that caused the lack of defense-in-depth. We should learn lessons on these points and reflect them in taking concrete measures for safety enhancement.

Firstly, we must be concerned more about preparedness for all-hazards and multiple disasters than had been. The safety barriers against tsunamis were very fragile, because people in the nuclear industry were so concerned about seismic motion that less attention was paid to the risk of tsunamis. Almost all of the equipment for emergency power supply and emergency water injection were located below the ground level, because the location is the best for protecting them from seismic motion. Such consideration, however, did harm for protecting them from tsunamis. We should have been more concerned about natural disasters other than seismic motion.

The backup systems against station blackout were insufficient, because the reliability of power grid is extremely high in Japan. The industry had made all efforts to maintain the reliability of power grid as high as possible, and they are too confident of it to think station blackout for a long period of time probable before Fukushima. The multiple disasters over a very wide area after Tohoku Earthquake easily denied such expectation and the external power supply from the grid became completely unavailable. Relying just on the quality of power grid is vulnerable in front of such multiple disasters.

Preparedness for all-hazards, unrestricted to natural disasters, is now a critical issue of nuclear safety in Japan after Fukushima. Aircraft crashes and terrorists' attacks should be considered also. Progress of these events may easily exceed the conventional event scenarios, and it is difficult to take preventive countermeasures to achieve prescribed design bases, in particular by installing some hardware equipment. It is therefore unsuitable to cover all these hazards by safety regulation. Meteorite strikes are out of the scope of design bases at present, but some response scenario should be imagined as an unforced activity. What can we do if most of the plant staff are down due to pandemic? Such questions must be asked behind the nominal scene of regulation.

3.2 Administration of Emergency Response

Secondly, we should attend more to the administration of emergency response rather than preventive measures with hardware equipment. While no casualties from radiation exposure have been reported, many people died during or just after evacuation due to improper evacuation planning and operation in Fukushima.

An offsite center, which is expected to be the local headquarter of nuclear emergency response, was constructed in each area of major nuclear facility sites after JCO. But the offsite center in the Fukushima area did not function at all due to the blackout and a high radiation dose. The administrator failed to collect monitoring data of radiation dose and could not use SPEEDI (System for Prediction of Environmental Emergency Dose Information) for decision-making in evacuation planning, in particular for deciding which areas to be evacuated. It is because data necessary for operating SPEEDI could not be transferred from the Safety Parameter Display System (SPDS) at the plant site due to the loss of external power supply. The Nuclear Safety Technology Center, which is an organization under the regulatory body, calculated the likely atmospheric dispersion of radioactive materials using SPEEDI assuming a unit radioactivity release from the Fukushima site and reported the results to relevant organizations. It was recognized afterwards that the calculation results were useful for evacuation planning, but none of the organizations used them due to improper information strategy by the government.

In addition, information sharing was so poor between different organizations such as TEPCO, the central government, Self-Defense Force, police, and the local governments, that evacuation planning and operation were carried out on an ad hoc basis. The most symbolic and miserable case of the consequence from the poor administration was 19 deaths in the evacuee patients from Futaba Hospital. Following the evacuation order, 209 patients who could walk on their own and almost all hospital staff left the hospital boarding five busses dispatched by the town on March 12, but some 130 patients of Futaba Hospital, 98 people staying at the related nursing home, two facility staff, and the hospital director stayed behind. Okuma Town, however, misjudged that the evacuation from the hospital was completed. Two days later, a squadron of the Ground Self-Defense Force Liaison transferred all 98 people from the nursing home and 34 patients from Futaba Hospital to Iwaki-Koyo High School. It took around 11 h due to confusion in deciding the facility to accept these evacuees, and 8 patients died meanwhile. Transfer of the patients remaining at Futaba Hospital delayed for the reported critical situation of the nuclear reactor as well as poor information sharing between the relevant organizations, and the operation was completed early morning on March 17. The delay resulted in additional deaths of 11 patients.

The disaster described above would have been avoided if we had elaborated the administration of emergency response considering accident scenarios that really match the crisis. Different from engineering design of hardware equipment, however, no systematic or technical design methods have been established for the administration of emergency response. Techniques for optimal planning or normative decision-making have been developed in Operations Research and applied to emergency response problems such as evacuation planning. Most of them do not work in ill-structured situations of emergency, because they rely on complete and accurate information to set up mathematical models and obtain solutions. In addition, the conventional mathematical methods cannot deal with organizational interactions, which play a very important role in emergency response as described so far.

Some new approaches of administration design therefore are expected such as agent-based organizational simulation or application of bio-inspired design of complex social systems. Kanno, Morimoto, and Furuta proposed agent-based organizational simulation for emergency response planning [5]. Figure 2 illustrates the proposed simulation architecture of organizational emergency response. The simulation model consists of many agents representing various organizations relevant to emergency response. The scenario manager is a controller, which provides messages on the progress of disaster to the agents following a particular scenario. The simulation system outputs logs of communications, actions and resource consumptions for each agent. One can evaluate the total performance of emergency response by analysing these logs. The time required for executing some task, for example, can be a measure of the effectiveness and efficiency of the task execution.

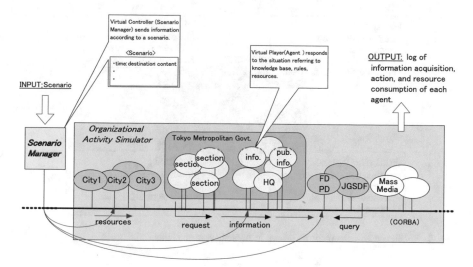

Fig. 2 Simulation architecture of organizational emergency response

Bottlenecks of the system in terms of the workload or information sharing can be identified as well by comparing the amount of executed tasks or processed information by each agent. Comaring simulation results by changing the action rules of agents and the disaster scenario will give us useful insights for rational emergency response planning.

4 Enhancing Resilience

Resilience, which is the ability of a system to absorb changes and to maintain its functionality, has attracted interests of experts in many areas after Tohoku Earthquake and Fukushima. While the conventional safety design of artifacts focuses just on the within design-basis region, resilience sheds light also on the beyond design-basis regions. Resilience of a system is often represented by the speed of recovery from a degraded state of system functionality after a crisis. It is, however, multifaceted features of a system, and Woods enumerated the following four essential characteristics of resilience [6]:

Buffering capacity   the size or kinds of disruptions the system can absorb or adapt to without a fundamental breakdown in performance or in the system's structure;

Flexibility   the system's ability to restructure itself in response to external changes or pressures;

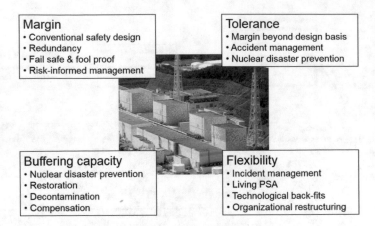

Fig. 3 Resilience enhancement measures in nuclear power plants

Margin   how closely or how precarious the system is currently operating relative to one or another kind of performance boundary;

Tolerance   how a system behaves near a boundary, whether the system gracefully degrades as stress/pressure increase, or collapses quickly when pressure exceeds adaptive capacity.

Following his proposal, Fig. 3 shows a summary of how the safety enhancement measures adopted in Nuclear Power Plants (NPPs) before and after *Fukushima* contribute to enhancing resilience. The conventional within design-basis approaches of safety design contribute to enhancing margin. Accident management is a typical enhancement measure of tolerance for beyond design-basis events. Nuclear disaster prevention appears at two places in this figure, tolerance and buffering capacity. The scale of disasters targeted in the two differs and they correspond respectively to the 4th and 5th level of defense-in-depth. Buffering capacity is related to the recovery process from damaged conditions after a disaster, while flexibility contributes to the improvement above the previous performance level by organizational learning and reengineering. The remedial actions undertaken by the utility companies for satisfying the new regulatory standards can be classified in the same manner.

5   Three Categories of Unsafe Events

Though the basic strategy of nuclear safety has not changed even after Fukushima, now we are requested to deal with a wider scope of events including beyond design-basis. This situation is described in Table 1, where unsafe events that occur in NPPs are classified into three categories.

Table 1 Three categories of unsafe events

	Category 1	Category 2	Category 3
Manifestation	Work accidents or single failures	Systemic or organizational accidents	Design basis events or anticipated incidents
Frequency	Relatively high	Extremely low	Very low
Scale of damage	Local and limited	Medium ~ devastating	Devastating
Primary victims	Interested people	Interested people and/or third party	Interested people and/or third party
Complexity of scenarios	Simple	Complicated and non-linear	Complicated but linear
Variety of scenarios	Diverse but classifiable	Extremely diverse	Limited and finite
Quantitative risk assessment	Statistically possible	Impossible	Theoretically possible
Safety goal	ALARP[a]	ALARP[a]	Absolute risk limit
Countermeasures	Quality assurance and work management	Systems approach	Engineered safety features
Trade-off with economy	Compatible	Partly compatible	Conflicting
Status in nuclear industry	Already resolved	Unresolved	Already resolved

[a]*ALARA* As low as reasonably practicable

The author made the original version of this table shortly after JCO. Category 1 corresponds to unsafe events of relatively a high frequency and low consequence and they do not differ from work accidents in the ordinary industries. The risk of these events can be evaluated statistically and remedial actions are taken in ergonomics and work management. In contrast, Category 3 includes design basis events of a low frequency and high consequence, and they are unique to the nuclear industry. The countermeasures for this category are evaluating their risks theoretically and installing some engineered safety features. Category 2 is a new type of unsafe events that emerged in the past decades. This category includes complex events of systemic or organizational accidents, and they sometimes exceed the design bases.

A locus of interest will be the trade-off of safety measures with economy. Since NPPs are protected from Category 3 events with engineered safety features, most of which are out of service during the normal operation, their enhancement conflicts with the plant economy. In contrast, safety enhancement measures for Category 1 events often contribute also to the improvement of efficiency and productivity of works, and they can be compatible with the plant economy. It differs from the natural image that safety and economy are in a trade-off relationship. Those for Category 2 are located between the both, i.e., if safety enhancement measures using engineered safety features are necessary, the investments are costs. Otherwise, they are sometimes compatible with the plant economy through the improvement in work efficiency.

When this table was created, a term of resilience was unknown among the community of nuclear safety, but now it has become clear that enhancement of resilience contributes to solving a problem how to prevent and mitigate Category 2 events. General principles as well as practical methods, however, to do so are not yet enough established, and resilience engineering should challenge to solve this issue.

6 Conclusions

The Fukushima Daiichi accident was caused mainly by the breach of defense-in-depth against tsunamis, which is the very basis of nuclear safety, and it is unnecessary to substitute it with a new concept. The accident rather showed that defense-in-depth is effective even in unanticipated emergency conditions of beyond design-basis, and the remedial actions now undertaken by the utility companies in Japan are in line with the principle. It must be taken into consideration, however, that the breach occurred due to insufficient preparedness for all-hazards and multiple disasters. The administration of emergency response rather than preventive measures by hardware equipment should be more concerned about than before. Enhancing the resilience and renovating the defense-in-depth of NPPs are crucial and Category 2 unsafe events will be the targets of these efforts.

References

1. Final Investigation Report, *Investigation committee on the accident at the Fukushima nuclear power stations* (2012) http://www.cas.go.jp/jp/seisaku/icanps/eng/final-report.html. Accessed 23 July 2012
2. K. Furuta, K. Sasou, R. Kubota, H. Ujita, Y. Shuto, E. Yagi, Human factor analysis of JCO criticality accident. Int. J. Cogn. Technol. Work **2**(4), 182–203 (2000)
3. S. Kanaida, N. Itou, A. Ueno, Y. Takabayashi, Strategy of installation of permanent and mobile equipment for countermeasures against severe accident at Tokai-2, in *Proceedings of ICMST-Kobe 2014*, Kobe, Japan, Japan Society of Maintenology, 2–5 November 2014
4. H. Uehara, T. Kawamoto, S. Kaneda, K. Ishikawa, K. Numata, Safety measures at Tomari nuclear plant, in *Proceedings of ICMST-Kobe 2014*, Kobe, Japan, Japan Society of Maintenology, 2–5 November 2014
5. T. Kanno, Y. Morimoto, K. Furuta, A distributed multi-agent simulation system of emergency response in disasters. Int. J. Risk Assess. Manag. **6**(4/5/6), 528–554 (2006)
6. D.D. Woods, in *Resilience Engineering: Concepts and Precepts*, ed. by E. Hollnagel, D.D. Woods, N. Leveson (Ashgate, Aldershot, UK, 2006), p. 21–34

Consequences of Severe Nuclear Accidents on Social Regulations in Socio-Technical Organizations

Christophe Martin

Abstract Major nuclear accidents have generated an abundant literature in the social sciences. They are the source of many key concepts that have led to studies of the organization and its links to system safety. Social psychology and sociology have shown that such bodies have their own modes of organization; while resilience engineering has hypothesized that they have the capacity to learn from the past and anticipate potential causes of serious damage. This paper revisits some major contemporary accidents, notably the accident at the Fukushima Daiichi nuclear power plant, through an analysis of the resilience capacity of systems in terms of the sociology of organizations and especially, social regulation.

Keywords Industrial disaster · Social regulation · Resilience engineering · Negotiation · Independence

1 Introduction

The major industrial accidents of the 20th and 21st centuries have been the source of a variety of interpretations. Technological causes initially held the top spot. However, it was subsequently agreed that human factors were a major cause of disaster in these complex technological systems leading, ultimately, to the idea that the causes of accidents could be found at the organizational level. Here, we do not offer an exhaustive overview of work that has modernised performance factors at the organizational level in at-risk industries; instead we present arguments from the sociology of organizations in order to better understand how the day-to-day work of organizations complements or substitutes what is prescribed, either to adapt to operational necessities, or in an emergency. In other words, is post-accident management possible based on "safety in action", which finds its foundations the

C. Martin (✉)
Centre for Research on Risks and Crises (CRC), MINES ParisTech/PSL Research University, Paris, France
e-mail: Christophe.martin@mines-paristech.fr

© The Author(s) 2017
J. Ahn et al. (eds.), *Resilience: A New Paradigm of Nuclear Safety*, DOI 10.1007/978-3-319-58768-4_4

negotiation of prescribed regulations, what de Terssac [1] describes as a consequence of social regulation [2].

The arguments used in this chapter are based on a reinterpretation of major industrial accidents in terms of the sociology of organizations; in particular we aim to establish bridges between knowledge of the organization's operations, and the restructuring of organizational ecosystems during the management of a crisis. We argue that modes of social regulation that enable prescriptive orders to be adapted to the daily work of organizations can play a positive role in the capacity of systems to anticipate and adapt, which in turn creates resilience.

This paper begins with a brief review of some major industrial accidents in order to highlight the main phases of research in the social sciences. It discusses the contribution of the sociology of organizations, particularly the French school of strategic analysis and social regulation, which examines *in fine* the role of social regulation in understanding both operational systems and the post-accident period.

2   Major Industrial Accidents and Changing Paradigms

The major accidents that have occurred over the past four decades have changed the research paradigms used in risk management. They have influenced industry practice both in terms of analytical tools and management culture. The engineering culture that dominated safety decisions opened a door to the humanities and led to the development of cross-cutting approaches that could address system complexity. This section presents a brief history of this evolution.

The industrial accident at Three Mile Island (TMI) was the origin for a profound examination of the organizational dimension of accidents (although it did not lead to work on prescriptive organizational design). Perrow [3] describes complex systems with a high potential for disaster and highlights the systemic dimension of accidents in tightly coupled systems where trivial errors can interact and lead to an unwanted event. However, according to Perrow [3] these systems only concern the 'normal' accident. This unacceptable sociological approach, in a society where risk management is a corollary to technology, was nevertheless, the starting point for the growing interest of sociologists in at-risk organizations.

This appeal to the sociologists of organizations would be reiterated by Reason [4]. Having observed the limits of engineering and cognitive science in understanding the Chernobyl accident, he used theories from sociology in order to understand and track the latent errors that hide at all levels of the system and (using the cancer model), interact with one last operator, resulting in disaster. Reason's well-known 'Swiss Cheese' model would lead to the development of many audit methods that aimed to detect weaknesses in the system. The Tripod method [5] is one example.

Moreover, the Chernobyl accident was the origin of the concept of safety culture [6] and would lead to further work on its definition in both high-risk organizations and industry in general. The importance of the safety culture concept would be widely discussed and the source of many industrial initiatives. This was the case in France, where the creation of the *Institut pour une Culture de Sécurité Industrielle* [Institute for an Industrial Safety Culture] followed the AZF accident on 21 September 2001.

These wide-ranging conceptual developments, which attempted to limit major disasters, are marked by the creation of methodologies for the observation of high operational reliability in organizations, in order to understand their characteristics and eventually design prescriptive operational principles. The emergence of High Reliability Organizations (HRO) [7] in the 1990s was a major advance on Perrow's work and the fatalistic vision of the 'normal' accident. However, despite an unprecedented observation methodology, researchers themselves were forced to admit that it was not possible to develop a theory of HROs, although they constitute an important set of case studies on high-risk, high reliability organizations. Nevertheless, this work has served as the basis for many industrial studies by organizations that want to change and improve their level of safety culture, for example in the oil sector.

In the 2000s, resilience engineering would once again change perceptions of safety systems. Hollnagel [8] argued for the understanding of the day-to-day operation of systems, through the study of system successes rather than failures. This understanding of the capacity of a system to anticipate an accident and to react to adverse events constituted an important development in the management of at-risk systems and major accidents.

The aim of these various currents of research was to provide a better under-standing of at-risk systems during both routine operations and in times of crisis. The work of sociologists would lead to the emergence of established concepts from the sociology of organizations. The next section presents a summary of French research, in particular the school of social regulation, which emphasizes the negotiated dimension of safety systems.

3 The Sociology of Organizations and at-Risk Industries

In the late 1990s, Bourrier [9] carried out a study of American and French nuclear plants. This study would conclude that, far from being HROs, nuclear power plants were normal organizations, given what was known about the sociology of orga-nizations. Specifically, normal organizations are the result of the negotiations and strategies undertaken by their actors. Such organizations may be the source of virtuous ecosystems, although their managers may not be aware of it.

We also found, in our study of the decommissioning of a nuclear plant, that the plant's informal organization may be relevant driver of safety [10].

In the French school of the sociology of organizations, this dimension of an organization that does not fully meet the prescribed, formal requirements of man-agers is well-known. Crozier and Friedberg developed and demonstrated a the-ory concerning the strategies of actors and power relations in organizations [11].

The work of de Terssac [1], particularly following the explosion of the AZF factory in France, refers to the negotiated dimension of safety in relation to social regulation theory. This theory argues that rules can be revisited and that they are the result of negotiations between actors. They structure collective action, while independent initiatives can create conflict with external controls. How the system is regulated becomes the result of compromise and negotiations between these two forms of regulation.

de Terssac [1] highlights the development of everyday safety in a factory, beginning with negotiations between workers and supervisors. He clarifies what he calls "safety manufacture", which does not depend on prescriptive procedures that explain what safe behaviour is, but is the result of rules that are supplemented and negotiated by users. For the author, "safety in action" is the ability to decide whether (or not) to apply a safety rule and adapt it to the context. Different actors in the organization will have different ideas of safety that are linked to their role in the company. Safety culture results from the comparison of these different ideas.

An at-risk organization is not therefore fundamentally different to a normal organization, although it has its own characteristics. The study of such organizations simply considers that during normal operations, what is prescribed has been negotiated and adapted to the situation on the ground, and that these adjustments are part of the daily life of the organization.

It therefore seems appropriate to ask whether maintaining this shared safety culture after a major accident is an element of system resilience. Specifically, do negotiated rules make the organisation better able to anticipate and adapt or, on the contrary, must the organization resort to extremely strict procedures to manage a major disaster?

The Fukushima Daiichi accident required rules to be adapted to the realities of the situation regardless of the procedures to be followed in an emergency. The next section highlights the decisions taken by the plant's Director in the application of the venting procedure and cooling the reactors with seawater. We show that in a post-accident situation, assessments of procedures are a function of the context, notably with respect to the positions occupied by actors.

4   Following the Rules, Post-Accident

In a crisis, where nothing corresponds to any previous situation, it seems foolish to guide behaviour with reference to known procedures. The management of the Fukushima Daiichi crisis showed that certain actions taken by the plant's crisis unit and its Director were taken in the light of their knowledge of the status of the system—and that their understanding was different to that of governmental authorities and TEPCO [12]. During the hearings that followed the accident, the plant's Director stated that technical problems were encountered during the venting procedure that even he was not able to grasp because the crisis unit was too far away from where the action was happening. Therefore, he initially tried to follow instructions from headquarters, given the difficulty of the situation and delays in executing procedures.

> Yes, but at that moment, it was the first time for me as well that I found myself confronted with such a situation, and, to be very honest, I didn't even understand it myself. We didn't yet know the details of the situation on the ground. And in that, we were in the same position as the people at headquarters. Of course, on the ground, they couldn't see the indicators in the control room any more – they were in the dark, all the main instruments were off, but we were under the impression that if they were set to vent, this could happen. Of course, there was no electrical power supply, or air supply, but bizarrely, we were completely convinced that in order to vent all we had to do was open a valve, that if we could open this valve, it would work. We only understood afterwards. The AOV had no air. Naturally, the, MOV did not work either. We wondered if we could do it manually. But there was too much radioactivity for us to go in. And that's where we finally realized how difficult it was. But we could not get the message across to the head office or Tokyo, get them to see how difficult this venting was [12].

Although the order to vent would be repeated by the government, it would be repeatedly delayed because the levels of radioactivity made it impossible to access the valves. The Director then realised the differences between the people at head office and the situation on the ground, and that the order could not be executed. He therefore adapted the procedure, taking into account the state of the system at the time. Later in the hearing, he spoke of the distance that was created between headquarters and plant staff. The same problem also existed at the plant itself— between the crisis unit, the control centre and shift teams who had to manually carry out the venting and who would be exposed to the high levels of radioactivity. It was this distance that led the Director and his team to take important decisions without the approval or authorization of headquarters. These actions included the decision to cool reactors with seawater.

The hearing indicates that preparations were carried out much further upstream than the strict chronology of events would suggest. Knowledge of the system status necessitated the use of a cooling source that was available in large quantities. The only option was the on-site seawater. Independent of any discussions with head-quarters, the plant's staff prepared to execute the order.

> Here, it's not really a case of 'continue'. To be really precise, we began preparations for this seawater injection well before 2:54 p.m. This means that the order to prepare the injection was given well before then. But it was at that time that the preparations were completed and the injection became possible. This is why I gave this order, which was more like an order to implement that an order to prepare, if I remember correctly. Except, this is when the explosion occurred. We could not move to implementation and we ended up back at the beginning. What is clear is that the order to look at how to inject seawater was given at an earlier stage. [12, p. 169].

While TEPCO's management were aware of the intentions of the plant's Director and the crisis unit, they did not take part in any discussions or decisions about pumping procedures or water transport. Only on-the-ground personnel knew what resources were available and how to adapt them to the situation. Furthermore, after an initial attempt, the order was given to suspend the manoeuvre; the Director decided to continue, but did not reveal his decision to headquarters.

So we had ended the test and we were going to stop. It had been decided to stop. It was only me, arriving at this point, I had no intention of stopping the injection of water. Furthermore, they were talking about stopping, but we didn't even know how long it would go on for. They could have said thirty minutes, or more. But stopping with no guarantee of recovery. For me, there was no question of following such an order. I decided to do it my way. So I announced to the people at the crisis table that we would stop, but I quietly took the 'safety' group leader to one side, XXXXX, which was in charge of the injection and I told him that I was going to announce to anyone who would listen that we would stop the injection, but that he, at all costs, must not stop sending water. Then I prepared a report for headquarters to say that we'd stopped [12, p. 188].

This manoeuvre was also hidden from certain members of the crisis unit. This suggests that amongst the network of actors in the field, there were some who would execute orders from the Director, which were not in line with the instructions issued by headquarters. This indicates that the internal authority of the Director was such that members of the safety group would follow his orders rather than instructions from headquarters.

The procedure implemented at this time was therefore based on the capacity to find technical solutions in an emergency situation and networks of actors who shared the Director's beliefs. These networks of actors were responsible for the production of the rules that were applied at the time.

In a crisis, social regulation takes places in compressed time; it is the result of negotiations between headquarters, supervisory and government authorities and independent regulators. The decisions of the Director could only be translated into action with the consent of his team, through a process of negotiation. This is reflected in both the venting procedure (that would be delayed several times for technical and human reasons) and the decision to inject seawater, which was the subject of an internal search for technical solutions and led to the decision to carry on with the action against the orders from headquarters.

5 Discussion

From the perspective of the sociology of organizations, the reinterpretation of major accidents and particularly the accident at the Fukushima Daiichi nuclear power plant leads to questions about respect for rules and procedure in crisis management. We argue that a crisis should not cause the strict application of control regulations that are the result of procedures that were established in advance. Decision-making and the rules that apply should be the result of negotiations between decisions taken by headquarters and independent, on-the-ground regulation that takes into account the context.

An analysis of the in-depth feedback from the Fukushima Daiichi accident suggests that the capacity of the plant's teams to find new solutions to deal with the various problems is wholly characteristic of the HRO as described by Weick and Sutcliffe [13]. In other words, a such organization is able to identify and anticipate failure, overcome *a priori* assumptions, and comply with (or defer to) authority and expertise based on experience and intuition. While all of this may be true, it also seems necessary to understand the negotiation processes and power relations internal and external to the group in order to understand its actions. We argue that the social regulation dimension in a constrained timeframe exists, and is the result of negotiations that enable collective action.

Moreover, it appears that there was a significant bias in the analysis of a decision that was temporarily successful. de Terssac's [1] safety paradox states that it is possible to act safely and still not avoid disaster. This leads us to believe, given the limited rationality of actors, that rules that are negotiated in periods of normal operation or crisis may also lead to disaster (which was the case for the AZF accident in particular).

6 Conclusions

The aftermath of accidents does not prevent social regulation processes, which appear to be constrained by time and the emergency. Negotiations between actors occur despite conflicting interests and value systems—in this case, protecting the population, making decisions in line with international expectations, and protecting equipment and the workforce.

All of these interests are the subject of negotiations that create cooperation (or in some cases conflict) between actors in the system. We are therefore far from the situation where safety in a crisis is governed by universal basic procedures, or the intervention of a providential hero. The resilience capacity of a system is based on its capacity to adapt, and therefore knowledge of the dynamics governing the relationships between its actors.

References

1. G. de Terssac, J. Mignard. Les paradoxes de la sécurité, le cas d'AZF, Coll. *Le Travail Humain, Paris, Presses Universitaires de France* (2011)
2. J.D. Reynaud, Les régulations dans les organisations: régulation de contrôle et régulation autonome. *Revue française de sociologie* (1988)
3. C. Perrow, *Normal Accidents: Living with High Risk Technologies*. Princeton University Press (2011)
4. J. Reason, *Human Error*. Cambridge University Press (1990)
5. J. Reason, A systems approach to organizational error. Ergonomics **38**(8), 1708–1721 (1995)
6. International Atomic Energy Agency, Safety Culture, No. 75-INSAG-4 (1991)
7. K.H Roberts, Some characteristics of one type of high reliability organization. Organ. Sci. **1** (2), 160–176 (1990)
8. E. Hollnagel, D. Woods, N. Leveson, *Resilience Engineering: Concepts and Precepts*. Ashgate Publishing, Ltd. (2007)
9. M. Bourrier, *Le nucléaire à l'épreuve de l'organisation*. Presses universitaires de France (1999)

10. M. de Borde, C. Martin, D. Besnard, F. Guarnieri, Decision to reorganise or reorganising decisions? A first-hand account of the decommissioning of the Phénix nuclear power plant, in *Decommissioning Challenges: An Industrial Reality and Prospects, 5th International Conference* (2013)
11. M. Crozier, E Friedberg, *Actors and Systems: The Politics of Collective Action.* University of Chicago Press (1980)
12. F. Guarnieri, S. Travadel, C. Martin, A. Portelli, A. Afrouss, *L'accident de Fukushima Daiichi. Le récit du directeur de la centrale.* Volume 1: L'anéantissement. Presses des Mines (2015)
13. K. Weick, K. Sutcliffe, *Managing the Unexpected: Resilient Performance in an Age of Uncertainty.* Wiley (2011)

Part II
Measurement of Damages

A Multiscale Bayesian Data Integration Approach for Mapping Radionuclide Contamination

Haruko Murakami Wainwright, Masahiko Okumura and Kimiaki Saito

Abstract This chapter presents a multiscale data fusion method to estimate the spatial distribution of radiation dose rates at regional scale around the Fukushima Daiichi nuclear power plant. We integrate various types of radiation measurements, such as ground-based hand-held monitors, car-borne surveys, and airborne surveys, all of which have different resolutions, spatial coverage, and accuracy. This method is based on geostatistics to represent spatial heterogeneous structures, and also on Bayesian hierarchical models to integrate multiscale, multitype datasets in a consistent manner. Although this approach is primarily data-driven, it has great flexibility, enabling it to include mechanistic models for representing radiation transport or other complex processes and correlations. As a first demonstration, we show a simple case study in which we integrate two datasets over Fukushima City, Japan: (1) coarse-resolution airborne survey data covering the entire city and (2) high-resolution ground-based car-borne data along major roads. Results show that the method can successfully integrate two datasets in a consistent manner and generate an integrated map of air dose rates over the domain in high resolution. A further advantage of this method is that it can quantify estimation errors and estimate confidence intervals, which are necessary for modeling and for robust policy planning. In addition, evaluating correlations among different datasets provides us with various insights into the characteristics of each dataset, as well as radionuclide transport and distribution. The resulted maps have started being used by local governments to plan the residents' return, and they are expected to be used for additional policy decisions in the future such as decontamination planning.

Keywords Radiation dose rate mapping · Bayesian hierarchical models · Geostatistics

H.M. Wainwright (✉) · K. Saito
Lawrence Berkeley National Laboratory, 1 Cyclotron Road, Berkeley, CA 94720, USA
e-mail: hmwainwright@lbl.gov

M. Okumura
Japan Atomic Energy Agency, 2-2-2 Uchisawai-Cho, Chiyoda, Tokyo 100-0011, Japan

© The Author(s) 2017
J. Ahn et al. (eds.), *Resilience: A New Paradigm of Nuclear Safety*,
DOI 10.1007/978-3-319-58768-4_5

1 Introduction

Radiation measurements and monitoring in the region around the Fukushima Daiichi nuclear power plant (NPP) have been performed continuously since the accident [1, 2]. Such mapping is essential for protecting the public, guiding decontamination efforts, estimating the amount of decontamination waste, and also in planning the return of evacuated residents. Radiation measurements have been conducted using various techniques such as portable hand-held monitors, car-borne surveys, and airborne surveys. Soil samples have been collected to assess the extent of contamination in the terrestrial environment [3].

Despite such large-scale and continuous efforts, there are still significant challenges in mapping the radiation dose rates and radionuclide contamination. Detailed ground-based measurements have revealed that the radiation dose rates and contamination are both quite heterogeneous, often with many hotspots [4]. Although many datasets are becoming available, it has been difficult to integrate those datasets, since each type of data has a different level of accuracy and represents a different support scale (i.e., spatial coverage and resolution). For example, although ground-based car-borne data provide high-resolution air dose rates, car-borne data are limited to the locations along roads [5]. Airborne surveys have been extensively used to map dose rates in the regional spatial coverage (e.g., 100 km radius) [6]; such data are, however, known to exhibit some discrepancies with co-located ground-based measurements. These discrepancies result mainly from the differences in support volume, since airborne measurements represent the average dose rate over a much larger area (typically a several-hundred-meter radius) than ground-based measurements ($\sim$ several tens of meters).

In environmental science, monitoring and spatial-temporal mapping of various properties—such as CO_2 concentration, wind velocity or reactive transport properties in subsurface—have been the focus of extensive research. Although many traditional datasets have been sparse in time and space, more recently available datasets can cover large areas, such as remote sensing data in atmospheric/terrestrial sciences and data from geophysical techniques in subsurface science. Such datasets, however, are known to have some discrepancy with traditional point measurements, because they tend to have a larger support volume (or lower resolution), such that each pixel represents the average of heterogeneous properties in the vicinity. Various approaches have been proposed to integrate remote-sensing or geophysical datasets with traditional point measurements [7–9]. Many of them are based on geostatistics, a powerful tool for characterizing spatial heterogeneity (or correlation) structure based on available datasets [8–10]. In addition, a Bayesian framework is often used to integrate different datasets consistently and also to quantify the uncertainty associated with the estimated maps [7, 9].

In this study, we develop a Bayesian data-integration approach to estimate the spatial distribution of air dose rates and radionuclide contamination in high resolution across the regional scale (several kilometers to several tens of kilometers). We integrate various radiation measurements, with particular focus on airborne and ground-based measurements. Geostatistical approaches are used to identify spatial correlations and represent small-scale heterogeneity that is not resolved in the coarse-resolution airborne data. We employ a Bayesian hierarchical model for integrating the low-resolution airborne data and sparse ground-based data. We demonstrate our approach using the airborne and car-borne datasets collected in Fukushima City, Japan, in 2012. This integration aims to provide a more resolved and integrated dose-rate map, and also quantify the uncertainty associated with the map for modeling and for policy planning. This approach is now being used to map the radiation dose rate in the Fukushima region, particularly in the evacuation zone [11, 12].

2 Methodology

Our approach is based on a Bayesian hierarchical model [7, 8], which typically consists of three types of statistical submodels: (1) data models, (2) process models, and (3) prior models. The process models describe the spatial pattern (or map) of dose rates within the domain, given the parameters defined by the prior models. Geostatistical models are often used as process models based on the spatial heterogeneity structure identified by available datasets. The data models connect this pattern and the actual data, given measurement errors. These data models can represent, for example, a direct ground-based measurement or a function of the pattern—for example, spatial averaging over a certain area for a low-resolution airborne dataset. The prior models determine the distributions or ranges of the parameters based on pre-existing information. The overall model—a series of statistical submodels—is flexible and expandable, able to include complex correlations (such as correlations with land use, soil texture, or topography) or various observations. Once all the submodels are developed, we can estimate the parameters, as well as the radiation map and its confidence interval, using sampling methods or optimization methods. To fully quantify the uncertainty, here we will use the Markov-chain Monte-Carlo method. When the domain size and the number of pixels are large, optimization-based methods will be used to obtain the mean estimate and its asymptotic confidence intervals.

In this chapter, we show one simple example; the integration of airborne and car-borne radiation measurements. We assume that each point of the airborne measurements is the weighted average of radiations from radionuclides distributed over the ground. To develop an integrated map, we denote the radiation dose rate at i-th pixel by y_i, where $i = 1,...,n$. We also denote two vectors, representing the airborne data z_A (each data point is represented by $z_{A,j}$, where $j = 1,...,m_A$) and car-borne data z_V (each data point is represented by $z_{V,j}$, where $j = 1,...,m_V$). The goal is to estimate the posterior distribution of the radiation dose-rate map y (i.e., the vector representing the radiation dose rate at all the pixels) conditioned on two datasets (z_A and z_V), written as $p(y|z_A, z_V)$. By applying Bayes' rule, we can rewrite this posterior distribution as:

$$p(\mathbf{y}|\mathbf{z}_A, \mathbf{z}_V) \propto p(\mathbf{z}_A \mid \mathbf{y}) p(\mathbf{y} \mid \mathbf{z}_V). \tag{1}$$

The first conditional distribution $p(z_A \mid \mathbf{y})$ is a data model representing the airborne data as a function of the dose-rate distribution $\mathbf{y}$. We assume a spatial weighted averaging function of the dose rate map:

$$z_{A,j} = \sum_{i \in C_j} w_{i,j} y_i + \varepsilon_j. \tag{2}$$

where C_j represents the pixels within the spatial averaging range, $w_{i,j}$ is the weight determined by the distance between i-th pixel and j-th airborne data point, and ε_j is an error associated with each data point. We assume an inverse square distance function for the weights, $w_{i,j}$. The weight can be computed by radiation transport equations [11].

The range is typically considered to be equal to the flight height, following Torii et al. [6]. We assume that the error ε_j includes not only measurement errors associated with hardware (such as instrument noises) but also the uncertainty associated with other factors (such as the variable height of buildings, small-scale spatial variability). In this example, we assume that ε_j follows an independent normal distribution with zero-mean and the error variance σ_A can be determined from the correlation analysis between two datasets.

As a process model, we assume that $\mathbf{y}$ is a multivariate Gaussian field described by geostatistical parameters. For simplicity, we assume that the ground-based measurements have insignificant errors compared to the airborne measurements, and hence we use the ground measurements as conditional points to constrain the distribution of $\mathbf{y}$ as $p(\mathbf{y}|\mathbf{z}_V)$. Since two conditional distributions are both multivariate Gaussian, we can derive an analytical form of this posterior distribution as a multivariate normal distribution with mean $Q^{-1}g$ and covariance Q^{-1}, where $Q = \Sigma_c^{-1} + A^T D^{-1} A$ and $\mu = \mu_c + A^T D^{-1} z_A$ [7]. In Q and g, μ_c and Σ_c are the conditional mean and covariance given the ground-based data (z_V) and geostatistical parameters. D is the data-error covariance matrix; each of the diagonal components is σ_A. The matrix A is n-by-m_A sparse matrices, where $A_{ij} = w_{ij}$ if i-th pixel is within the range C_j; otherwise A_{ij} is 0. We may directly sample $\mathbf{y}$ or estimate the mean (or expected) dose map by $Q^{-1}g$. Although the example shown here is quite simple, we may add more complexity, such as, for example, physics-based radiation transport models (instead of weighted averaging) to represent the airborne data or data-derived correlations between dose rates and land use or topography [2].

3 Demonstration

In this example, we applied the developed method to the air-dose-rate data collected in Fukushima, Japan. We used the ground-based car-borne data from the second car-borne survey (December 5–28, 2011; http://radioactivity.nsr.go.jp/en/contents/

5000/4688/24/255_0321_ek.pdf), as well as the airborne data from the fourth airborne monitoring survey (October 22–November 5, 2011; http://radioactivity.nsr.go.jp/en/contents/4000/3179/24/1270_1216.pdf). We assume that the effect of radiocesium decay is negligible between the two surveys. The airborne data were processed and converted to the values equivalent to the dose rate one meter above the ground surface.

We first compare the co-located data values of the car-borne datasets to the airborne datasets. Direct comparison (Fig. 1a) shows significant scatters in the higher dose region, although the datasets are clearly correlated (the correlation coefficient is 0.78). The car-borne data have larger variability (larger variance), suggesting that small-scale variability is averaged out in the airborne data. When we take into account the weighted spatial average for the airborne data (Fig. 1b), the correlation improves significantly (the correlation coefficient is 0.84). We would note that the airborne data values are systematically higher than the car-borne ones, with and without spatial averaging. There are several possible reasons for such a shift: (1) there could be calibration issues in the airborne data, and (2) the center of roads (where the car-borne data are collected) is known to have lower contamination than the side of the roads or undisturbed land. For demonstration purposes, we assume that the car-borne data are still accurate representation of the radiation map in this example.

Using the correlation between the airborne and car-borne datasets that we found in Fig. 1, we determined the error variance σ_A a well as the shift factor. In addition, geostatistical parameters were estimated based on the variogram analysis of the car-borne datasets, representing the spatial correlation structure of small-scale heterogeneity. Figure 2a shows the airborne survey data in the part of Fukushima; the eastern portion of the domain has higher dose rates, possibly because the area

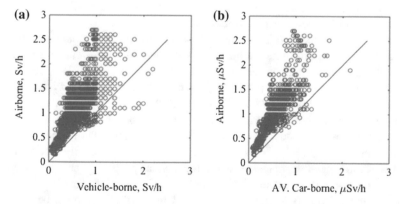

Fig. 1 Comparison between the car-borne data and airborne data: **a** direct comparison of data values and **b** including spatial averaging for the airborne data

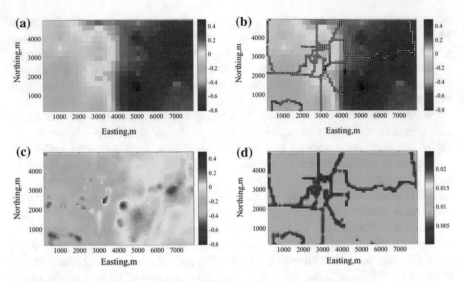

Fig. 2 **a** Airborne dose-rate data over Fukushima City (December 2011), **b** car-borne data (colored circles) over the airborne data (colored map), **c** the estimated integrated dose-rate map (mean field) based on the developed data integration, and **d** the estimation variance. In all the plots, the data values are log-transformed

lies along the initial plume direction and is also a forested area. By overlaying the car-borne and airborne data (Fig. 2b), we see that the car-borne data show smaller-scale variability than the airborne data, and that the airborne data overestimates the air dose rate. The estimated map (mean field) from the data integration in Fig. 2c shows more detailed and finer-resolution heterogeneity than the original airborne data (Fig. 2a). The systematic shift in airborne data was also corrected. As shown in Fig. 2d, the estimation variance is smaller near the car-borne data points, since the model includes spatial correlation.

Figure 3 shows the validation result to evaluate the performance of the data integration and the dose-rate estimation. One hundred of the car-borne data are excluded from the estimation, and used for validation purposes. Without the data integration, the airborne data (blue dots) have large scatters and a systematic shift compared to the car-borne measured data. After the data integration, the predicted values (based on the both airborne and car-borne data at other locations) are tightly distributed around the one-to-one line and are mostly included in the 95% confidence interval. Figure 3 shows that this method successfully estimates the fine-resolution dose-rate map based on the spatially sparse car-borne data and coarse-resolution airborne data. Having such a confidence interval would be useful for practical applications, such as estimating the range of the potential health effects or estimating the decontamination waste volume.

Fig. 3 Comparison between the predicted and measured air dose rates (log-transformed) at the car-borne data locations not used for the estimation. The *red dots* represent the predicted values based on the data integration method; the *blue dots* are the airborne data before the integration. The *blue line* is the one-to-one line; the *red lines* are the 95% confidence intervals

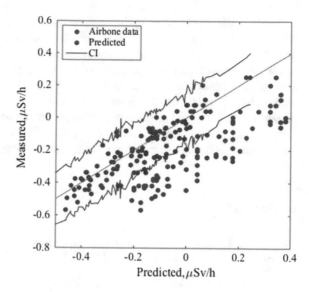

4 Summary and Future Work

In this chapter, we described a multiscale hierarchical Bayesian method for integrating multiscale, multitype dose-rate measurements. As an example, we illustrated how this method could be used to integrate coarse-resolution airborne data and fine-resolution (but sparse) car-borne data in a consistent manner, with the estimation uncertainty quantified. Although the current example model is still simple, results have suggested that the effective combination of ground-based data and airborne data could provide detailed and integrated maps of radiation air dose rates at regional scale around the Fukushima Daiichi NPP. In addition, this method could quantify estimation errors or confidence intervals, representing the uncertainty associated with the integrated maps. We also showed that statistical analyses could provide various insights into both the characteristics of each dataset and the spatial trend of contamination, which would be useful for predicting future radiation levels at the regional scale.

Further improvement was made to improve the estimation approach by including other information, such as the correlations between dose rates and land use and/or topography [11]. Physics-based radiation transport models are used to replace the spatial averaging function to accurately represent airborne data [11]. In the future, spatiotemporal integration—by integrating spatially sparse but continuous-time monitoring data and temporally sparse but spatially extensive data, such as airborne data—will be carried out to provide a detailed map of the air dose rate and radionuclide contamination at regional scale, at any given location and time, including their confidence interval.

These integrated maps of radiation dose rates has started being used in the local governments and agencies to plan the return of residents [12]. These maps can be used to provide additional important measures with the estimates of upper and lower bounds such as the expected cost and waste amount from decontamination. Having the confidence intervals or the upper/lower bounds of those measures will be helpful for planning the decontamination in more robust manner and also preparing for the worst-case scenarios. In addition, the detailed map and its confidence interval are critical to inform the public of the contamination level as well as the progress of decontamination for better decision-making.

References

1. K. Saito, Y. Onda. Outline of the national mapping projects implemented after the Fukushima accident. J. Environ. Radioactivity **139**(C), 240–249 (2015). doi:10.1016/j.jenvrad.2014.10.009
2. S. Mikami, T. Maeyama, Y. Hoshide, R. Sakamoto, S. Sato, N. Okuda et al. The air dose rate around the Fukushima Dai-ichi Nuclear Power Plant: its spatial characteristics and temporal changes until December 2012. J. Environ. Radioactivity **139**(C), 250–259. (2015). doi:10.1016/j.jenvrad.2014.08.020
3. K. Saito, I. Tanihata, M. Fujiwara, T. Saito, S. Shimoura, T. Otsuka et al, Detailed deposition density maps constructed by large-scale soil sampling for gamma-ray emitting radioactive nuclides from the Fukushima Dai-ichi nuclear power plant accident. J. Environ. Radioactivity **139**(C), 308–319. (2015). doi:10.1016/j.jenvrad.2014.02.014
4. Japan Atomic Energy Agency, Establishing the methodology to understand the long-term impact of radionuclides released during the Fukushima Daiichi nuclear power plant accident, JAEA Report (2012), http://fukushima.jaea.go.jp/initiatives/cat03/entry05.html. (in Japanese)
5. S. Tsuda, T. Yoshida, M. Tsutsumi, K. Saito, Characteristics and verification of a car-borne survey system for dose rates in air: KURAMA-II. J. Environ. Radioactivity, **139**(C), 260–265 (2015). doi:10.1016/j.jenvrad.2014.02.028
6. T. Torii, Y. Sanada, T. Sugita, A. Kondo, Y. Shikaze, Y. Urabe, Investigation of radionuclide distribution using aircraft for surrounding environmental survey from Fukushima Daiichi nuclear power plant. JAEA-Technology 2012-036, (2012) (in Japanese)
7. C.K. Wikle, R.F. Milliff, D. Nychka, L.M. Berliner, Spatiotemporal hierarchical Bayesian modeling: tropical ocean surface winds. J. Am. Stat. Assoc. **96**(454), 382–397 (2001)
8. H.M. Wainwright, D. Sassen, J. Chen, S.S. Hubbard, Bayesian hierarchical approach for estimation of reactive facies over plume-scales using geophysical datasets. Water Resour. Res. (2014)
9. Y. Zhou, A.M. Michalak, Characterizing attribute distributions in water sediments by geostatistical downscaling. Environ. Sci. Technol. **43**(24), 9267–9273 (2009)
10. P. Diggle, P.J. Ribeiro, *Model-Based Geostatistics*. Springer Science & Business Media (2007)
11. H.M. Wainwright, J. Chen, A. Seki, K. Saito, A multiscale Bayesian data integration approach for mapping radionuclide contamination in the regional scale. J Environ Radioact. **167**, 62–69 (2017a)
12. H.M. Wainwright, A. Seki, J. Chen, K. Saito, A multiscale Bayesian data integration approach for mapping air dose rates around the Fukushima Daiichi Nuclear Power Plant. Proceeding of WM2017 Conference, March 5–9, 2017, Poenix, Arisona, USA (2017b)

Challenges for Nuclear Safety from the Viewpoint of Natural Hazard Risk Management

Tatsuya Itoi and Naoto Sekimura

Abstract Lessons learned from the Fukushima Daiichi NPP accident and challenges for enhancement of the concept of nuclear safety are summarized from the viewpoint of risk management as well as the concept of defense in depth, for the protection against natural hazards, i.e., design against natural hazards and emergency response combined with regional disaster prevention and mitigation. The concept of resilience is also discussed, as a means for refining the fundamental concept of nuclear safety.

Keywords Fukushima Daiichi NPP accident · Earthquake · Tsunami · Nuclear safety

1 Introduction

It is well understood among the public as well as engineers that the use of nuclear energy involves potential risk associated with accidents. This is clearly recognized by experiences in which the potential risk has become obvious, e.g., during the Fukushima Daiichi NPP accident. Analysis of the experiences before, during, and after the accident is considered essential to discuss the safety of nuclear power in the future. In this chapter, future challenges addressed by several reports on the Fukushima Daiichi NPP accident are summarized from the viewpoint of natural hazard risk management. It is also discussed that the fundamental concept of nuclear safety, i.e. defense in depth, can be enhanced by introducing the concept of risk as well as resilience.

T. Itoi (✉) · N. Sekimura
The University of Tokyo, Tokyo, Japan
e-mail: itoi@n.t.u-tokyo.ac.jp

N. Sekimura
e-mail: sekimura@n.t.u-tokyo.ac.jp

© The Author(s) 2017
J. Ahn et al. (eds.), *Resilience: A New Paradigm of Nuclear Safety*,
DOI 10.1007/978-3-319-58768-4_6

67

2   The Great East Japan Earthquake Disaster and the Fukushima Daiichi NPP Accident

The 2011 Tohoku earthquake and tsunami caused 19,335 deaths and 2,600 people are still missing [1]. Spatially distributed damage is also the characteristic of the earthquake and tsunami. Almost a hundred thousand people were forced to evacuate as a result of the Fukushima Daiichi NPP accident, whereas almost 300,000 people in total were evacuated in the aftermath of the earthquake and tsunami. In terms of fatalities, a number of people died related to the Fukushima Daiichi NPP accident, e.g., two plant workers due to tsunami, and more than 20 hospital patients during and after evacuation, though no people died because of radiation effects due to the release of radioactive material. Large negative impacts to society have also been caused by the nuclear accident as well as by the earthquake and tsunami. Some have pointed out that the evacuation orders following the nuclear accident prevented rescue activities for people under collapsed houses in the areas surrounding the Fukushima Daiichi and Daini NPPs, which may have caused more fatalities.

3   Challenges Identified in Light of the Fukushima Daiichi NPP Accident

Risk management is a process that consists of identification, analysis, evaluation, treatment, and monitoring of risk. If risk is evaluated to be high, it has to be reduced by introducing a countermeasure to be retained. A contingency plan is also needed to prepare for the retained risk, if it should be realized. Risk analysis can be effectively used only if it is organically integrated into a risk management process. In our society, however, risk analysis of nuclear power plants, i.e., estimation of the probability that an accident will occur, tends to be used only to judge whether the risk is acceptable or not. Afterwards, simply speaking, nuclear plant operators together with regulators may fail to prepare a contingency plan in the case when the risk becomes obvious, and may also fail to implement an effective mechanism to continuously manage risk.

3.1   Risks of Nuclear Power Plant Accidents

Conventionally, risk R has been defined as the mean value of the possible adverse consequences, i.e., consequence times its frequency, as follows:

$$R = \sum_i C_i P_i \tag{1}$$

where, C_i is the consequence and P_i is the probability of occurrence of C_i for the i-th scenario. This definition of risk, however, represents only one aspect of risk, which is complex in nature.

It is described in ISO 31000 [2] that organizations face internal or external factors and influences that make it uncertain whether and when they will achieve their objectives. The effect this uncertainty has on an organization's objectives is defined as "risk." The objective of nuclear safety is to protect people, individually and collectively, as well as the environment from the harmful effects of ionizing radiations [3]. Therefore, the risks of nuclear power plant accidents are the effect of uncertainty in the various predecessors of accidents, e.g., earthquakes, fires, flooding and human errors, etc., on the objective of nuclear safety.

Analysis of risk is to attempt to envision what will happen if a certain course of action, including inaction, is taken [4]. Therefore, risk is defined as an answer to the following questions [4]:

- What can go wrong? (Scenario)
- How likely is it? (Likelihood)
- What might its consequences be? (Consequence)

The importance of the scenario, in addition to the consequence and the likelihood (or frequency), is more emphasized in this definition to describe the characteristics of risk. It is also emphasized that the risk information should include information about what is within/out of scope and how uncertain the result of the risk assessment is. The consequences that are assessed by risk assessment are not limited to fatalities due to radiation exposure, but include other consequences related to quality of life, e.g., environmental damage. It should also be noted that the actual fatalities related to nuclear power plant accidents are not always limited to radiation effects but may include other factors as discussed above, which should be included in the scope of the risk assessment of nuclear power plants.

3.2 Risk-Informed Decision Making

The purpose of probabilistic risk assessment is not limited to discussions whether a certain nuclear power plant is safe enough based on the estimated value of accident occurrence probability. We can also identify the weak points of the plant and contribute to an improved quality of decision making related to the introduction of safety-enhancement measures. Examples of required decision-making qualities are reasonability, accountability, openness, and transparency.

On the other hand, assessment of the risk due to external factors, e.g., earthquakes, fires, flooding, and aircraft impact, is hampered by an inherently large uncertainty. Therefore, an integrated framework to deal with these kinds of risks, i.e. a decision-making process under large uncertainty [5], is essential to ensure the safety of nuclear power plants now and in the future. Key elements for the integrated decision-making process include deterministic consideration, e.g., defense in depth, good practices, operating experiences, and organizational considerations [5], some of which are discussed briefly below.

Uncertainty due to lack of knowledge, e.g. uncertainty related to modeling or insufficient amount of data, is called "epistemic uncertainty", which is distinguished from inherent randomness, i.e. aleatory uncertainty. Appropriate methodologies to analyze risk depend on the magnitude of epistemic uncertainty. Epistemic uncertainty is related to the amount of knowledge about accident scenarios and their consequences as well as their probability.

It has been discussed, e.g. by Stirling [6], that decision makers who receive the results of risk assessment as well as engineers who provide the results of risk assessment have a tendency to simplify and trivialize the characteristics of risk so as to make them easier to analyze and evaluate. However, when considering the basic purpose of risk analysis, i.e. to contribute to rational decision making, the result of such simplified risk assessment may be one of reasons that the decision is wrongly distorted. To avoid this distortion, it is recommended to use various methodologies in addition to probabilistic risk analysis, e.g., sensitivity analysis, deliberation among experts, and enhancement of diversity as well as resilience capacity, to tackle the entire range of risk including ignorance.

3.3 Defense in Depth

"Defense in depth" is an approach to designing and operating nuclear facilities that prevents the occurrence and mitigates the consequences of accidents. The approach consists of multiple levels of defense to compensate for potential failures [7]. Table 1 shows the objectives and essential means for each of five levels of defense defined by IAEA INSAG [8]. Defense in depth is considered to be the fundamental concept to achieve nuclear safety under uncertainty, especially for installation of nuclear power plants, and is regarded still important after the Fukushima Daiichi NPP accident. Meanwhile, it is sometimes said that the Fukushima Daiichi NPP accident, which occurred due to a severe natural event, is a challenge to the defense in depth concept. It was conventionally considered that natural hazard risks can be avoided through appropriate siting criteria and conservative design, i.e. up to the first level of defense in depth, without much consideration of higher levels of defense in depth, though it is discussed in IAEA INSAG-25 that the concept of defense in depth includes consideration of external hazards. A typical example is that the accident management introduced at NPPs in Japan was prepared mainly for accidents from internal event, i.e., random failure of components, and consideration of external events in case of accident management was not discussed before the Fukushima Daiichi NPP accident. It can be said that there was no effective mechanism to evaluate and reduce the risk from both internal and external events continuously in Japan.

Table 1 Objectives and essential means of defense in depth (IAEA, 1996)

Level of defense in depth	Objective	Essential means
Level 1	Prevention of abnormal operation and failures	Conservative design and high quality in construction and operation
Level 2	Control of abnormal operation and detection of failures	Control, limiting and protection systems and other surveillance features
Level 3	Control of accidents within design basis	Engineered safety features and accident procedures
Level 4	Control of severe plant conditions, including prevention of accident progression and mitigation of the consequences of severe accidents	Complementary measures and accident management
Level 5	Mitigation of radiological consequences of significant releases of radioactive materials	Off-site emergency response

It is natural and reasonable, even if the Fukushima Daiichi NPP accident had not occurred, that external events including natural events should be considered among the main factors for nuclear power plant accidents. In case of natural events, e.g., earthquakes and tsunamis, simultaneous occurrence of damage to many structures, systems and components, both on-site and off-site, must be considered to prepare for accident management and for off-site emergency response to the conditions that are most likely to occur. Probabilistic risk assessment is considered one of the effective approaches for estimating reasonable and realistic plant conditions.

The fourth and fifth levels of defense in depth are the on-site and off-site response, respectively, to mitigate the impact of the accident. Though these measures are not hardware-oriented but management-based, it should be assumed that structures, systems, and components, e.g., mobile equipment for accident management or access routes to a nuclear power plant from off-site, related to the fourth and fifth levels of defense in depth, may become unavailable due to prior damage to these components related to the second and third levels of defense in depth. This is because usually the on-site and off-site facilities other than the reactor building are designed to resist natural events at a smaller scale than the reactor itself; this is discussed in the following section. Accident management and emergency response plans should be prepared to be effective under these kinds of severe conditions.

3.4 Design Against External Hazards Such as Earthquakes

Hardware design of nuclear power plants plays a central role in their systems safety. Major direct factors of the Fukushima Daiichi NPP accident are considered to be lack of preparation for beyond-design events as well as underestimation of design tsunami height, i.e., both lack of contingency planning and deficiency in design. The former issue is related to the importance of identification and resolution of the "cliff edge effect."

There are two background factors that led to the underestimation of design tsunami height. One is an insufficient understanding of the importance of following the most up-to-date scientific knowledge, where paradigms occasionally shift dramatically, as our knowledge of natural hazards is expanded; the other is an over-insistence on valid historical evidence to take preventive action. Design earthquake ground motion and tsunami are evaluated for supposed active faults and subduction zone earthquakes. For subduction zone earthquakes, the supposed earthquake characteristics are determined based on the historical records for only several hundreds of years, while geological data over a longer period of time can be available for active faults.

Table 2 shows a reference probability level of design ground motions for different categories of structure, such as NPP and ordinal civil structure. For the earthquake resistant design of NPP, an earthquake that has not been experienced in history has to be assumed in some cases, because the reference probability level for NPP design is quite small, one-tenth of that for ordinary civil structures and sometimes smaller than that for disaster preparation for a nation.

It may also be added that nuclear power plant design against external events such as earthquakes tends to focus on prevention of component failure. From the viewpoint of implementation of defense in depth, however, the concept of systems

Table 2 Examples of reference probability level for earthquake-resistant design

		Annual probability of exceedance	Cf. Exceedance probability in 50 years (%)
Design ground motion of NPP	Level 1	10^{-2} (mean) (IAEA)	40 (mean) (IAEA)
	Level 2	10^{-4}–10^{-3} (mean) (IAEA) 10^{-5}–10^{-4} (median) (IAEA)	0.5–5 (mean) (IAEA) 0.05–0.5 (median) (IAEA)
Design ground motion for ordinal civil structure	Service ability limit state	1/500-1/25 (AS/NZ) 1/50-1/20 (Japan)	5–86 (AS/NZ) 63–92 (Japan)
	Ultimate limit state	1/2500 (US) 1/2500-1/250 (AS/NZ) 1/500-1/1000 (Japan)	2 (US) 2–20 (AS/NZ) 5–10 (Japan)
Cf. Regional disaster prevention & mitigation		$<10^{-3}$ (Japan)	<5 (Japan)

design, i.e. performance-based design, related to the second and third levels of defense in depth should be more emphasized, to control abnormal operations or accidents effectively when external events occur.

3.5 Accident Management

Provisions for management of severe accidents are required for the fourth level of defense in depth. For nuclear power plants in Japan, accident management was planned and introduced as a countermeasure for severe accidents around 2000 by operators as voluntary basis without regulatory requirements. Reports on probabilistic risk assessment for internal events were published in 2002 to confirm the effectiveness of introducing severe accident countermeasures. However, it has been recognized among experts that the main source of risk is not from internal events but from external events such as earthquakes. Therefore, a standardized method was prepared and published as AESJ (Atomic Energy Society of Japan) standard for seismic risk assessment by 2007. Risk assessment due to natural hazards, i.e. individual plant examination for external events, was not published for each specific plant in Japan before the Fukushima Daiichi NPP accident in 2011. Accident management was not yet reinforced to suppose natural hazards by continuous efforts.

3.6 Regional Disaster Prevention/Mitigation

It is of significant concern for the public whether they can survive, e.g. by successfully evacuating in case of a nuclear accident. For this purpose, we need to provide information on the likelihood, timing, and possible amount of radioactive material released for possible accidents. Because all units may suffer from identical external events, multi-unit risk assessment is necessary considering the disturbance of on-site and off-site activities related to mitigate the consequences of the accident, as was observed in the hydrogen explosion in the Fukushima Daiichi accident.

Considering off-site emergency response, it is critically important that we recognize that off-site facilities may suffer damage from a nuclear accident due to natural events occurring simultaneously. There are many kinds of possible interactions. A first point of consideration is the difficulty for local residents to evacuate and also the difficulty for external organizations to provide support to the nuclear site, due to the spatial distribution of damage to the infrastructure (see the examples of damages for surface transportation around the Fukushima Daiichi NPP shown in Figs. 1 and 2) and other factors. A second point of consideration is that the rescue activity for people affected by the severe natural event may be disturbed because of the forced evacuation due to nuclear accident, such as when a rescue team is forced to leave the site. Additionally, people are discouraged from multiple damages repeatedly by the natural event as well as by the nuclear accident.

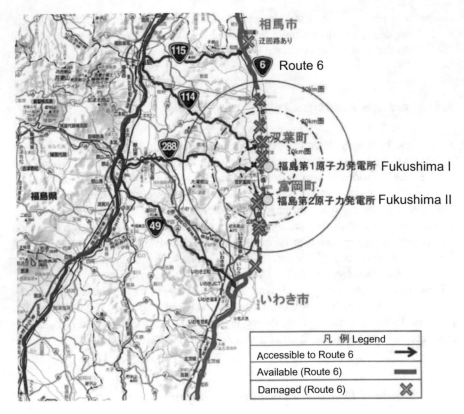

Fig. 1 Location of damage along Route 6. http://www.thr.mlit.go.jp/road/jisinkannrenjouhou_110311/dourohisaijyoukyou.pdf

Fig. 2 Collapse of road surface due to ground motion (Route 6). http://www.thr. mlit.go.jp/road/ jisinkannrenjouhou_110311/ dourohisaijyoukyou.pdf

4 Resilience in the Field of Nuclear Safety Engineering

The term "resilience" is understood among nuclear engineers as a concept for enhancing nuclear safety. It is, however, used to describe a wide range of perspectives. Consensus is required about the meaning and the role of the term in the field of nuclear safety engineering. In this chapter, the role of resilience is discussed from the viewpoint of enhancing nuclear safety.

The concept of resilience is considered important when dealing with risk under large uncertainty as discussed above. The concept of resilience is not introduced when risk is simply regarded as the possibility that something untoward may happen, but when it is regarded as something whose occurrence is rare but inevitable, to be managed when it in fact does occur. In such case, the importance of understanding the characteristics of the scenario when the risk becomes obvious is more emphasized, including the temporal sequence. Conventionally, as discussed heretofore, defense in depth is fundamental to nuclear safety, and was gradually refined to include lessons learned from the Three Mile Island accident as well as the Chernobyl accident. Resilience is considered to be a concept that can further refine and enhance the concept of defense in depth.

4.1 Resilience Engineering for Possible Future Nuclear Accident

Resilience can be defined as the ability to prepare for and plan for, absorb, recover from, or more successfully adapt to actual or potential adverse events [10]. In this sense, resilience is a concept that is relevant in the context of emergencies, such as nuclear accidents. Figure 3 schematically shows the accident sequence with respect to time, from occurrence to conclusion of nuclear accidents, all of which are in the scope of resilience engineering. The vertical axis of the figure shows the function, i.e. malfunction of barrier in each level of defense in depth, while the horizontal axis represents time. The temporal sequence to deal with abnormal and accidental conditions until recovery, e.g., accident management, off-site emergency response, decontamination and decommissioning, is illustrated in the figure.

To protect both the public and the workers, defense in depth is a widely accepted approach combining both prevention of incidents and accidents, and mitigation of their consequences, as discussed above. The safety barriers and procedures installed based on the concept of defense in depth are to prepare for, mitigate and respond to the accident, which are within the scope of resilience engineering.

In other words, from the viewpoint of nuclear safety engineering, resilience is a concept that expands the concept of defense in depth by enlarging the scope of nuclear safety engineering from only preventing accidents and mitigating consequences to responding to and recovering from accidents in the medium and long term.

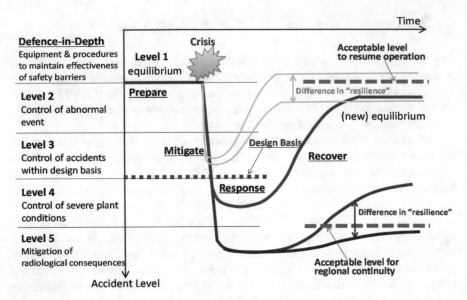

Fig. 3 Relationship between defense-in-depth and resilience from the viewpoint of nuclear safety engineering [11]

4.2 *Resilience Engineering for Integrated Risk-Informed Decision-Making Process*

It has been discussed that resilience is not restricted to a systems' response to crisis, but also includes adapting to slow and long-term changes. In other words, resilience can be defined also as the adaptive capacity of systems to maintain, through proactive maintenance, functions that deteriorate over time, the capacity of systems to evolve by remaking itself in order to adapt to existing and future environmental challenges, and the capacity of a system to become more robust by learning from past failures or from recent findings. In other words, maintenance science and technology should be effectively integrated into the decision-making process associated with continuous improvement of nuclear safety. This type of resilience is related to the first level of "defense in depth."

Hollnagel [13] suggests that the conventional safety perspective that defines safety as the condition where the number of adverse outcomes is as low as possible (i.e. safety-I) has some limitations, and a new safety perspective should also be emphasized, which defines safety as the condition where the number of intended outcomes is as high as possible (i.e. safety-II). This framework is considered to be important when the decision-making process with continuous improvement is discussed. Resilience engineering needs to include resilience for both safety-I and safety-II.

5 Summary

In this chapter, the challenges for nuclear safety with respect to natural hazard risk management were summarized from the viewpoint of a risk-informed framework, defense in depth, design, and regional disaster prevention/mitigation.

The application of the resilience concept to nuclear safety and maintenance science and technology was discussed. Resilience engineering is considered to be a discipline that broadly applies theories and technologies related to safety, which can especially refine and enhance the concept of defense in depth, the fundamental concept of nuclear safety. Roles of stakeholders (public, utilities, vendors, regulatory bodies, government, academia, etc.) need to be discussed in this context, to reconsider the conventional procedures for nuclear safety.

References

1. http://www.fdma.go.jp/bn/higaihou_new.html (in Japanese) Accessed 20 Sept 2015
2. International Organization for Standardization: ISO 31000:2009, Risk management—Principles and guidelines, (2009)
3. International Atomic Energy Agency: Fundamental Safety Principles, Safety Fundamentals, No.SF-1 (2006)
4. S. Kaplan, B.J. Garrick, On the quantitative definition of risk. Risk Anal. 1(1), 11–27 (1981)
5. International Atomic Energy Agency: A Framework for an Integrated Risk Informed Decision Making Process, INSAG-25, A report by the International Nuclear Safety Group (2011)
6. A. Stirling, Keep it complex. Nature 468, 1029–1031 (2010)
7. United States Nuclear Regulatory Commission, http://www.nrc.gov/reading-rm/basic-ref/glossary/defense-in-depth.html. Accessed 20 Sept 2015
8. International Atomic Energy Agency, Defence in Depth in Nuclear Safety, INSAG-10, A report by the international safety advisory group (1996)
9. Joint Editorial Committee for the Report on the Great East Japan Earthquake Disaster, Report on the Great East Japan Earthquake Disaster Nuclear Engineering Volume (in Japanese with English abstract) (2015)
10. The National Academies: Disaster Resilience: A National Imperative (2001)
11. N. Sekimura, H. Miyano, T. Itoi, Resilience Engineering: New Discipline for Enhancement of Nuclear Safety, Proceedings of ICMST-Kobe 2014 (2014)
12. Committee on Increasing National Resilience to Hazards and Disasters; Committee on Science, Engineering, and Public Policy; The National Academies: Disaster Resilience: A National Imperative (2012)
13. E. Hollnagel, Safety-I and Safety-II The Past and Future of Safety Management (Ashgate, 2014)

The Economic Assessment of the Cost of Nuclear Accidents

Romain Bizet and François Lévêque

Abstract This paper discusses the obstacles that hamper robust estimations of the cost of nuclear accidents. From an economic standpoint, risks of accidents are often quantified by the assessment of an expected cost; that is the product of a monetary loss by its probability of occurrence. In the case of nuclear power, this definition is inadapted. Estimating the probability of a nuclear disaster is subject to high uncertainties, and so is the assessment of its monetary equivalent. This paper first discusses the two specific challenges in estimating the probabilities of nuclear accidents: these accidents are too sparse to identify frequencies of occurrence with probabilities; they are also dreadful, which makes risk-aversion a complex phenomenon. This paper then focuses on the assessment of nuclear damage. The large discrepancies exhibited in the existing literature arise from three sources: the scope of the assessments, the conflict between the use of past data or PSA studies in the assessment of radiation damage, and the methods that are used to quantify non-monetary welfare losses.

Keywords Probabilistic analysis · Dreadful event · Cost assessment · Monetary valuation · Nuclear countermeasures

1 Introduction

Resilience has been defined as the ability of a society facing an extreme event to react, adapt to its new environment and recover from the damage incurred. If economics is not a priori well suited to discuss the short-term reaction or decision-making processes of people facing extreme events, it nonetheless offers

R. Bizet (✉) · F. Lévêque
CERNA-Centre for Industrial Economics, MINES ParisTech/PSL-Research University, Paris, France
e-mail: romain.bizet@mines-paristech.fr

F. Lévêque
e-mail: francois.leveque@mines-paristech.fr

J. Ahn et al. (eds.), *Resilience: A New Paradigm of Nuclear Safety*,
DOI 10.1007/978-3-319-58768-4_7

79

tools to describe how the risks of nuclear accidents can be anticipated, prepared for, and somehow mitigated in case an accident occurs. This paper will thus focus on the economic assessment of the risks of nuclear accidents. We will emphasize on its major drawbacks and on the policy guidelines this assessment entails.

Risks of accidents are often quantified by the assessment of their expected costs [1]; that is the product of damage by the probability of occurrence. In the case of nuclear power, this definition is particularly inconvenient. Estimating the probability of a nuclear disaster is subject to high uncertainties, and so is the assessment of its damage. Yet, this assessment serves two purposes. First, *ex ante* cost assessments provide policy-makers with guidelines regarding hazardous activities; such as choosing among various technologies for electricity production, or setting efficient safety standards. *Ex ante* assessments are thus *economically* driven. Second, *ex post* cost assessments allow victims to be compensated according to their losses. This assessment takes place after the accident. It focuses on the evaluation of the different damage to individuals, to society and to nature. Such *ex post* assessments are usually *legally* driven (see BP Deepwater Horizon or TEPCO Fukushima-Daiichi payouts).

One major difference between the *ex ante* and *ex post* approaches is that the former is confronted with much more uncertainties. The future is less known that the past and the accident is only one possible outcome. The cost to consider is thus an expected cost. Generally speaking, the probability of an event is derived from its observed frequency. Some authors also increase this product by a coefficient to account for risk-aversion. This canonical method in the economics of accident is not easy to apply to nuclear catastrophes. Probabilities cannot be derived from observed frequencies and psychological biases regarding dreadful events are not simply risk-aversion. These issues will be addressed in the first part of this paper, which is divided into two parts. The first part deals with the limitations of observed frequencies and of the so-called Probabilistic Safety Assessment (hereafter, PSA) as a method to estimate the overall probability of nuclear accidents. It argues that knowledge from observed occurrence of accidents and knowledge from safety engineers and experts have to be combined. The second part focuses on the gap between the probabilities of nuclear accidents as calculated by experts and the probabilities of nuclear accident as perceived by individuals. It shows how psychological biases in estimating probabilities are amplifying the perceived risk of nuclear accidents.

Numerous assessments of the cost of a nuclear accident have been performed over the years. This paper will review some of these assessments and present their very different results. We argue that these differences are not peculiar from an economic standpoint. Similar differences are also observed in other hazardous activities, such as car accidents, oil spills or climate change. We then try to identify the origins of the discrepancies in the assessments. The uncertainty characterizing these results stems from three sources: existing studies have a very different scope; they rely on either past data or probabilistic safety assessments (PSA); and they assess the consequences of the accident and their monetary costs with different methodologies. Second, we highlight the fact that most studies focus on damage and try to produce assessments that account for as many consequences of an accident as possible. Even if this is necessary for the aforementioned goals, it fails to provide insights into how the aftermath of nuclear accidents might be mitigated. Indeed, numerous countermeasures are available and can help reduce the impact of a nuclear accident on the economy. Yet, few economic studies try to assess the impact of mitigation policies on the cost of nuclear accidents. Those who do reduce their scope to a specific nuclear countermeasure—such as land decontamination—often yield less uncertain results or clearer insights for policy-makers. Therefore, we stress the need for future research in the economics of nuclear countermeasures, which could provide guidelines for mitigation policies.

2 Evaluating the Expected Cost of Nuclear Power

2.1 Limitations in Estimating the Probabilities of Nuclear Accidents

The expected costs of nuclear accidents seen as car crashes

How to compare different technologies to produce electricity? What is the optimal mix of power generation? To answer these questions economists use the levelized cost of electricity (hereafter, LCOE), defined as the price of electricity required to balance discounted costs and benefits throughout a power plant's service life. From a welfare perspective, the costs and benefits must include both the private costs the operators will incur (e.g., fuel costs) and the external costs society will have to pay for (e.g., polluting emissions). For instance, the UK department of energy and climate change estimates the LCOE of a Combined Cycle Gas Turbine and of a nuclear power plant (hereafter, NPP) to respectively 80 and 90£/MWh for a project starting in 2013 with a 10% discount rate.[1]

As far as nuclear generation is concerned, accidents have to be included in the LCOE. However, this inclusion has a negligible impact because the huge amount of damage is multiplied by an infinitesimal probability. Let's consider a simple back-of-the-envelope calculation: 1 billion euros of damage and a large early release frequency of 10^{-5} per reactor-year. This leads to 1 €/MWh, that is about one hundredth of the total cost to produce one MWh from a nuclear plant. According to a recent study on nuclear costs based on a comprehensive literature survey [1], the order of magnitude of external costs due to nuclear accident is between 0.3 and .

Such an approach to compute the social external cost of nuclear power generation is far from satisfactory. It views nuclear accidents as transport crashes. It is

---

[1]DECC, Electricity Generation Costs 2013, https://www.gov.uk/government/uploads/system/uploads/attachment_data/file/223940/DECC_Electricity_Generation_Costs_for_publication_-_24_07_13.pdf.

reasonable to make cost comparison of different means of transportation because their probabilities can be derived from observed frequencies. For instance, there have been worldwide 90 to 118 annual airplane accidents between 2009 and 2013 (including 9–13 annual fatal accidents with 173–655 fatalities per year).[2] Based on the frequencies observed during this period, the probability for a passenger to have an airplane accident when embarking at the airport is approximately 3×10^{-6}. Road traffic accounts for more than 1 million deaths per year worldwide, mainly pedestrians and motorcyclists.[3] In a small country like New Zealand, the number of fatalities due to car crashes has been 254 in 2013. It corresponds to a frequency of 0.8 deaths per 10,000 vehicles.[4] On the basis of 2011–2013 statistics the probability for a driver to be killed is about 3×10^{-9} per km. In 2013, the social cost of fatal car accident is estimated to NZ\$ 4.5 million. The expected cost of fatal accident per km can be estimated to 0.03 NZ\$, that is about one tenth of gas price. Given the observed frequencies of transport accidents (and assuming the value of loss life is the same and neglecting the other damage), one can easily compare the social accident costs of rail, air, maritime and road transportation per km or per travel. Moreover, data on car accidents can generally be broken down by local area, models of car, types of roads, age categories of the driver, etc. As a result precise probability can be estimated according to different situations.

2.2   Nuclear Accident Are no Car Crashes

Estimating probabilities of nuclear accident from frequencies is a non-sense. Since the first grid-connection of nuclear power plant in 1956 there have been 12 core-meltdowns of reactors, including very limited ones [2]. According to the INES classification there have been 2 major, or level 7, accidents (Chernobyl and Fukushima-Daiichi) and 21 accidents with a level equal or higher to 4. Knowing that since the end 1950s 14,500 reactor-years have passed worldwide the observed frequencies are 1.6×10^{-3} per reactor-year for INES >3; 8.3×10^{-4} per reactor-year for core-meltdown; and 2.7×10^{-4} per reactor-year for INES = 7. Is it sound to infer probabilities from these values? For instance, using a Poisson distribution and knowing that the worldwide nuclear fleet amounts to 435, is it relevant to say that the probability of an INES 7 accident in 2015 on the planet is 0,11 (i.e., $[1-(1-2.7 \times 10^{-4})^{435}]$?

No! The reasons are twofold. The obvious one is that the number of observations is too small. The observed events cannot be assumed as representative. Reactors are neither identical, nor exposed to the same locational risk (e.g. earthquake, flooding).

---

[2]International Civil Aviation Organization, 2014 Safety Report.
[3]World Health Organization, Global Safety Support on Road Safety 2013.
[4]Ministry of Transport, Motor Vehicle Crashed in New Zealand 2013.

In addition, nuclear accidents are not independent—the current nuclear fleet is close to the 1980s fleet for more than three quarters of reactor is over 25 years old. Moreover, the evolution of safety performances and standards makes heroic to assume that safety is time-invariant.

A second reason is that assessing the risk of nuclear accidents exclusively on data from past observations implicitly assumes that no other knowledge is available on nuclear safety. It ignores all the works carried out over the past 50 years by thousands of nuclear scientists and engineers on safety. This knowledge has partly crystallized in PSAs. The first large-scale probabilistic assessment was carried out in the US in the 1970s. It was led by Norman Rasmussen, then head of the nuclear engineering department at MIT. PSAs have now been carried out on all nuclear power plants (hereafter, NPP) in the US and many others worldwide. Similarly reactor vendors carry out such studies for each reactor model while it is still in the design stage. For instance, the calculated core meltdown frequency for the UK EPR is 10^{-6} per year and the core damage with early containment failure is estimated to 3.9×10^{-8} per year.

As for observed frequencies, assessing the risk of nuclear accident based exclusively on PSAs would be unsound. The use of PSAs has strong limitations, too. Firstly, they are not mainly designed to provide a final single number. They are designed to detect exactly what may go wrong, to identify the weakest links in the process and to understand the failures which most contribute to the risk of an accident. Secondly, PSAs have a limited scope. They study known initiating events such as seism or loss of coolant but not all the possible states of the world because the list of all causes and failures is unknown. Thirdly, PSAs assumes perfect compliance with safety standards and regulatory requirements. An implicit assumption is that safety standards are enforced thanks to an independent, competent and powerful safety regulatory authority. All these limitations can explain in part why PSAs figures are much lower than observed frequencies.[5]

If we want to make progress in estimating probabilities of nuclear accident, we have to use all the current available quantitative knowledge and therefore to combine information from PSAs and observed accidents. Escobar-Rangel and Lévêque have made such an attempt [4]. The issue addressed in their paper is to compute the post Fukushima-Daiichi global probability of a core-meltdown. Different models are used including a Poisson Exponentially Weighted Average model to capture the idea that recent accidents are more informative than past ones and to introduce some inertia in the safety performances of the fleet. This model shows that the Fukushima Daiichi accident results in a huge increase in the probability of an accident. The arrival rate in 2011 is similar to the arrival rate computed in 1980s. To put it another way, this catastrophe has increased the probability of an

---

[5]For a detailed discussion on the discrepancy between observed frequencies and calculated frequencies in PSA models see [3], chapter "Consequences of Severe Nuclear Accidents on Social Regulations in Socio-Technical Organizations".

accident for the near future in the same extent it has decreased over the past 30 years owing to safety improvements. This huge effect of Fukushima Daiichi in revising the global estimation of a core meltdown can be interpreted as evidence that besides the design, the location and the operating of reactors the probability of an accident also depends on institutional factors like the strength and ability of nuclear safety authorities, a factor which is not taken into account in probabilistic assessments. In fact, like in Japan there are a lot of countries wherein nuclear safety authorities are captured by operators and fail to enforce safety standards.

As a conclusion, uncertainties prevail. There is no overarching probability of nuclear accident to use to make a rational decision for society to invest in or to phase out nuclear power generation, to determine the right level of nuclear safety expenditures, not to say to identify the economically optimal level of nuclear safety. Unlike transport crashes no means can be inferred from observed frequencies. Moreover, the probability of a nuclear accident differs according to the design and the location of reactors but also according to institutional characteristics (independent regulator, liability rules, experience of operators, etc.). We do not know the probability distribution of nuclear accidents, even for a given reactor design and location. Last but not least, one has always to keep in mind that probabilistic analysis requires knowing all the states of the world. A probability cannot be assigned to an unknown event, or to put it another ways to black swans and unknown unknowns.

2.3   Perception of Probabilities

Utility function and human behavior

It is well known that many people are risk-averse: they would rather, for instance, a certain gain of 100 to an expected gain of 110. Since Bernoulli [5], this psychological trait is represented by a concave utility function. The Swiss mathematician opened the way for progress towards decision theory[6] through a back-and-forth between economic modeling and psychological experimentation. The latter would, for instance, pick up an anomaly—in a particular instance people's behavior did not conform to what theory predicted—and the former would repair it, altering the mathematical properties of the utility function or the weighting of probabilities. The works by Allais and Ellsberg were two key moments in this achievement. Following an experiment showing that people with good knowledge of the theory of probability were violating an axiom of expected utility theory, Allais [7] proposed to weight probabilities depending on their value, with high coefficient for low probabilities, and vice versa. Putting it another way, preferences assigned to probabilities are not linear. This is more than just a technical response. It makes

---

[6]For a comprehensive panorama on decision theory under uncertainties see [6].

allowances for a psychological trait, which has been confirmed by a large body of experimental study: people overestimate low probabilities and underestimate high probabilities.

Another anomaly well known to economists is ambiguity aversion. This characteristic was suggested by Keynes and latter demonstrated by Ellsberg [8] in the form of a paradox. In his treatise on probabilities Keynes [9] posited that greater weight is given to a probability that is certain to one that is imprecise. Ellsberg has shown that just as there is a premium for taking risks, some compensation must be awarded to individuals for them to be indifferent to gain (or lose) with a one-in-two probability or an unknown probability with an expected value of one-in-two. Recent developments in economic theory offer several solutions to tackle this problem, in particular by specifying new types of utility functions, yet again [10]. What is important to keep in mind here is that individuals usually prefer the exposure to a hazard associated with a clearly defined probability—because experts are in agreement—rather than the exposure to a hazard characterized by uncertain or fuzzy probabilities—because experts may disagree. Putting it another way, in the second instance people side with the expert predicting the worst-case scenario.

More recently, Kahneman's work followed on that of Bernouilli, Allais and Ellsberg. He and his fellow author, Tversky, introduced loss-aversion: individual are more affected by loss than gain [11]. Kahneman also diverged from his predecessors in adopting a more positive approach. Observing the distortion of probabilities is a way to understand how our brain works rather than to build a theory where the decision-maker optimizes or maximizes the outcome. Kahneman's line of research is comparable to subjecting participants to optical illusions to gain a better understanding how our brain functions. For example, a 0.0001 probability of loss will be perceived as lower than a 1/10,000 probability. Our brain seems to be misled by the presentation of figures, much as our eyes are confused by an optical effect which distorts an object's size or perspective. This bias seems to suggest than our brain takes a short cut and disregards the denominator, focusing only on the numerator.

2.4   The Effects of Perception Biases on Nuclear Accidents

The overall biases in our perception of probabilities, briefly discussed above, amplify the risk of a nuclear accident in our minds. A nuclear accident is a rare event, so its probability is overestimated. The risk of a nuclear accident is ambiguous. As expert appraisals diverge, people are therefore inclined to opt for the worst-case scenario. The highest probability of accident prevails. Along with plane crashes or terrorist attacks targeting markets, hotels or buses, a nuclear accident is a dreadful event. Rather than acknowledging the probability of the accident, attention focuses exclusively on the accident itself, disregarding the denominator. Moreover, several other common routines or heuristics which have been identified by experimental psychologists distort the probability of a nuclear accident[7] and increase our aversion vis-à-vis such a disaster.

As a consequence, public decision exclusively based on perceived risk entails a series of drawbacks. Firstly, it tends to an over-investment in nuclear safety. The perceived risk of a nuclear accident being amplified, the benefits to decrease it seems higher and therefore efforts to reduce it seems more worth to be undertaken. Secondly, the choice of technology is distorted in favor of ways generating electricity which are not less hazardous. Coal is perceived as less dangerous whereas according to data on fatalities it is more [12]. Thirdly, public decisions exclusively based on perceived probabilities could lead to costly premature phase-outs. After Fukushima-Daiichi, the German government decided to accelerate the decommissioning of NPPs. It entails an economic loss estimated to a 100 billion of euros in comparison to the previous more progressive nuclear exit as enacted in the Atomic law passed a few months before the accident [13].

However, public decisions ignoring perception biases can also result in wasting a lot of money. It could be costly to treat the attitude of the general public as the expression of fleeting fears which can quickly be allayed, through call to reason or the reassuring communication of the 'true' facts and figures. The reality test, in the form of hostile demonstration or electoral reversals, may substantially add to the cost for society of going back on past decisions ignoring public perception. Nuclear power history is full of cases of abandoned projects after several years of construction. In France, for instance, about 10 billion of euros have been spent to build the fast breeder commercial reactor Superphénix for nearly nothing. It had only produced a modest quantity of electricity when it was shut down.

In short, public decision-making must avoid two pitfalls: ignoring how probabilities of nuclear accident are perceived and exclusively taking them into account.

3 Economic Assessment of Nuclear Damage and Their Insights into Mitigation Policies

3.1 Existing Assessments of the Cost of a Nuclear Accident

A review of existing studies

Assessments of the cost of nuclear accidents have been carried out since the mid-seventies and the beginning of probabilistic safety assessments. Since then, numerous studies have been published, and several reviews of these studies exist. Namely, in 2000, the Nuclear Energy Agency published a methodological review in which several cost assessments were described [14]. In 2011, after the Fukushima-Daiichi accident, the German Renewable Energy Foundation performed a

---

[7]See [3] pp. 115–117.

calculation of the adequate insurance premium that the nuclear industry would need to pay to fully cover the accident risk. This study also reviewed some existing assessments [15]. The D'Haeseleer report for the European commission also provides a comprehensive review of studies that assess the external cost of nuclear accidents [1]. Finally, the IRSN[8] published in 2013 an assessment of the cost of severe and major accidents in which other studies were reviewed [16]. Those four reviews reference numerous studies and give a thorough overview of the state of the art literature on the evaluation of the costs of nuclear accidents. As we do not wish to tackle here the question of the probability of nuclear accidents, Table 1 below only presents the studies that assess the cost of nuclear accidents before weighting.[9]

Table 1 shows high discrepancies. How can one assess the cost of nuclear accidents at approximately €10 billion [17], while others announce a cost of more than a trillion Euros [21]? First, it can be noticed that all studies do not assess the same cost. Some only focus on the damage to the population (health and food costs), while others try to assess the total impact of the accident on the economy. Yet, this cannot be the only cause of these differences. Indeed, even within cost sections (health, food...), there is little consensus as to which cost section represents the highest share of the total cost. The comparison between the "IRSN-major" [16] and the assessment from the German Renewable Energy Federation [15] embodies this observation: even though it only assesses health, food and production costs, the German study calculates a total cost ten times superior to the IRSN figure, which accounts for a larger panel of consequences.

3.2   The Assessments of the Costs of Other Hazards Exhibit Similar Discrepancies

This is not specific to nuclear power: other hazardous activities exhibit the same kind of discrepancies. In 1995, Elvik studied the assessments of the cost of car accidents in twenty countries. This work was motivated by the observation of large disparities in the evaluation of this cost: while the Netherlands were evaluating the total cost of a car accident at U.S. $0.12 million, Switzerland estimated it at U.S. $2.5 million [25]. This study argued that the deviation was caused by the lack of common methodology in the assessment of the cost and the consideration by only a limited number of countries of the value of the lost quality of life.

The estimations of the damage caused by oil spills are also prone to large disparities. In 1995, Cohen assessed the damage of the Exxon Valdez oil spill. She claimed that the upper bound of the estimation of the damage caused by the oil spill in the first two years following the disaster was U.S. $155 million [26]. In 2003,

---

[8]The IRSN is France's technical support organization for the Nuclear Safety Authority (ASN).

[9]The cost of a nuclear accident can be weighted by an electrical output to yield an external cost or by a probability to yield an expected cost.

Table 1 A review of existing assessments of the cost of nuclear accidents

References		Year	Health cost	Food cost	Loss of land, production and cost of mitigation actions	On-site cost	Image cost	Fleet cost	Cost of a nuclear accident (b€)
[17]	WASH 1400	1975	x	x	x	–	–	–	14
[18]	CRAC-2	1982	x	x	x	–	–	–	314
[19]	Hohmeyer	1988	1370	–	–	–	–	–	1370
[20]	Ottinger	1990	629	38	–	–	–	–	667
[21]	Ewers-Rennings 1	1991	2740	38	828	–	–	–	3606
[21]	Ewers-Rennings 2	1992	7815.6	307.4	179.1	–	–	–	8302
[22]	ExternE	1995	74.3		37,9	–	–	–	112.2
[23]	Eeckhoudt	2000	10.85	6.162	0.098	–	–	–	342
[15]	German Renewable Energy Federation	2011	x	x	x	–	–	–	5900
[24]	Rabl-Low	2012	10	5	100	50	–	–	165
[24]	Rabl-Central	2012	18.8	75	250	78	–	–	354
[24]	Rabl-High	2012	50	50	1000	290	–	–	1390
[16]	IRSN-severe	2013	0	9	11	10	50	44	124
[16]	IRSN-major	2013	27	14	110	28	180	88	447

"x" signifies that the cost section is at least partly assessed. "–" signifies that the cost section is not assessed

another study assessed the cost of this oil spill at approximately U.S. $2.8 billion [27]. In these studies, Cohen limited her assessment to the costs incurred by southcentral Alaska's fisheries, while Carson assessed the population's willingness-to-pay to restore the lost passive use of the damaged environment.

Finally, the climate change literature also exhibits large discrepancies. In a review published in 2009, Tol shows that there is little agreement on the long term effects of climate change. While some authors [28] predict an overall small positive effect due to the heating of cold regions, others predict dramatic consequences [29]. This overview highlights the fact that the uncertainty pertaining to cost estimates can originate from several sources.

4 Uncertainties and Mitigation Policies

If these discrepancies are not peculiar from an economic standpoint, it is nonetheless interesting to try and understand why they occur. In basic economic theory, costs are often defined as anything that causes a loss of welfare [30]. The cost of a nuclear accident can thus be defined as the gap between the welfare levels obtained with and without its occurrence. This theoretical definition induces divergence in the assessments: the consequences of an accident—direct or induced by mitigation countermeasures—are so numerous and intricate that it is impossible to be sure that all consequences have been accounted for properly. Studies differ first in their assessment of the consequences of the accidents, and then on the monetary valuation of these consequences.

4.1 Cost Assessments Do not Speak the Same Language

First, it seems obvious that results will be different if the type and location of accidents assessed are different. Nuclear plants are highly sophisticated, so there is a large panel of possible accidents which do not have the same consequences. Likewise, nuclear plants are located in areas which are not equally densely populated [31]. As an example from Table 1, Hohmeyer's study calculates the external cost of a hypothetical Chernobyl-like accident in the Biblis nuclear plant (Germany) in 1990. The IRSN study calculates the social cost[10] of a hypothetical DCH[11] nuclear accident in France in 2025. The scopes of these two studies are radically

---

[10]The private cost of an accident is the cost incurred by the utility; the external cost is the cost that will not be incurred by the firm because either a legal liability threshold exists, or the firm is limited by the total worth of its assets. The social cost of the accident is the sum of private and external costs.

[11]The IRSN describes a Direct Containment Heating accident, which consists in a direct heating of gases within the containment vessel.

different. More generally, the comparison of the studies presented in Table 1 is impossible because they do not stand on common definitions.

Similarly, the boundaries of a cost assessment also need to be clearly defined. In 2006, two reports on the consequences of the Chernobyl accidents were published. Their assessments of the number of radio-induced cancers differed by a factor ten. The IAEA/WHO report [32] focused on the consequences of the accident in Belarus, Ukraine and Russia, while the TORCH report accounted for all consequences across Western Europe [33]. Nuclear accidents can have cross borders consequences, so it is paramount to define clearly the boundaries of cost assessment studies to fully understand their implications for public policies. As an example, Hayashi and Hughes have shown that the Fukushima-Daiichi accident had an impact on the electricity bill of households in gas-intensive countries such as the United-Kingdom or South-Korea [34]. How and by whom should these impacts be accounted for?

Finally, the statistical choices in the presentation of the cost are also crucial for their comparison or their use in policy making. There is for example no consensus as to whether the cost of a nuclear accident should be presented as a distribution function or as a single number. The IRSN decided to produce a median cost so that decision makers know that there is a 50% chance for the cost of an accident to be above or below the result. Conversely, the 2011 study from the German Renewable Energy Federation provided an average maximum value in order to calculate an "adequate" insurance premium.

4.2 The Aftermath of a Nuclear Disaster: PSA or Past Events?

The consequences of a nuclear accident are numerous and intricate. An accident has on-site consequences, such as casualties, highly-irradiated workers or material losses in adjacent reactors. It also causes off-site consequences, such as the release of radioactive materials in the atmosphere, the collective absorbed dose, the area of contaminated lands or the quantity of crops and cattle contaminated. The negative consequences of the countermeasures, such as psychological distress, also have to be estimated. Yet, these numerous consequences have to be assessed in order to derive their monetary value.

The source of divergence in the assessment of the consequences is twofold. First, all studies do not assess the same range of consequences. Some studies argue that health effects dominate all other effects [15, 19, 20]. They thus focus on the collective absorbed dose and neglect other consequences. Other studies focus on a wider panel of effects, such as land exclusion, or image effects (tourism, regulatory changes…). Second, studies also differ in their assessment strategies. Physical consequences can be modelled by dedicated programs (MACCS, COSYMA… [35]) that rely on level-three probabilistic safety assessments; or assessed by adapting the figures derived from past catastrophes. Most studies performed in the early nineties were based on Chernobyl's figures, and find particularly high values for the total cost of the accident [19–21]. More recently, another very high cost was assessed by the German Renewable Energy Foundation which happens to be also based on Chernobyl's figures. This observation raises an important question. Can we assess future accidents solely by using the consequences of past catastrophes? A preliminary answer is that we cannot. Relying on past figures fails to account for the learning from past consequences, the enhancement of safety standards, and the progress in available mitigation technologies.

4.3   Converting Consequences into Costs Requires Various Hypotheses and Assessment Methodologies

Once the consequences of a nuclear accident have been assessed, they have to be given a monetary value. Indeed, a cost is the monetary valuation of foregone welfare. Among the consequences discussed previously, some welfare losses are easily derived (cost of material losses). For other physical consequences, various hypotheses are required to bridge the gaps in our limited knowledge. Regarding health issues, we do not know precisely the effect of exposure to low doses on cancer or hereditary diseases probabilities. Regarding the environmental impact of an accident, the size of lost lands depends on the geographical spread of the radioactive materials and on the acceptable radioactivity threshold that a population can bear. The consequences on food are also uncertain since the population can react to food-bans by boycotting healthy products. The harm caused by nuclear countermeasures, such as psychological distress due to relocations, is also hard to assess. Some hypotheses substantially differ from one study to another. As an example, the excess rate of radio-induced cancer varies from 5 to 10% in the assessments presented in Table 1.

Some of these welfare losses such as reduced tourism, strengthened safety standards for nuclear plants, or higher energy prices, can easily be given a monetary value. They are assessed through macroeconomic methods such as the IO-table method. Yet, all welfare losses caused by nuclear accidents are not necessarily monetary. Therefore, some methodologies have been developed in health and environmental economics in order to give monetary values to non-monetary losses. Environmental losses can be assessed by the evaluation of individuals' willingness to pay (WTP) to avoid these losses. Two families of methods allow the assessment of this WTP: the revealed-preference methods and the stated-preference methods. *Revealed-preference* methods such as the travel cost method or the hedonic pricing method, use past individual behaviors to infer the value of environmental losses. These are hard to apply to nuclear accidents because they rely on past behaviors and thus require data [36–38]. *Stated-preference* methods are based on surveys that try to elicit the willingness to pay of people to restore the environment. The contingent valuation method is often used to value the environmental consequences of rare disasters.

Regarding health costs, the human capital method calculates the economic value of fatalities or impairments by assessing the number of lost years of production and multiplying it by the average yearly production of a human being. Other methods, such as the friction cost method, exist and have very different ways of calculating those health costs [39–42]. This variety of methods is responsible for some of the discrepancies observed in Table 1. First, a consequence can be assessed by different methods. Second, even if a cost is assessed by two studies with the same method, some aspects of the evaluation remain quite arbitrary. In the human capital method applied to the cost of radio-induced cancers, Hohmeyer assesses the cost of a death at $1 million while Ottinger assesses it at $4 million [15].

4.4 Drawbacks

Table 1 shows that the tendency over the last twenty years has been to provide an estimation of the cost of nuclear accidents which would account for as many consequences as possible. This emphasis on completeness, which is particularly stressed in the IRSN study, is indeed necessary for the goals mentioned in the introduction of this paper. *Ex ante* policy making and *ex post* compensations both need to rely on a complete assessment of the consequences of a nuclear accident, since an incomplete assessment might lead to an underestimation of the cost and entail an underinvestment in nuclear safety, a disproportionate share of nuclear power in the electricity mix, or an inadequate compensation of victims [43, 44].

This quest for completeness also has its drawbacks. First, it fosters the aggregation of numbers that differ by nature. As we have seen, all costs of consequences are not assessed using the same methodologies, and are thus not subject to the same uncertainties. Summing them up to provide a global cost of nuclear accidents propagates the highest uncertainty to the final result. Second, completeness can be detrimental to scientific rigor. Some costs currently have no corresponding assessment methods. It is the case for food bans; which are estimated in the most recent study by comparison with recent non-nuclear food bans [16]. Their integration in cost assessments is thus parochial, since they don't stand on robust economic grounds. Finally, to achieve completeness, existing studies have focused on damage and consequences, and tried to identify new consequences, or "lines of cost". By doing so, most studies overlook the impact of nuclear countermeasures, which is the object of the next part of this paper.

4.5   Cost Assessment Fails to Provide Guidelines into Mitigation Policies

Current research on cost assessment focuses on providing complete assessments by identifying more and more consequences of an accident. This trend is necessary, but is not adapted to mitigation policies. First, the theory of "sunk costs" [45] explains that once a cost has been incurred, it is no longer relevant for decision making regarding the future. In the case of mitigation policies, the capital losses due to the destruction of a power plant are incurred at the time of the accident. Those losses are an example of sunk costs, and should thus not enter the mitigation policy decisions. Current estimates, as they account for all kinds of losses regardless of the time at which they are incurred, cannot be used in the determination of mitigation policies. This observation raises one question: can we expect cost assessments to provide useful guidelines for mitigation policies?

We believe it can. Cost-benefit analysis (CBA) of countermeasures could provide at least three useful insights regarding mitigation policies. First, it was shown by the report on the consequences of Chernobyl that countermeasures are costly [32]. Cost-benefit analysis could thus help determine which countermeasures are most efficient by comparing their costs to society with the valuation of the prevented damage. Second, there are numerous countermeasures that address the same harmful consequences. Some measures are substitutes (emergency relocation and confinement), while others are complements (iodine prophylaxis and confinement). Hence the assessment of their costs and benefits could help policy-makers identify tradeoffs or synergies when implementing several countermeasures. Finally, the consequences of a nuclear accident do not happen all at once. Cost-benefit analysis is thus a good tool to search for the optimal inter-temporal allocation of mitigation resources.

This kind of assessment is already carried out in other hazardous activities such as car accidents or biosecurity [46, 47]. In the case of nuclear power, Munro studied the tradeoff between long-term relocation and land decontamination. As radioactive decay reduces the cost of land decontamination over time, he calculated the optimal decontamination date which occurs approximately ten years after the accident [48]. Other studies also focus on particular tradeoffs between countermeasures, namely land decontamination and food restrictions [49, 50]. Yet, these studies focus on multi-criteria decision making rather than on performing a CBA of countermeasures.

Existing studies that deal with mitigation only focus on long-term countermeasures. Being able to deal with emergency countermeasures is a barrier that needs to be overcome if CBA is to provide guidelines for mitigation policies. Indeed, an important tradeoff has to be solved right after the accident, and concerns the confinement or the emergency relocation of populations. A question for future research is whether CBA can deal with this emergency. Indeed, the optimal mitigation scheme cannot be determined ex ante, as it requires ex post data such as the plant impacted or the weather and its impact on the path of the radioactive materials dispersed in the atmosphere.

5 Conclusions

Regarding the estimation of the probabilities of nuclear accidents, two directions of research could be worthwhile. First, more research seems necessary to get a better knowledge on the uncertainties related to these probabilities. It includes uncertainties propagation in PSAs event trees, combinations of observed frequencies and PSAs and the use of new probability axiomatic such as imprecise probability theory. Second, more research is needed on methodologies that could help law-makers to make decision based both on probabilities as perceived by individuals and probabilities as calculated by experts.

This paper also raises two research questions regarding the cost of nuclear damage. The first is whether assessing the cost of nuclear accidents using the figures derived from past events is a robust method. As it fails to account for safety enhancements, progress in mitigation technologies, and learning from past catastrophes; it can drive cost assessments upwards, provide pessimistic numbers and entail overinvestments in safety or an unbalanced electricity technology mix. The second question is whether cost assessments should only focus on *ex ante* policy making and *ex post* compensations. We believe that cost assessments should also be used in order to improve mitigation policies.

References

1. W.D. d'Haeseleer, *Synthesis on the Economics of Nuclear Energy* (Study for the European Commission, DG Energy, 2013)
2. T.B. Cochran, M.G. McKinzie, *World Federation of Scientists' International Seminars on Planetary Emergencies* (Ettore Majorana Centre, Erice, 2011)
3. François Lévêque, *The Economics and Uncertainties of Nuclear Power* (Cambridge University Press, Cambridge, 2014)
4. L.E. Rangel, F. Lévêque, How Fukushima Dai-ichi core meltdown changed the probability of nuclear accidents? Saf. Sci. **64**, 90–98 (December 21, 2013)
5. D. Bernouilli, Exposition of a new theory on the measurement of risk. Econometrica **22**, 23–36 (1954)
6. Itzhak Gilboa, *Theory of Decision under Uncertainties* (Cambridge University Press, Cambridge, 2009)
7. M. Allais, *Fondements d'une théorie positive des choix comportant un risque et critique des postulats et axiomes de l'École américaine*, vol. XL (Colloques internationaux du CNRS, Paris, 1953), pp. 257–332.
8. D. Ellsberg, Risk, ambiguity, and the savage axioms. Q. J. Econ. **75**, 4 (1961)
9. J.M. Keynes, *A Treatise on Probability*. (Macmillan, London, 1921)
10 Massimo Marinacci, Model uncertainty. J. Eur. Econ. Assoc. **13**(6), 1022–1100 (2015)
11. Daniel Kahnemann, Amos Tversky, Prospect theory: an analysis of decision under risk. Econometrica **47**, 263–291 (1979)
12. Nuclear Energy Agency, *Comparing nuclear accident risks with those from other sources* (Organization for Economic Co-operation and Development, Paris, 2010)
13. Jan Horst Keppler, The economic cost of the nuclear phase-out in Germany. NEA News **30**, 8–14 (2012)

14. Nuclear Energy Agency, *Methodologies for Assessing the Economic Consequences of Nuclear Reactor Accidents* (Organization for Economic Co-operation and Development, Paris, 2000)
15. German Renewable Energy Federation (BEE). *Calculating a risk-appropriate insurance premium to cover third-party liability risks that result from operation of nuclear power plants.* (Versicherunsforen, Leipzig, 2011)
16. IRSN. *Méthodologie appliquée par l'IRSN pour l'estimation des coûts d'accidents nucléaires en France.* PRP-CRI/SESUC/2013-00261 (2013)
17. Rasmussen, N. *WASH-1400 (Reactor Safety Study). An assessment of accident risks in US Commercial Nuclear Power Plants*, vol. 8 (1975)
18. U.S.N.R.C. *CRAC-2 Report NUREG/CR-2239* (U.S.N.R.C. & Sandia National Lab, Rockville, 1982
19. O. Hohmeyer, *Latest Results of the International Discussion on the Social Costs of Energy— How Does Wind Compare Today?* (The European Commission's Wind Energy Conference, Madrid, 1990)
20. R.L. Ottinger et al., *Environmental Costs of Electricity. Pace University Center for Environmental Studies* (Oceana Publications Inc, 1990)
21. H.-J. Ewers, K. Rennings, *Abschätzung der Schäden durch einen sogenannten Super-GAU. Prognos-Schriftenreihe Identifizierung und Internalisierung externer Kosten der Energieversorgung* (Prognos, Bâle, 1992)
22. CEPN. *ExternE : Externalities of Energy—Volume 5 : nuclear* (European Commission Directorate-General XII, Luxembourg, 1995)
23. L. Eeckhoudt, C. Schieber, T. Schneider, Risk aversion and the external cost of a nuclear accident. J. Environ. Manage. **58**, 109–117 (2000)
24. A. Rabl, V.A. Rabl, External costs of nuclear: greater or less than the alternatives? Energy Policy **57**, 575–584 (2013)
25. R. Elvik, An analysis of official economic valuations of traffic accident fatalities in 20 motorized countries. Accid. Anal. Prev. **27**(2), 237–247 (1995)
26. M.J. Cohen, Technological disasters and natural resource damage assessment: an evaluation of the exxon valdez oil spill. Land Econ, 65–82 (1995)
27. R.T. Carson et al., Contingent valuation and lost passive use: damages from the exxon valdez oil spill. Environ. Resource Econ. **25**, 257–286 (2003)
28. Robert Mendelsohn et al., Country-specific market impacts of climate change. Clim. Change **45**(3–4), 553–569 (2000)
29. Richard S.J. Tol, The economic effects of climate change. J. Econ. Perspect. **23**(2), 29–51 (2009)
30. M. Shechter, Valuing the environment, in *Principles of Environmental and Resource Economics: A Guide for Students and Decision-Makers*, H. Folmer, H.L. Gabel and Hans Opschoor (Edward Elgar Publishing Ltd., Aldershot, 1995)
31. Declan Butler, Reactors, residents and risk. Nature **474**, 36 (2011)
32. AIEA. *Chernobyl's Legacy: Health Environmental and Socio-Economic Impacts and Recommendations to the Governments of Belarus, the Russian Federation and Ukraine,* 2nd edn. (The Chernobyl Forum, Vienna, 2003–2005, 2006)
33. I. Fairlie, D. Summer, *The other report on Chernobyl.* (MEP Greens/EFA, Berlin, 2006)
34. Masatsugu Hayashi, Larry Hughes, The Fukushima nuclear accident and its effect on global energy security. Energy Policy **59**, 102–111 (2013)
35. J.A. Jones, P.A. Mansfield, Irmgard Hasemann, *PC COSYMA Version 2: An Accident Consequence Assessment Package for use on a PC* (European Commission, Brussels, 1996)
36. W. Adamowicz, J. Louviere, M. Williams, Combining revealed and stated preference methods for valuing environmental amenities. J. Environ. Econ. Manage. **26**, 271–292 (1994)
37. W.M. Hanemann, Willingness to pay and willingness to accept: how much can they differ? Am. Econ. Rev. **81**(3), 635–647 (1991)
38. J.L. Knetsch, J.A. Sinden, Willingness to pay and compensation demanded: experimental evidence of an unexpected disparity in measures of value. Q. J. Econ. 507–521 (1984)

39. Bruno Fautrel et al., Costs of rheumatoid arthritis: new estimates from the human-capital method and comparison to the willingness-to-pay method. Med. Decis. Making **27**(2), 138–150 (2007)
40. W.Kip Viscusi, The value of risks to life and health. J. Econ. Lit. **31**, 1912–1946 (1993)
41. W.B. Van den Hout, The value of productivity: human-capital versus friction-cost method. Ann. Rheum. Dis. **69**(Suppl 1), i89–i91 (2010)
42. W. Kip Viscusi, E. Aldy Joseph, The value of a statistical life: a critical review of market estimates throughout the world. J. Risk Uncertainty **27**(1), 5–76 (2003)
43. Michael G. Faure, Göran Skogh, Compensation for damages caused by nuclear accidents: a convention as insurance. The Geneva Pap. on Risk and Insur. **17**(65), 499–513 (1992)
44. Nigel Evans, Chris Hope, Costs of nuclear accidents: implications for reactor choice. Energy Policy **10**(4), 295–304 (1982)
45. Hal R. Arkes, Catherine Blumer, The psychology of sunk cost. Organ. Behav. Hum. Decis. Process. **35**, 124–140 (1985)
46. J. Matheny, M. Mair, B. Smith, Cost/success projections for US biodefense countermeasure development. Nat. Biotechnol. **26**(9), 981–983 (2008)
47. McFarland, William F, et al., *Assessment of techniques for cost-effectiveness of highway accident countermeasures* (Federal Highway Administration, Environmental Design and Control Division, 1979)
48. Alistair Munro, The economics of nuclear decontamination: assessing policy options for the management of land around Fukshima Dai-ichi. Environ. Sci. Policy **33**, 63–75 (2013)
49. Patrice Perny, Daniel Vanderpooten, An interactive multiobjective procedure for selecting medium-term countermeasures after nuclear accidents. J. Multi-criteria Decis. Anal. **7**, 48–60 (1998)
50. Raimo P. Hämäläinen, Mats R.K. Lindstedt, Sinkko Kari, Multiattribute risk analysis in nuclear emergency management. Risk Anal. **20**(4), 455–468 (2000)

Considering Nuclear Accident in Energy Modeling Analysis

Ryoichi Komiyama

Abstract After Fukushima nuclear accident, alternative energy sources show a dramatic growth such as natural gas, petroleum and solar photovoltaic to compensate the loss of nuclear energy supply in Japan, and in the latest national energy policy, the government plans to promote renewable energy at a scale larger than the one aimed in the previous policy. Hence, the Fukushima accident can be regarded as the tipping point for the country to pursue alternative energy and environmental policy adjusting into the social circumstance after the Fukushima. So far, energy model has been developed to discuss long-term energy scenario in a consistent way and to analyze the effectiveness of energy policy. However, the most of the model developed until now does not explicitly consider the impact of nuclear accident on the long-term pathway of energy portfolio, in spite of the fact that the Fukushima accident is actually observed to dramatically change the situation of energy demand and supply in Japan. This manuscript aims to overview the transition of energy supply and demand in Japan after the Fukushima and to discuss the possibility of considering nuclear accident in energy modeling analysis by applying stochastic dynamic programming.

Keywords Nuclear energy · Nuclear accident · Energy mix · Energy modeling · Stochastic dynamic programming

1 Introduction

Energy supply serves as a basis to maintain socio-economic activity. For Japan which highly relies on the import of energy resource from other countries, the reinforcement of energy security is considered as an important challenge to be tackled with. In the global energy market, Japan is one of the big energy consumers

R. Komiyama (✉)
Resilience Engineering Research Center, The University of Tokyo, Tokyo, Japan
e-mail: komiyama@n.t.u-tokyo.ac.jp

© The Author(s) 2017
J. Ahn et al. (eds.), *Resilience: A New Paradigm of Nuclear Safety*,
DOI 10.1007/978-3-319-58768-4_8

and importers, becoming fifth in primary energy consumption, third in both pet-roleum imports and consumption, and first in liquified natural gas (LNG) imports. In addition, petroleum holds the largest fraction in the primary energy supply mix (44%), followed by coal (27%) and natural gas (22%), and the ratio of fossil fuels in total energy supply amounts to 93% in 2013 [1]. Moreover, the Japanese energy self-sufficiency ratio shows a considerably lower level, since the energy supply in Japan depends on imports of almost all fossil fuels; the Japanese energy self-sufficiency ratio is only 7% in 2013 [1], which exhibits a level below that in other developed countries. In addition, Japan is heavily dependent on the Middle East for about 80% of its domestic crude oil supply. Thus, nuclear power has traditionally played an essential role to ensure domestic energy supply in Japan where the situation of energy balance is considerably vulnerable as explained.

However, the impact of Fukushima nuclear accident, caused by Great East Japan Earthquake in Japan, is quite influential on the Japanese energy mix and socio-economy, and has caused intensive discussion for rethinking energy policy thereafter which strongly supported nuclear energy. Elaborate political and tech-nical effort has been dedicated to replace the loss of nuclear power supply, an important base-load technology contributing to energy security and environmental sustainability before the Fukushima. Actually, after the Fukushima accident, alternative energy sources compensating nuclear have shown a dramatic increase such as natural gas, petroleum and solar photovoltaic (PV) as well as electricity conservation. Therefore, the severe nuclear accident is considered to be one of driving force which might change the pathway of the country's energy mix, and the Fukushima can be understood as the tipping point for the country to pursue energy, environmental and nuclear policy adjusting into the socio-economic circumstance after the Fukushima.

Until now, a lot of academic effort has been dedicated to the development of energy system model which allows us to yield long-term energy scenario in a consistent way and to analyze the effectiveness of energy and environmental political instrument such as carbon tax, regulation or subsidization. However, the majority of existing analysis does not explicitly assess the impact of disruptive nuclear accident and its successive shutdown on the long-term pathway of energy portfolio, although the accident is actually observed to dramatically alter the situ-ation of energy balance in Japan.

The objective of this chapter is to overview the transition of energy supply and demand in Japan after the Fukushima, and, based on that, to discuss the possibility of considering nuclear accident as contingency risk in energy modeling analysis by applying the methodology of risk analysis such as stochastic dynamic program-ming. The chapter is organized as follows: Sect. 2 describes the energy balance situation in Japan before and after the Fukushima accident; Sect. 3 discusses the possible methodology to consider a nuclear accident in energy model and Sect. 4 depicts the concluding remark.

2 Impact of Fukushima Nuclear Accident on Japanese Energy Market

The Great East Japan Earthquake and following tsunami attack caused serious damage to nuclear power plants together with other critical energy infrastructures such as thermal power plants, oil refineries, and LNG import terminals in the broad area of eastern Japan. Particularly the significant damage was put on the power sector. Due to the sudden loss of 27 GW of power generation capacity in Japan where the total capacity is 231 GW, power supply shortage became a profound problem shortly after the earthquake. To resolve the power shortage, enormous efforts were concentrated on incrementing the power generation capacity by restarting aging and idle petroleum-fired power plant, while mandatory power saving was implemented to cope with insufficient power supply. In order to cover the electricity supply loss caused by the earthquake, urgent measures were conducted to maximize the utilization of natural gas (LNG)-fired and petroleum-fired power plants, including the aging idled power plants. In the demand side, rolling blackouts were implemented from March to April in Tokyo/Kanto area in 2011, and then the government ordered compulsory curtailment in power usage from July, 2011 to treat the peak power load in summer for the first time since the oil crisis in 1970s. Due to those measures, no critical unplanned blackouts were implemented during the summer period immediately after the Fukushima accident. However, thereafter, all of 50 nuclear power reactors shut down in Japan due to their regulatory inspection. After the Fukushima, there has remained strong concerns over the safety of nuclear power plant in the public, and the local authority which holds nuclear power plants in their sites does not easily provide an official permissions for those restarts. At last, on August 11, 2015, Sendai nuclear power reactor in Kyushu Electric Power Company restarted its operation, which is the first case of the restart since 2013. However, the nuclear shutdown has still influenced the power supply mix in Japan.

2.1 Trend of Energy Mix After Fukushima

After Fukushima, national concerted efforts to maximize the usage of thermal power generation as well as electricity saving have contributed to prevent the occurrence of power shortage. LNG-fired power plants and petroleum-fired power plants, which serves regularly as middle-peak power generator, operate so as to compensate the loss of nuclear power which is responsible for base load in power load profile. It should be also noted that power utility companies attempt to newly build thermal power capacity as well as to bring very aging power plants back on line.

The Fukushima nuclear accident triggered the shutdown of the country's entire nuclear power plants, which accounted for 30 percent of the country's electricity

Fig. 1 Annual power
generation mix in Japan [2, 3]

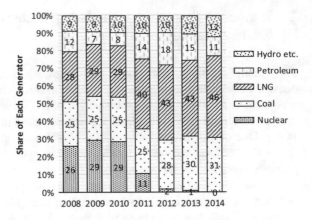

Fig. 2 Monthly power
generation mix in Japan [3]

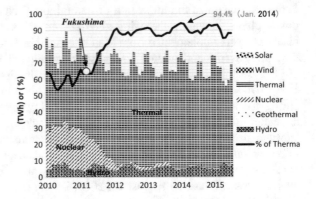

supply before the Fukushima (Fig. 1) [2, 3]. Due to the declined utilization of
nuclear power since the Fukushima accident, the fraction of fossil fuel over total
power generation reached at the highest level (94.4% in January 2014) in the last
three decades (Fig. 2) [3].

Substantial increase in LNG and petroleum is attributed to power generation to
offset the decline in nuclear power generation after the accident. In particular, a
radical shift to LNG occurred to compensate the loss of nuclear. Currently,
LNG-fired power accounts for almost half of the power generation mix in Japan
(Fig. 1) and LNG becomes an essential fuel in the Japanese energy supply portfolio.
Actually, Japan's LNG imports increased from 70.5 million tons (MT) in FY 2010
to 83.2 MT in FY 2011, 86.9 MT in FY 2012, 87.7 MT in FY 2013 and 89.1 MT in
FY 2014 [1]. Growing LNG consumption has caused an increase in import payments for LNG (Fig. 3) [4, 5], electricity supply cost, dependence on Middle East
for LNG supply (Fig. 4) and CO_2 emissions, and has provided a decline in the
country's energy self-sufficiency. Figure 3 shows Japan's annual imports of LNG.
Japan's import payments for LNG in 2014 were a record high at 7.9 trillion yen,

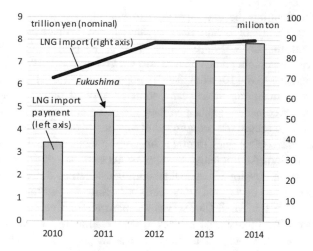

Fig. 3 Annual payment for LNG import in Japan [4, 5]

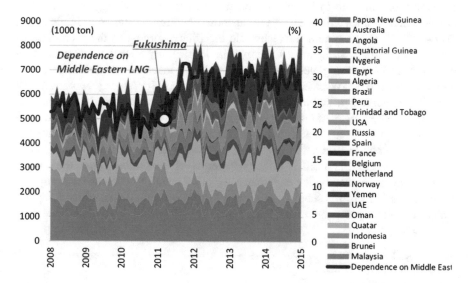

Fig. 4 Monthly LNG import by country in Japan [5]

more than doubling from FY 2010. By this soaring LNG import, the balance of payment turned to negative in fiscal year 2011 for the first time since 1980 [4].

Meanwhile, coal holds an important advantage of being economically competitive cost, although it has an environmental disadvantage of its higher intensity of CO_2 emissions. Thus, coal-fired plant plays an important role for an economical power supply as base load power generator. As nuclear power plants in Japan remain shut down since the Fukushima nuclear accident, Japanese utility companies actually began planning the new operation of coal-fired power plants in order to enhance the capability of economical power supply, influenced by the soaring import payment for LNG. For example, Tokyo, Kansai, Chubu, Kyushu and Tohoku electric power companies plan to add 1.0 GW [6], 1.2 GW [7], 1.0 GW [8], 1.2 GW [9] and 0.6 GW [10] of coal-fired power plant in their power generation mix respectively. Including those plans, total 13 GW of newly building plan of coal-fired is under consideration. Japan needs to place coal as an important energy resource in order to secure the economical competitiveness of power supply and to effectively purchase other fuel through the employment of coal as a bargaining power in negotiation. Additionally, to suppress its external environmental impact, Japan is expected to develop clean coal technologies such as IGCC (integrated coal gasification combined cycle) and CCS (carbon capture and storage).

As explained so far, fossil fuel serves as main alternative energy substituting nuclear energy after Fukushima. However, expectation has been currently concentrated on renewable energy sources such as solar photovoltaic and wind power systems, since those are domestic and carbon-free energy sources that will contribute to create advanced electricity business. In addition, renewables are socially preferable options due to its reliance on natural sources of energy as distributed power sources. Since July 1, 2012, the Japanese government started to implement a Feed-in Tariff (FIT) system for renewable electricity, aiming at the promotion of renewable energy in the country's energy mix. Particularly after the implementation of FIT in 2012 by the government, the cumulative installed PV capacity rapidly increased from 6.6 GW in 2012 to 23.3 GW in 2014 (Fig. 5) [11]. Moreover, PV capacity, which is certified to be built for the future and now under consideration, amounts to 68.9 GW as of October 2014 against 231 GW of total utility capacity in Japan (Fig. 5) [11], suggesting that the effects of FIT have been very powerful, although the power generation cost of PV is still more expensive (Fig. 6) [12]. However, a set of FIT tariff is difficult in terms of integrating renewables in an optimal way as implied by the experiences of other countries such as Germany and Spain. If the FIT tariff is set to be profitable for PV owners, PV investment rapidly expands well beyond the managerial capability of power grid, and increasing

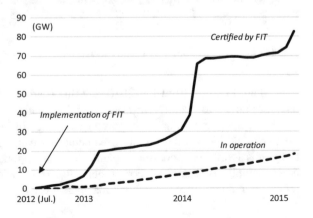

Fig. 5 Installed solar PV capacity (certified by FIT & in operation) in Japan [11]

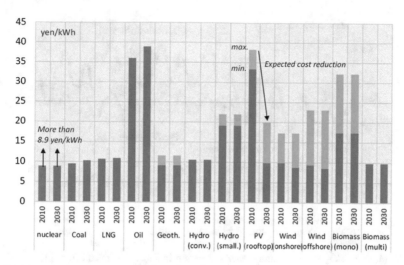

Fig. 6 Assessment of power generation cost in Japan [12]

electricity price through a surcharge by FIT puts a financial burden on end-users. It should be carefully considered that the massive integration of renewables poses a lot of challenges in technical and financial aspects, derived from their output intermittency and higher generation costs. Technical efforts are required to stabilize the power supply system through flexible resources such as rechargeable battery, back-up generators and demand response, and massive investments to enhance power grid capability will be indispensable. Meanwhile, if the tariff is set at a lower level, the installation of renewable power will not be successfully promoted. Paying cautious attention on the status of electricity market such as electricity price, it is important to promote renewable energy and improve the grid capability for the sustainable development of energy supply.

Increased dependence on fossil fuel supply and the associated growing import payments for the fossil fuels have eventually resulted in higher power generation cost and soaring electricity retail price. The increase in import payments for fossil fuel for power generation is projected to still cause an increase in average unit costs by 3.0 yen/kWh in FY 2015 from their FY 2010 levels [13]. Higher power generation cost has serious economic impact, causing more than 30% escalation in electricity bill for the average household in Tokyo after the Fukushima [14]. Actually, Kansai Electric Power Company, around half of which power generation mix is derived from nuclear energy, submitted applications for approval to the government about a 10.23% increase in power rates for the customers from FY2015. In those surroundings, rising electricity bills will pose a serious challenge for Japanese economy. Particularly for Japanese manufacturing sectors which come up against challenges in their international competitiveness, a further energy cost escalation is considered to be economically harmful. Since electricity price in Japan

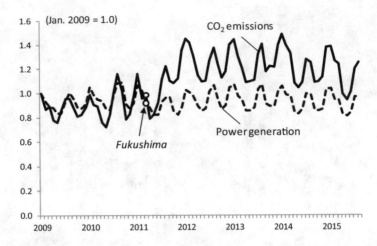

Fig. 7 Monthly CO_2 emissions and power generation in Japan (estimated from [3])

is relatively expensive in the context of an international comparison, its further increase can be understood as an additional financial burden.

CO_2 emissions will increase as well, because of increased usage of fossil fuel to replace the loss of nuclear power after Fukushima. The utility power companies accounted for 439 million tons of CO_2 for the year 2011 (after Fukushima), up 17 percent from 374 million tons in the year 2010 (before the Fukushima) (Fig. 7). Kansai Electric Power Company, which relies most on nuclear power, produced 65.7 million tons, a 40 percent increase in CO_2 emissions. Japan's total CO_2 emissions increased from 1.124 billion tons (6.1% higher than that in FY 1990) in FY 2010 to 1.173 billion tons (11% higher) in FY 2011, 1.208 billion tons (14% higher) in FY 2012 and 1.224 billion tons (16% higher) in FY 2013. Growing CO_2 emissions due to higher fossil fuel usage has imposed new challenges for Japan to strategically consider post-Kyoto climate change policy. As discussed so far, it is important to note that Japan currently faces multiple difficulties concerning energy security and environmental conservation together with sustainable economic growth.

2.2 Energy Policy Overview in Japan Before and After Fukushima

Before Fukushima, the Japanese government, in its Strategic Energy Plan officially approved in June 2010 [15], assigned the target of increasing the energy self-sufficiency ratio from the present level of 38 to 70% by 2030 and of mitigating CO_2 emissions by 30% compared with the 1990 level. For achieving those targets, an important measure was to expand the contribution of nuclear power supply in future energy mix. The Energy Plan assumes that carbon-free energy sources, that is, nuclear and renewable, account for 70% in power generation mix in 2030. The fraction of nuclear power itself amounts to more than 50% in the mix as indigenous and emission-free sources. The Japanese government planned to build additional 14 reactors, thereby boosting the installed nuclear capacity from 49 GW in 2010 to 68 GW by 2030, while maintaining all the existing plants. In addition, the government aimed to raise the capacity factor of nuclear power plants to average 90%. However, the Fukushima accident required a rethink of this Plan in terms of nuclear energy development.

The changing circumstances after Fukushima necessitate the fundamental review of the energy policy which was formulated before Fukushima. The new Strategic Energy Plan of Japan [16], the first national energy policy after Fukushima, was officially approved in April 2014, and the new governmental energy outlook to 2030 was presented in July 2015 [17]. The plan and the outlook discuss the normative view of desired energy mix to 2030 by reviewing the properties of nuclear power and other alternative energy sources, and proposes the optimal power generation mix to satisfy the requirement of energy demand and supply in terms of 3Es: energy security, environmental conservation and economic efficiency. The plan places nuclear power as an important base load power generator which satisfies all of the agenda in 3Es. The new energy outlook of Japan assumes the appropriate fraction of total electric power generation in 2030: 20–22% for nuclear, 27% for liquefied natural gas (LNG)-fired thermal, 26% for coal-fired thermal, 3% for oil-fired thermal, and 22–24% for renewables (Fig. 8). The plan describes that nuclear has advantage in economic efficiency based on its higher density of power output and in environmental compatibility as carbon-free source. Meanwhile, the plan describes that the enhancement of nuclear safety has higher priority than the assurance of 3Es and the dependence on nuclear in the energy supply portfolio should be decreased. It is therefore indispensable that nuclear operations strictly comply with the new safety standards, regarded as the world's most stringent code by the Nuclear Regulation Authority (NRA). According to the basic direction in the

Fig. 8 Power generation mix of Japan to 2030 [17]

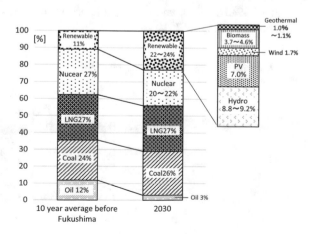

plan, the new outlook sets the desired fraction of nuclear in 2030 as 20–22%, a level below that in the plan before Fukushima (more than 50%). Regarding the appropriate fraction of nuclear (20–22%), the possibility of supplying power at that designated ratio by nuclear needs to be cautiously discussed. Because the existing capacity of nuclear power plants will decline by half to 2030, if the regulation of 40-year operational lifetime is assigned on all the existing plants [17]. However, the 20–22% nuclear fraction could be realized if a 20-year lifetime extension is endorsed by NRA through the additional investment to comply with the new safety standards. Other challenge for maintaining nuclear at the fraction is electricity market reform in Japan to 2020 where new investment for nuclear becomes difficult because nuclear faces competition to other alternative power resources. The new outlook describes that the desired fraction of nuclear energy could play an indispensable role to satisfy the requirement of 3Es in Japan, although multiple challenges remain to maintain the status of nuclear in the future power supply portfolio.

The plan also suggests to maximize the ratio of renewables at a level higher than the one pursued in the previous plan. In the new outlook, renewables will account for 22–24% of total power generation in 2030 (Fig. 8). A special consideration was conducted concerning the cost of purchasing renewable energy under the FIT system which was introduced in 2012. FIT has played a crucial role to increase the installation of renewable energy in Japan, while the financial burden of purchasing renewable energy at a specific rate from 10 to 20 years is regarded as one of the challenge in energy policy of Japan like Germany. The new outlook forecasts that annual purchasing cost of renewable will increase from 0.5 trillion yen in 2013 to 3.7–4.0 trillion yen in 2030. However, the outlook estimates that total electricity costs in 2030 will decrease by 3–5% from 9.7 trillion yen in 2013, since renewables, nuclear and energy efficiency reduce the consumption of fossil fuels, even if the purchasing cost for renewables will grow significantly.

In addition, the fraction of energy source in total primary energy supply in 2030 is estimated as well in the new outlook: oil 32% (40% in 2010, before Fukushima), coal 25% (22%), natural gas 19% (19%), nuclear 10–11% (11%), and 13–14% for renewables (7%) (Fig. 9). The energy self-sufficiency in Japan will increase to around 24% by 2030, and the dependence ratio on fossil fuels show a significant decline from 81% in 2010 to 76% in 2030.

Fig. 9 Primary energy mix in Japan to 2030 [17]

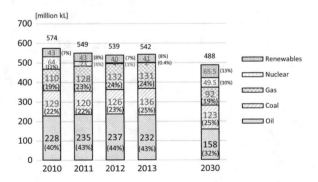

The new outlook describes the vision to satisfy the 3E targets. However, keeping the nuclear fraction at 20–22% and expanding renewable energy in power grid are considered to be big challenges, and political supports, technical advancement and gaining the public understanding are indispensable.

3 Attempt for Energy Modeling Considering Nuclear Accident

After the Fukushima nuclear accident, Japan recognizes itself as the country where the energy planning considering the risk of nuclear disruption is required to enhance the country's energy security, experiencing the unprecedented increase in fossil fuels imports to guarantee the loss of nuclear energy as explained so far. The previous section introduces the official energy outlook of Japan to 2030. However, generally speaking, those kinds of long-term projection do not consider the possible disruptive events such as nuclear accidents and the subsequent suspension of all nuclear power plants which causes the increase in power supply cost by massive fuel import and deep electricity saving. Moreover, the huge cost of decontaminating radioactive material needs to be considered in the case of nuclear accidents, based on the Fukushima experience. In addition, after Fukushima, the government started feed-in-tariff to promote renewable energy which has pushed up electricity price and the Japanese power generation mix has shown the radical change. Thus, if the nuclear accident is explicitly considered which possibly assigns severe socio-economic burden, the long-term outlook of the country's energy mix will take the alternative pathway such that the deployment of nuclear energy is suppressed and the alternative sources are more introduced against the nuclear accident.

Based on the author's existing literature [18], this section attempts to discuss a potential methodology of considering the nuclear accident in energy modeling analysis. The literature [18] regards nuclear accidents as phenomena to cause a shutdown of nuclear power supply and tries to specify the optimal operation of fuel stockpiles, corresponding to SPR (strategic petroleum reserve) in the U.S., as control variables, considering the possible occurrence of a nuclear accident. Other than nuclear accident, in the analysis, the closures of Hormuz Straits and Straits of Malacca are explicitly considered as the extreme events, which cause the increase in energy supply cost. The literature [18] attempts to formulate the mathematical optimization which minimizes the country's total energy system cost, adopting stochastic dynamic programming, taking into account nuclear power shutdown. As an estimation period, the literature assumes a short-term such as nearly one year where the minimum unit of the estimation period is each day of the year, and eventually energy supply capacity is assumed to be fixed variable. Through the calculation, the optimal charge and discharge strategy of fuel stockpile is endogenously determined under the risk of the nuclear power supply disruption.

Firstly, fuel price volatility can be modeled as stochastic process. The daily time profile of the fuel import price in the study is assumed to have a certain cyclic pattern in average. In the model, the fluctuating change of the fuel import price P_t is expressed with Winner process dZ_t. As the specific stochastic process, the mean reverting process was assumed where price recurs to the average price for the long term. The stochastic process of fuel price is shown as follows.

$$dlogP_t = \frac{dP_t}{P_t} = \alpha(log\theta_t - logP_t)dt + \sigma dZ_t \tag{1}$$

$$log\theta_t = \frac{1}{\alpha}\frac{\partial logF(0,t)}{\partial t} + logF(0,t) + \frac{\sigma^2}{4\alpha}\left(1 - e^{-2\alpha t}\right) - \frac{1}{2\alpha}\sigma^2 \tag{2}$$

where P_t: fuel price at t [yen/specific unit], θ_t: equilibrium fuel price at t [yen/specific unit], α: reversion rate, σ: volatility, dZ_t: winner process, F: future price.

The expected total energy system cost (the total cost that is necessary from t to the expiration of an analytical period) in the state of s and i at t is defined as $V_i(P_t, s, t)$. By stochastic dynamic programming, $V_i(P_t, s, t)$ becomes the minimum of the sum of the expected total energy system cost after the time of $t + dt$, $\sum_j \Pr(i \to j) \cdot E[V_j(P_t + dP_t, s + ds, t + dt)]$ and the total energy system cost in each unit of time, $TC(P_t, Av(u, Im_i), F)dt$, as follows.

$$V_i(P_t, s, t) = \min_u \Big\{ TC(P_t, Av(u, Im_i), F)dt + Stk(P_t, u, s)dt$$
$$+ e^{-rdt} \sum_j \Pr(i \to j) \cdot E[V_j(P_t + dP_t, s + ds, t + dt)] \Big\} \tag{3}$$

where i, j: state of fuel supply and nuclear, t: time step, $V_i(P_t, s, t)$: discounted total energy system cost, u: daily change of fuel stockpiles, s: stockpiles of crude oil and LNG, Im_i: Import of crude oil and LNG, $Av(u, Im_i)$: oil and LNG available in a day, F: power generation capacity, dt: differential of time(=1 day), $Stk(P_t, u, s)$: daily O&M cost of fuel stockpile, r: discount rate, $\Pr(\cdot)$: State transition probability, $TC(P_t, Av(u, Im_i), F)$: daily total system cost.

Through the computational simulation of the above Eq. (3), the optimal operation of crude oil or LNG stockpile can be theoretically identified under the risk of nuclear supply disruption assumed in certain probability $\Pr(\cdot)$ (state transition probability). Thus, the author so far analyzes the short-term impact of nuclear supply disruption on the operation of fuel stockpile, because the energy supply capacity is assumed to be fixed variable. The analysis reveals that the arrangement of adequate scale of energy stockpile such as LNG significantly decreases the expected cost of the country's energy supply against those extreme events. However, this is not the analysis regarding the long-term impact of disruptive events on the desirable pathway of power generation mix. As a future challenge to be considered, the author attempts to conduct the evaluation of the optimal installed capacity of fuel stockpiles and individual power generator in a long-term perspective.

4 Conclusions

The Fukushima nuclear accident has necessitated the fundamental review of Japan's energy supply portfolio and is a turning point for the country to rethink energy, environmental and nuclear policy acclimating to the socio-economic situation after Fukushima. For developing effective long-term energy policy, it is important to have a firm vision optimizing the country's energy mix, considering contingency risk such as nuclear severe accident which has actually and dramatically changed the country's energy balances. This chapter reviews the transition of energy balance in Japan after Fukushima and discusses the potential energy modeling approach to consider nuclear accident by applying stochastic dynamic programming.

To revise appropriately the long-term energy policy, various uncertainties need to be considered, and the discussion over long-term energy planning should be closely rooted in consistent and quantitative analysis based on certain mathematical tool. In this context, the simulation tool such as stochastic dynamic programming is expected to provide insight for developing effective energy policies. Actually, however, constructing the optimal energy portfolio is not merely a discussion about identifying the optimal combination of energy source. For the adequate formulation of energy policy against various risks, comprehensive argument should be conducted in various viewpoints such as sustainable economic growth, energy resource diplomacy and national technical competitiveness.

References

1. IEEJ (The Institute of Energy Economics Japan), EDMC *Handbook of Energy & Economic Statistics in Japan* (2014)
2. FEPC (Federation of Electric Power Companies of Japan), *Electric Power Statistics* (2014)
3. METI (Ministry of Economy, Trade and Industry), *Monthly Electricity Report* (2014)
4. MOF (Ministry of Finance), *Monthly Trade Report* (2014)
5. METI (Ministry of Economy, Trade and Industry), *Monthly Resource and Energy Report* (2014)
6. Tokyo Electric Power Company (TEPCO) (2014), http://www.tepco.co.jp/cc/press/2014/1236420_5851.html
7. Kansai Electric Power Company (KEPCO) (2014), http://www.kepco.co.jp/corporate/pr/2014/1128_1j.html
8. Chubu Electric Power Company (CHUDEN) (2014), http://www.chuden.co.jp/corporate/publicity/pub_release/press/3254876_19386.html

9. Kyushu Electric Power Company (KYUDEN) (2014), http://www.kyuden.co.jp/press_ h141119-1.html (in Japanese)
10. Tohoku Electric Power Company (2014), http://www.tohoku-epco.co.jp/news/normal/ 1188477_1049.html
11. METI (Ministry of Economy, Trade and Industry), *Present Status and Promotion Measures for the Introduction of Renewable Energy in Japan* (2014), http://www.meti.go.jp/english/ policy/energy_environment/renewable/
12. NPU (National Policy Unit), *Final Report, Cost Verification Committee* (2011)
13. IEEJ (The Institute of Energy Economics Japan), *Short-term Energy Outlook in Japan* (2014)
14. Tokyo Electric Power Company (TEPCO), *Electricity Rate of the Average Model, Electricity Rates and Rate Systems* (2014)
15. METI (Ministry of Economy, Trade and Industry), *The Strategic Energy Plan of Japan— Meeting Global Challenges and Securing Energy Futures* (2010)
16. METI (Ministry of Economy, Trade and Industry), *The Strategic Energy Plan of Japan* (2014)
17. METI (Ministry of Economy, Trade and Industry), *Long-term Energy Demand and Supply Outlook of Japan* (2015)
18. Y. Kawakami, R. Komiyama, Y. Fujii, Development of energy security evaluation method using mathematical programming and analysis of optimal strategy for fuel stockpile operation. J. Jpn. Soc. Energy Resour. **34**(5), 21–30 (2013)

Deprivation of Media Attention by Fukushima Daiichi Nuclear Accident: Comparison Between National and Local Newspapers

Ryuma Shineha and Mikihito Tanaka

Abstract On March 11th in 2011, a huge earthquake and tsunami struck Japan and caused severe accidents at the Fukushima first nuclear power plant (NPP). The impact and damages of these triple disasters, called "3.11," continue to this day. There was a diversity of damages and social conditions among devastated areas. This means that this disaster struck so broad area that it brought many kinds of "realities" to different areas. Therefore, we cannot treat the various regions that were affected uniformly. At the same time, the attention given to the 3.11 based on location has ultimately been covered differently between various media sources. The aim of this paper is to share basic descriptions and media analysis of the 3.11 disasters for future discussions. Through our analysis, it is showed that there is a different framing of the 3.11 between national and local media. The difference implicates that it is deprived of social interest in the national newspaper by the NPP accident, and on the other hand, the local newspaper kept their perspectives reflecting damages from the earthquake and tsunami.

Keywords 3.11 · Media analysis · Newspaper · Local media

1 Introduction

On March 11th in 2011, a huge earthquake and tsunami struck Japan and resulted in many victims. The earthquake and tsunami caused severe accidents at the Fukushima first nuclear power plant (NPP). The impact and damages of these triple disasters, called "Higashi-Nihon-Daishinsai" or "3.11," continue to this day.

R. Shineha (✉)
Seijo University, Tokyo, Japan
e-mail: r_shineha@seijo.ac.jp

M. Tanaka
Waseda University, Tokyo, Japan
e-mail: steman@waseda.jp

© The Author(s) 2017
J. Ahn et al. (eds.), *Resilience: A New Paradigm of Nuclear Safety*,
DOI 10.1007/978-3-319-58768-4_9

111

To consider various issues resulting from the 3.11 disasters, we must understand the continuing damages, social-structural issues behind the 3.11, and the complex effects of triple disasters. In addition, how "realities" of damaged sites were overlooked in the society is another important point. This is rephrased to, "how were social attentions deprived?" In response to this question, we conducted quantitative media analysis between national and local newspapers: *Asahi-shimbun* and *Kahoku-shimpo*.

1.1 Basic Description of the 3.11 Disasters

The aim of this section is to share basic descriptions of the 3.11 triple disasters. For this purpose, we would like to use a description-oriented method. Through descriptions of damaged sites, we will know the complexity of damages brought by the earthquake, tsunami, and the NPP accident.

For the horrible triple disasters, there were over 20,000 deaths and missing persons. Approximately 170,000 people evacuated from their homes in Japan. Three prefectures in particular—Miyagi, Iwate, and Fukushima—were the most affected (for example, deaths in the Miyagi prefecture reached over 10,000). We summarize the number of victims, focusing on these three prefectures, in Table 1.[1] The tsunami wiped out several hundred kilometers of coastline in towns. Concerning causes of death, drowning was responsible for over 90% of lost victims, and about 65% of the victims were over the age of 60 [3].

Looking at the press release by the national government on June 24, 2011, it was estimated that total losses for the building, life-line, social infrastructure, agriculture, and others reached approximately 16.9 trillion yen. Over 25 million tons of rubble and general waste were generated. Of course, there continues to be other serious problems, such as the stress of disaster victims, discrimination, etc.

In addition, the social conditions were different between damaged areas. Previous studies showed that local towns that were seriously injured were considered to be an aging society, farming and fisheries society, and generally poorer than when compared to metropolitan society [1, 2]. And of course, such negative conditions compound upon each other. Generally speaking, the municipalities that were seriously injured faced deeply-indented coastlines; this geographical disadvantage is not only vulnerable to tsunamis but also reinforced aging and economically vulnerable society. Furthermore, these vulnerabilities (such as an aging

---

[1]Previous data of the Table 1 were represented in Shineha [1] and Tanaka et al. [2]. This data is updated, using statistics published by governments until May 8th 2015: Prefectural governments of Miyagi, Iwate, Fukushima, and the Reconstruction Agency. Concerning the death toll, the associated number of deaths from the 3.11 was included. In addition, the number of the disappeared is the number of those who never received death notifications of their family members. There are 224 persons who were registered in this type of situation.

Table 1 Breakdown of damages

		Iwate	Miyagi	Fukushima
Human damages	Fatalities	5,124	10,534	3,727
	Missing	1,129	1,246	3
	Evacuees (to other Prefecture)	28,242 (1557)	67,510 (7055)	69,208 (46,170)
Collapse of buildings/houses	(Complete or half destroyed)	26,163	238,123	92,905

society) have been reinforced more and more in some areas of damaged sites after the 3.11.

In addition to the terrible natural disasters, we also faced the severe accident of NPPs. In "Risk Society," Ulrich Beck discussed the globalization of risk and inequality on the distribution of risk and benefit [4]. Such discussions regarding risk society point out a collision of risk receiver and benefit receiver as well as risk gap according to social stratum and area. Generally speaking, the weak members of society tend to become risk receivers.[2]

1.2 The Effect of the Fukushima-Daiichi Nuclear Power Plant Accident

In addition to the huge damage of the earthquake and tsunami, evacuation zones were set up after the NPP accident and many people were forced to leave their towns, particularly in the Futaba area of the Fukushima prefecture. At the first, the evacuation zone was set within a radius of 20 km from the NPP. However, after April 1, 2013, the evacuation zone was redefined in three categories according to the amount of radiation: Difficult-to-return zone (Kitaku-Konnan Kuiki); No-residence zone (Kyojyu-Seigen Kuiki); and zone being prepared for lifting of evacuation order (Hinan-Shiji Kaijo Junbi Kuiki).[3]

Figure 1 shows the current situation of Namie, a town that was established as a mandatory evacuation zone after the NPP accidents. However, after the re-categorization of evacuation areas on April 1, 2013, a part of Namie became available for short-time stays with the local government's permission. At the current status, Namie has three different types of evacuation areas.

---

[2]In a model of disaster studies, disaster is regarded as an opportunity in which social vulnerability comes up to the surface [5]. Thus, when we think disaster and its hazard, we also have to think about the social structure and vulnerability behind the disaster.

[3]Since the 1st April 2016, zone being prepared for lifting of evacuation order (Hinan-Shiji Kaijo Junbi Kuiki) was opened. As the result, the center of the town can be accessed without permission.

Fig. 1 The town Namie collapsed by earthquake and tsunami. *Left* Main street of the town (April 11th, 2013). *Right* The seaside area (Uketo area; March 2nd, 2014). They show the effects of isolation from the NPP accident

At the time of taking these pictures on April 11, 2013, destroyed houses were left as-is for more than 2 years after the NPP accident. In Namie, there were many houses and buildings destroyed by the earthquake. The level of collapse has progressed according to time. The inside walls became moldy and eventually collapsed, because rains invaded through the holes of damaged ceilings. There are few figures in the street.

In addition, the coastal area was devastated by the tsunami. The horrible destruction and traces of the tsunami were left as-is for more than 4 years. In other areas, removing of houses, cars, and ships destroyed by the tsunami has been developed. However, in the coastal area of Namie, those remained for more than 4 years, because of the recovery and reconstruction process, which was prevented, and continued the isolation from the NPP accident. As a result, the level of damage varied among the affected regions and still remains.

The continuous problems of these affected situations make it difficult for residents to come back. According to the current report of a questionnaire conducted by the research group of Waseda University collaborated with Namie local office, the answer of "I will not come back to Namie" (34.8%) and "I don't know" (45.6%) were most chosen responses. On the other hand, the answer related to "I will come back" constituted just 17.0% of responses.[4] The popular reason given for why they will not come back was, "if we come back, it's difficult (or impossible) to live as we used to" [6].

Over 4 years have already passed since the 3.11. Some evacuees succeeded in making new lives in other areas. The relationship between their children and friends are also points of consideration to stay there. Moreover, if they come back to Namie in its current state, serious damages in terms of infrastructure, community, and

---

[4]Total score of "I will come back to the Namie after the completion of de-contamination process by the national government" and "I would like to come back to the Namie if de-contamination process takes a long time."

various capitals (cultural, social, economic, and human) have still not been recovered. Although the Namie local office planned for the returning process to start from 2017, many issues and hurdles remains. Officers recognize their difficult tasks still ahead. When we think of the effects of the NPP accident, we should consider the complex influences of the NPP accident on the reconstruction process.

1.3 The 3.11 and Information Ecology

Keeping in mind the variety of continued damages and issues described above, we should start to consider problems of social structure behind the 3.11 from the viewpoints of "information" and "media." Huge amounts of information and discourses concerning the 3.11 have persisted, and in our opinion, this is one of the most significant features of the 3.11.

We surmise that many readers can imagine the effect of the Internet and social network sites (SNSs) in the 3.11 contexts. Actually, the spread of information via the Internet enables us to create and use large amounts of information. This situation has been reinforced by SNSs such as Twitter, Facebook, and so on. These convenient informational tools set the base of information for volunteers in a disaster.

However, we should take a point of view that there is an information gap according to social stratum, income, age, etc. Looking at the status of Japanese Internet usage just before the 3.11, the reports published by the Ministry of Internal Affairs and Communications (MIC) in 2011, we can extract three important aspects:

- Generally speaking, the coefficient of use of Internet for elderly people is lower than for younger people (e.g., 60–64 age, 70.1%; 65–69 age, 57.0%; 70–79, 39.2%; over 80, 20.3%).
- Families with low annual incomes have a low coefficient of use of Internet (e.g., under 2 million, 63.1%; 2 million to 4 million, 68.6%; over 6 million, over 80%).
- Use of Internet depends on the geographical area. The Tohoku region has a low score compared to other metropolitan areas.

When we think of 3.11, we should take such conditions of the Japanese information society into account, in addition to destruction of information infrastructure. Generally, the damaged areas from 3.11 are aging and low income communities. Moreover, in the 1970s, Japanese society attempted to be a society where "all of the citizens are middleclass." As a result, many upper-middleclass people appeared. However, looking at the media usage, the relationship between social class and usage and thoughts regarding media has been discussed and investigated since the 2000s. Several gaps according to social stratum have been recognized, such as the term "digital divide." For example, lower social capital cultures use mobile phones but not smartphones; they are social gamers but do not have PCs; and they are light Internet users. Another example is that TV audiences are part-time workers. On the other hand, higher social capital clusters have smartphones, they are users of tablet PC and SNSs, and so on. In summary, current gaps of information and media usage have appeared.

In any event, it seems that the majority of Internet users regarding the 3.11 were in the non-damaged metropolitan areas such as Tokyo, Yokohama, etc. For the gap of information environment, victims could not collect and transmit information effectively, compared to a majority in the information society. It seems that this "gap of information environment" is one of the background factors for the gap of "information, which is focused and taken up by and in media."

Considering this situation, we should review different information sources used in local sites to understand "realities" of information ecology of the 3.11. Rausch discussed the particular and significant role of local newspapers in local society in Japan [7]. Also, in damaged areas of the 3.11, they have their key local newspapers. Thus, we have to focus on the contents and framings of local newspapers in addition to national-level media. From the point of view that media have played various roles in agenda-building and bringing attention to issues through framing [8–11], analysis of news media should be conducted. What was the focus in national and local newspapers? What kinds of differences existed between national and local newspapers in the 3.11 context? To answer these questions, we conducted comparative quantitative analysis of *Asahi-shimbun* and *Kahoku-shimpo*.

2 Materials and Methods

2.1 Data Collection of National and Local Newspapers

In this chapter, we focus on the *Asahi-shimbun* as an example of a national newspaper, and also *Kahoku-shimpo* as an example of a local newspaper. *Asahi-shimbun* is one of the famous national newspapers of Japan. News articles of *Asahi-shimbun* were collected from the database *KikuzoII*. *Kahoku-shimpo* is the most famous local newspaper, which has top share in the Miyagi Prefecture. Text data of *Kahoku-shimpo* were collected from the *Kahoku-shimpo* database. As for the results, we collected 5,405 articles of *Asahi-shimbun* from March 11 to April 7. Concerning *Kahoku-shimpo*, 2,049 articles from March 12 to April 8, 2011 were collected.[5]

---

[5]Straight news from news service companies were excluded from the Kahoku-shimpo data set. In addition, at the first several days, "Kanto-Tohoku Great Earthquake" was used in Kahoku-shimpo, and these words were replaced with "Higashi-Nihon Daishinsai" in this analysis.

2.2 Seeking Specific Keywords and Co-word Mapping Comparison Between National and Local Newspapers

Since March 11, there has been a large amount of information while discourse related to 3.11 has been rampant. Thus, we conducted quantitative analysis of discourses regarding 3.11 in legacy and social media. We conducted frequency analysis, analysis of specific keywords of each week, and network analysis focusing on the co-occurrence of keywords (co-word).

For analysis of frequency and specific keywords ranking[6] of each week, we used KH coder (http://khc.sourceforge.net/) developed by Koichi Higuchi [12, 13].

We then searched for the target keywords in each text. We employed co-word analysis to describe Japanese media trends on the 3.11. The analysis employed in the present study relied on previous research by Leydesdorff and Hellsten [14, 15]. They expressed co-words such as "the carriers of meaning across different domains" ([15]: 232). The method of analysis here basically depended on previous studies by Leydesdorff and Hellsten [14, 15]. Salton's cosine was calculated as the similarity index [16–18]. In all analysis of this chapter, the cosine threshold was settled at 0.375. Networks have been described by Pajek [19]. For the configuration algorithm for nodes, the Kamada and Kawai algorithm was used [20]. As the index of centrality, "Betweenness centrality" was used [21].

3 Results

We would like to discuss another gap: "gap of information, which is focused and taken up by and in media". Figure 2 shows the time-lined change of article number and appearance ratio of keywords of *Asahi-shimbun* and *Kahoku-Shinpo*. *Asahi-shimbun* is one of the most prestigious newspapers in Japan. *Kahoku-Shinpo* is the most famous and important block newspaper in the Miyagi Prefecture. Looking at this line-chart of *Asahi-shimbun*, we analyzed the oscillation cycle of article numbers and the change of trend in keyword appearance. The appearance ratio of "Earthquake" gradually decreased, but on the other hand, the ratio of "NPP" increased. At the same time, "tsunami" marked a lower score than "NPP." Meanwhile, it is clear that *Kahoku-Shinpo* has different trends from *Asahi-shimbun*. In *Kahoku-Shinpo*, the appearance ratio of "tsunami" continued to remain equal to "NPP." The oscillation cycle of article number in *Asahi-shimbun* had the tendency to decrease, whereas, there are no such trends of decrease in article numbers in *Kahoku-Shinpo*.

---

[6]For making ranking of specific keywords of each time-tag, KH coder calculated Jaccard index between time-tag and keywords.

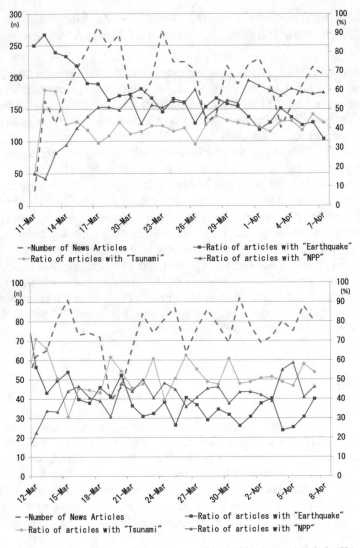

Fig. 2 Time-lined change of keywords appearance of *Asahi-Shimbun* and *Kahoku-Shimpo* during the first months. In these line charts, *left axis* shows the number of news articles concerning the 3.11, *right axis* shows the appearance ratio of each keyword

Tables 2 and 3 show the time-lined changes of top 10 characterized keywords of *Asahi-shimbun* and *Kahoku-Shinpo* from first to fourth week. In this tables, Jaccard index of time-tag (week 1, week 2, etc.) and keywords was used. From the Table 2 of *Asahi-shimbun*, we can find that keywords related to earthquake and tsunami were particular in the first week. However, this situation changed after the second

Table 2 Top 10 characterized keywords of *Asahi-Shimbun* from first week to fourth week analyzed by KH Coder

Week 1		Week 2		Week 3		Week 4	
Earthquake	0.242	Higashi-Nihon Daishinsai	0.232	NPP	0.208	NPP	0.192
Higashi-Nihon Daishinsai	0.202	Disaster-affected	0.225	Fukushima	0.198	The quake	0.188
Damage	0.173	Evacuation	0.214	Accident	0.179	Disaster-affected	0.182
Explanation	0.164	Local	0.208	The quake	0.157	Accident	0.17
Effect	0.16	Fukushima	0.207	Prefectural-level	0.156	Explanation	0.158
Tsunami	0.158	NPP	0.206	Area	0.129	Support	0.158
Occurrence	0.158	Earthquake	0.205	Sougou	0.128	Tsunami	0.156
Confirmation	0.153	Accident	0.175	Safety	0.127	Japan	0.148
Situation	0.151	Explanation	0.174	Life	0.122	Area	0.133
Disaster	0.148	Perfectural-level	0.173	Many	0.115	Reconstruction	0.124

Table 3 Top 10 characterized keywords of *Kahoku-Shimpo* from first week to fourth week analyzed by KH Coder

Week 1		Week 2		Week 3		Week 4	
Earthquake	0.243	evacuation	0.193	Higashi-Nihon Daishinsai	0.244	NPP	0.216
Higashi-Nihon Daishinsai	0.234	recovery	0.175	Tsunami	0.22	Fukushima	0.213
Occurrence	0.198	NPP	0.174	Fukushima	0.216	Accident	0.207
Evacuation	0.191	Life	0.172	Miyagi	0.211	Tsunami	0.206
Sendai	0.19	Iwate	0.138	Disaster-affected	0.211	Damage	0.199
Resident	0.178	Area	0.134	Damage	0.202	Disaster-affected	0.197
Explosion	0.174	condition	0.128	Life	0.199	The quake	0.186
Confirmation	0.168	Missing	0.127	Accident	0.199	Reconstruction	0.179
Information	0.162	Possible	0.125	NPP	0.199	Speak	0.177
Situation	0.154	Whereabouts	0.124	Evacuation	0.191	Support	0.16

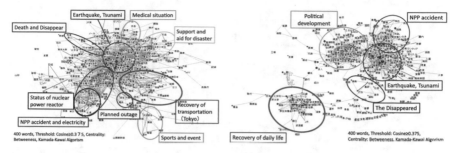

Fig. 3 Co-word map of *Asahi-Shimbun* (*left* week 1, *right* week 4)

week. Keywords concerning the Fukushima NPP accident appeared clearly in this ranking. This trend was strengthened according to time. On the other hand, Table 3 of *Kahoku-Shinpo* shows a different trend. For example, in week 2, "recovery" appears in the second position of the ranking. Simultaneously, "Iwate," "missing," and "whereabouts," are ranked. This is a different character from the results of *Asahi-shimbun*. Also in week 3, "tsunami" is ranked in the second position, higher than "NPP." The rank of "NPP" of *Kahoku-Shinpo* in week 3 is lower than the result of *Asahi-shimbun*.

Figure 3 describes the co-word network from March 11 to March 17 (first week after the 3.11), and from April 1 to April 7 (fourth week after the 3.11) of *Asahi-shimbun*. Clearly, we can find the difference of network structure among two co-word networks. This means the degree of organization of topics. The co-word network of the first week shows the confusion of topics and agendas in the after-math. The organization of topics and agendas progressed according to time (see Fig. 3). Furthermore, we can extract more implications from these co-word networks. The keyword cluster meaning "Earthquake and Tsunami" has a dense network with keyword cluster meaning "NPP accident and electricity." This means that the topics on "Earthquake and Tsunami" and topics on "NPP accident and electricity" were often connected in articles. This trend is seen in all co-word networks of each week (data not shown).

On the other hand, Fig. 4 shows the co-word network of *Kahoku-shimp*o in the first (March 12–March 18) and fourth week (April 2–April 8) after the 3.11. They have some different characters from the co-word networks of *Asahi-shimbun*. One of the features of *Kahoku-shimp*o is that there are relatively weaker connections between "Earthquake and Tsunami" and "NPP accident" than others. We can find this point from the result of the first week after the 3.11. The co-word of *Asahi-shimbun* in the fourth week showed clear co-word clusters on "Political develop-ment"[7] but on the other hand, "Medicine and human service" and "Support and aid for disaster" appeared in *Kahoku-shimpo* at that time.

---

[7]At that time, nationwide local elections were conducted, and this theme attracted interests of national media.

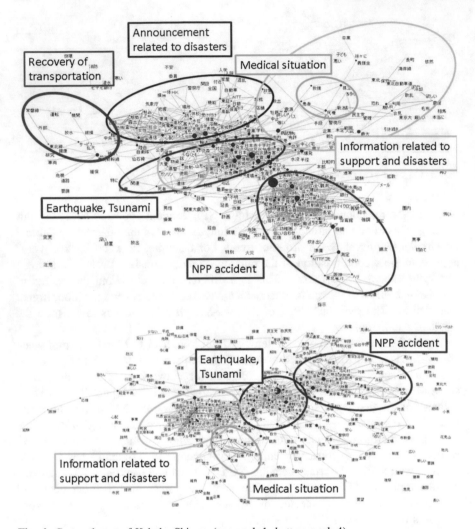

Fig. 4 Co-word map of Kahoku-Shimpo (*top* week 1, *bottom* week 4)

4 Discussion

Tanaka, Shineha, and Maruyama also showed that there are gaps of interests between newspapers and social media [2]. From their study, web news showed a rapid decrease in topics on the 3.11 disasters, but there were simultaneous increases in topics regarding "entertainment," "sports," "economics," and so on. Although this shift of interest is common for media, the difference of degree between media is illustrated. This result can be interpreted as newspapers continuing to take up 3.11 issues more actively than Web news and underpin public interests on the 3.11 disasters. However, we should not dismiss that topics of 3.11 in newspapers have also been dominated by topics regarding "NPP," and the topics of "earthquake and tsunami" were covered with them.

At the same time, blog entries with contents on "Earthquake" and "Tsunami" rapidly decreased. After the decrease of "Earthquake" and "Tsunami," "Radiation exposure" and "NPP accident" rapidly increased. However, interest on them were not sustained. In summary, interest on disasters and the NPP accident have rapidly faded out. The remained and stable interest was about "low radiation exposure" and "internal exposure." The rapid decrease of interest on NPP was also found on Twitter. In the analysis of Twitter, we categorized a variety of topics such as "NPP" such as "NPP accident," "Nuclear power policy," etc. In the aftermath of the NPP accident, tweets regarding "NPP" occupied in over 10% of all tweets, however, by the third month after the accident, the coverage decreased to fewer than 2%. Shortly thereafter, in the Internet sphere, it seemed that the interest concerning the 3.11 rapidly decreased [2]. Considering insights from the report on Japanese Internet use published by the Ministry of Internal Affairs and Communications [22], there is a gap of coefficient use of Internet according to different areas affected by the disasters. The Tohoku region has a low score compared to other metropolitan areas. This indicates that the decrease of interest in the earthquake and tsunami was driven by interests of urban residents with less damages. In other words, the majority of generator and consumer of discourses concerning the 3.11 on the Internet were people in the metropolitan areas with less damages.

Simultaneously, our current data indicated that there are different media interests between national and local newspapers. The national newspaper (*Asahi-shimbun*) tended to focus on the NPP accident rather than tsunami and earthquake, while on the other hand, the local newspaper (*Kahoku-shimpo*) tried to keep their attention placed toward the tsunami and earthquake. This indicates that these two newspapers faced different contexts or "realities." However, at least, it can be said that the "NPP" accident played a strong role in setting topics and engulfed interests on "Earthquake and Tsunami" in national and social media, and the gap of interests between national and local media became broader. We should discuss the effect of the NPP accident from the perspectives of "exploitation of interests" hereafter. Considering previous discussions of media studies, it can be said that gaps of attention between national/social media and local media will influence the distribution of capital and social interests during the reconstruction process through their agenda and frame building process [9–11].

5 Conclusion

There are continuous damages in the sites devastated by earthquake, tsunami, and the NPP accident. Damages show their complex characters according to areas. Particularly in areas where it has been regarded as evacuation areas for the NPP accident, various effects and hurdles for the reconstruction process appeared. Therefore, we cannot treat those effects uniformly. In addition, there are different trends of media attention between national/social and local media. "Realities," framings, and agendas of local media of damaged sites faced difficulties to spread at the national level. Although the rapid decrease of interest concerning 3.11 occurred in national and social media, local media continued to face their "realities" at each site.

The disaster strucked local sites but the discussion of the reconstruction process related to national level regulations. During dialogues with local journalists conducted by our research group, they said that they felt weak to set the agenda process at the national level. Briefly, the locals were "peripheralized" and agenda-setting was developed in the "center" without enough care for local contexts and diversity of "realities." While the situation continues, the gap of reconstructions has been spread gradually according to gaps of damages and social conditions of areas. However, attention continues to decrease.

Acknowledgements Authors were supported by grant-in-aid of SOKENDAI-CPIS, Seijo University (Tokubetsu-Kenkyu-Joseikin), Suntory Foundation, and the JSSTS-Kakiuchi-Kinen Support Award 2014.

References

1. R. Shineha, Society behind the 3.11 disasters, in *Science and Politics after the Disaster of March 11 in Japan*, ed. by M. Nakamura, pp. 179–224 (Nakanishiya Press, Kyoto, 2013)
2. M. Tanaka, R. Shineha, K. Maruyama, *Disaster Vulnerable and Information Vulnerable: What has been overlooked after the 3.11* (Chikuma Press, Tokyo, 2012)
3. Cabinet Office, White paper of disaster prevention 2011 (2011). http://www.bousai.go.jp/kaigirep/hakusho/h23/index.htm. Accessed 30 Aug 2015
4. U. Beck, Risikogesellschaft - Auf dem Weg in eine andere Moderne. Suhrkamp Verlag (1986)
5. B. Wisner, P. Blaikie, T. Cannon, I. Davis, At Risk: Natural Hazards, People's Vulnerability and Disasters, 2nd edn. (Routledge, 2003)
6. Waseda University, *Report on actual condition survey of damages of residents in the Namie-town from results of questionnaire* (2015). http://www.town.namie.fukushima.jp/uploaded/attachment/2040.pdf. Accessed 30 Aug 2015
7. A.S. Rausch, *Japan's Local Newspapers: Chihōshi and Revitalization Journalism* (Routledge, New York, 2012)
8. A. Downs, Up and Down with Ecology: The Issue Attention Cycle. The Public Interest no. **28**, 38–51 (1972)
9. R.M. Entman, Framing: Toward Clarification of a Fractured Paradigm. Journal of Communication no. **43**, 51–58 (1993)
10. M.E. McCombs, D.L. Shaw, The agenda-setting function of mass media. Public Opinion Quarterly no. **36**, 176–187 (1972)
11. D.A. Scheufele, Framing as a theory of media effects. International Communication Association no. **49**, 103–122 (1999)
12. K. Higuchi, Tekisutogata-data no keiryouteki bunseki: futatsu no apurochi no shunbetsu to tougou (Quantitative Analysis of Textual Data: Differentiation and Coordination of Two Approaches). Riron to Houhou (Sociological Theory and Methods) no. 19: 101–115 (2004)
13. K. Higuchi, *Syakaichousa no tameno keiryou tekisuto bunseki (Quantitative Analysis for Social Researchers: A Contribution to Content Analysis)* (Nakanishiya Press, Kyoto, 2014)

14. L. Leydesdorff, I. Hellsten, Metaphors and diaphors in science communication: mapping the case of "stem-cell research". Science Communication no. **27**, 64–99 (2005)
15. L. Leydesdorff, I. Hellsten, Measuring the meaning of words in contexts: an automated analysis of controversies about "monarch butterflies," "frankenfoods," and "stem cells." Scientometrics (67) 231–258 (2006)
16. W.P. Jones, G.W. Furnas, Pictures of relevance: a geometric analysis of similarity measures. Journal of the American Society for Information Science no. **36**, 420–442 (1987)
17. G. Salton, A. Wong, C.S. Yang, A vector space model for automatic indexing. Information Retrieval and Language Processing **18**, 613–620 (1975)
18. G. Salton, M.J. McGill, *Introduction to Modern Information Retrieval* (McGraw-Hill, New York, 1983)
19. V. Batagelj, A. Mrvar, Pajek: A Program for Large-Network Analysis. Connections no. **21**, 47–57 (1998)
20. T. Kamada, S. Kawai, An algorithm for drawing general undirected graphs. Information Processing Letters no. **31**, 7–15 (1989)
21. L.C. Freeman, Centrality in Social Networks Conceptual Clarification. Social Networks no. **1**, 215–239 (1979)
22. Ministry of Internal Affairs and Communications. (2011). *Report on Current Status of Information and Communications 2010* (2011). http://www.soumu.go.jp/main_content/000114508.pdf. Accessed 30 Aug 2015)

Development of a Knowledge Management System for Energy Driven by Public Feedback

Massimiliano Fratoni, Joonhong Ahn, Brandie Nonnecke,
Giorgio Locatelli and Ken Goldberg

Abstract The Nuclear Engineering Department at the University of California, Berkeley, in collaboration with the Industrial Engineering and Operations Research Department and the University of Lincoln in the United Kingdom, is proposing to create an open web platform that makes high-quality scientific data on energy sources readily available, assembles those data into metrics more suitable to the general public's knowledge and interest (e.g. impact on the family's budget or green house gas emission), and visually renders such information in a straightforward manner.

Keywords Knowlegement management · Web platform · Metrics · Nuclear energy

1 Introduction

In the era of information technology a large amount of data is readily available at everyone's fingertips. Energy and its implications, scarcity or abundance of resources, impact on climate change, emissions of pollutants, and more are topics of global interest that receive strong attention across all media. Opinions, official statements, and scientific data create a continuous flow of information. Nuclear energy among all sources is the subject of strong debates with cohorts of supporters and detractors ready to pinpoint its benefits or its drawbacks, respectively. In this large pool of information, it is of paramount difficulty even for field experts to isolate scientific data on energy, and to select reliable and coherent sources. Furthermore, higher quality data are often packaged in scientific jargon and are presented in forms and ways to which the general public does not relate

M. Fratoni (✉) · J. Ahn · B. Nonnecke · K. Goldberg
University of California, Berkeley, CA, USA
e-mail: maxfratoni@berkeley.edu

G. Locatelli
University of Lincoln, Lincoln, UK

© The Author(s) 2017
J. Ahn et al. (eds.), *Resilience: A New Paradigm of Nuclear Safety*,
DOI 10.1007/978-3-319-58768-4_10

127

(e.g. investment NPV, Sox produced, GDP impact, etc.). The Nuclear Engineering Department at the University of California, Berkeley, in collaboration with the Industrial Engineering and Operations Research Department and the University of Lincoln in the United Kingdom, is proposing to create an open web platform that (1) makes high-quality scientific data on energy sources readily available, (2) assembles those data into metrics more suitable to the general public's knowledge and interest (e.g. impact on the family's budget or green house gas emission), and (3) visually renders such information in a straightforward manner. Through this platform users will be able to create "energy portfolios" by mixing energy sources and evaluating how different choices impact the metrics they are interested in. Rather than a top-down approach, the platform will solicit feedback from the end-user on the prioritized topics as well as contribute additional topics with help of a knowledge management system.

2   Functionalities of the Envisioned Platform

The proposed web platform will include two major components: a user opinion component with working name "Energy Report Card" and an information component with working name "The Energy Challenge".

The "Energy Report Card" integrates elements from the Opinion Space project (http://opinion.berkeley.edu/) and the California Report Card project (http://californiareportcard.org) developed at the CITRIS Data and Democracy Initiative and informed by work done by the World Bank on the use of report cards as assessment tools of government performance. The Energy Report Card gathers feedback on users' perceptions toward environmental, social, and economic impacts of energy sources. Upon entering the system users will be asked to assign a value from 0 "Strongly Disagree" to 9 "Strongly Agree" on six quantitative assessment questions that will be used to gauge each user's preference for environmental, social, and/or economic impacts as high priority issues (Fig. 1). For example, participants will be asked whether they believe global warming (environmental impact) is a high priority issue, whether job creation (social impact) from energy production is a high priority issue, and whether energy cost stability (economic impact) is a high priority issue, among others.

Participants will then enter "The Energy Challenge" where they will be presented with an energy portfolio that matches their personal environmental, social, and economic interests. Participants will be able to adjust the different energy sources composing their energy portfolio. As they add and remove components to the portfolio they can observe how the selected metrics respond to each change. Additional text, graphics, videos, and links will also be provided through the page to explain the correlations between sources and metrics (similarly to what is done in the "California Budget Challenge"). Unrealistic scenarios, i.e. 100% nuclear energy or 100% solar energy, will prompt a warning message with an accurate and straightforward explanation of why such scenarios are unrealistic. A visual

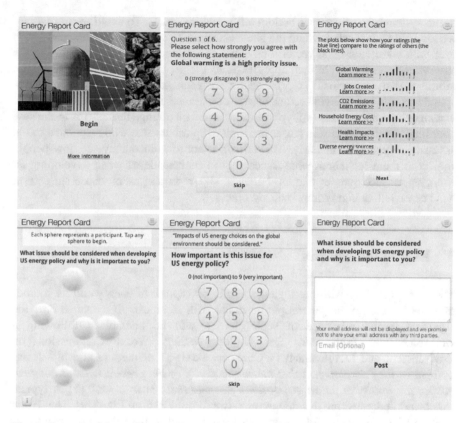

Fig. 1 Example of the structure and functionalities of the "Energy Report Card". The panels from *left* to *right*, *top* to *bottom* show: introductory panel; example of quantitative assessment; individual versus average assessment distribution; 2-D Principal Component Analysis display; assessment of opinions of other users; user input panel. This example was adapted from the "California Report Card" and actual name, content, metrics, functions, and graphics will be developed as part of the proposed project

rendering system will be developed to visualize the outcome of the users' choices in intuitive ways. For example, users could choose to visualize a comparison of the volume of waste created by each source, or visualize the fraction of US territory that needs to be used for each source on a US map. Users will finally have the option to share their personalized energy portfolio and metrics of choice through email and social media.

After completing "The Energy Challenge", participants will then enter the final portion of the "Energy Report Card" where they will be able to suggest additional issues they believe are important to consider when designing an energy portfolio. Participants will also rate the importance of others' suggestions, enabling crowd-sourced insights. We apply Principal Component Analysis (PCA) to display each participant's suggestion on a two-dimensional plane. Each user is represented in the system by a sphere (see bottom left panel in Fig. 1). To avoid overcrowding, we load only a few spheres onto the plane at a time. In a first step, we associate each user with a k-dimensional vector: one entry corresponding to each response to the assessment questions. We then apply PCA to the set of vectors and the algorithm returns a two dimensional (x, y) location for each participant. This point corresponds to the top 2 eigenvectors of the covariance matrix. We then center the visualization on the user's (xp, yp) position, and then arrange the spheres in the new coordinate space. Spheres in closer proximity represent users who responded to the assessment questions similarly. This allows users to immediately see how people similar to them feel about what issues should be considered when developing an energy portfolio. Spheres that are larger in size represent users whose suggestion has been rated as highly important by others.

3   Evaluation Metrics

The metrics that we will use to gauge public perception of energy and its sources must be familiar to the general public rather than technical. At the same time the significance and relevance of such metrics will be guaranteed following a well-established framework. The United Nations World Commission on Environment and Development (WCED) in the 1987 defined sustainable development as the *"development that meets the needs of the present without compromising the ability of future generations to meet their own needs."* [1] A typical framework, empowering this definition is the "Triple Bottom Line" [2]. The Triple Bottom Line (3BL) is a framework, well established in the scientific literature as well as public-oriented publications, with three key elements: social, environmental (or ecological) and economics. It provides a holistic perspective to assess the sustainability of several engineering solutions. A state-of-the-art framework to assess the sustainability of power plants and their life cycle (nuclear in particular) is provided in [3]. Regarding environmental indicators in particular, the US EPA has focused on determining and developing the best impact assessment tool for Life Cycle Impact Assessment (LCIA), Pollution Prevention (P2), and Sustainability Metrics for the US. This research led to the creation of TRACI—the Tool for the Reduction and Assessment of Chemical and other environmental Impacts. The methodology has been developed specifically for the US using input parameters consistent with US locations. Site specificity is available for many of the impact categories, but in all cases a US average value exists when the location is undetermined. The average values were implemented in the ecoinvent data. Further information is available at http://www.epa.gov/nrmrl/std/traci/traci.html. TRACI is therefore useful to compare different power plants and their life cycle. Unfortunately, these frameworks are hardly compressive for non-experts. In particular regarding the power sectors, people often have misconceptions that the tool envisaged by this research program will contribute to overcome. Some of the most relevant examples that we will address are:

(1) Thinking at technology level is inappropriate

- The same technology has different performances in different scenarios: e.g. technology X can have great performance in scenario A (desert with plenty of sun), poor performance in scenario B (north country with several rainy days).
- An electrical system to work in an efficient way (from technical and economical perspectives) needs the right mix of power plants: base load, peak load, ancillary services, etc.

(2) Energy cost is just one aspect of economics

- People need to distinguish between Production Cost (technology driven), Electricity Price (market driven) and Value (usage driven). Gas turbines working as spinning reserve are costly, get a high price, but are extremely valuable. A private company working in a market has, usually, the goal to maximize profits minimizing risk, not minimize production costs.
- Let us assume that technology A has an overall production cost (LUEC) of $100 per MWe and B $70 per MWe. Is B better than A? We need to include environmental issues, but also social. Let's think about social. Maybe B is not creating local national jobs, while B is more expensive, but the cost is boosting local/national economics.

(3) Global warming

- The majority of scientific publications say it is an issue. However, we still lack understanding of how citizens feel about global warming and their preferences for dealing with it. In a world (or nation) with limited resources it is important to prioritize budget allocations for important social, economic, and environmental issues. Identifying how citizens would allocate limited resources could provide insights into citizens' feelings toward global warming. For example, having $100 to invest—how much should be allocated to "cutting greenhouse gas emission", "funding cancer research", "paying for vaccinations in poor countries", "creating grants for student education", and "developing more sustainable food production techniques"?

This research leverages the state-of-the-art knowledge to create an innovative social engagement platform that will allow for key insights to emerge on public perception toward different energy sources, including perceptions toward different environmental, social, and economic impacts. The 3BL elements can be broken down into categories (and eventual sub-categories) and the categories in quantitative indicators. This framework, common for all the energy sources, differs for the specific values of each indicator, specific for the source considered. The key idea is to use indicators that are intuitive for the "average citizen". This indicator requires a "life cycle perspectives" and needs to be tuned from existing research and database (e.g. http://www.externe.info/). In this way the user can focus his/her attention on specific aspects.

We will give the option to the user to assign "weight" to different categories to obtain the "ideal ranking". For example, an "Environmentally sensitive user" can assign a high importance to the environmental indicators and/or categories and the system will return an energy portfolio that reflects these interests. There is precise set of mathematical methods to address in an exact way this issue, and they are built around the Multi Attribute Decision Making theories. The Analytic hierarchy process (AHP) is rather simple and straightforward [3], but if there are interactions between categories it is better to use the Analytic Network Process (ANP) [4]. The system, receiving the input from the user, will apply these methods for the ranking of different energy-mix alternatives.

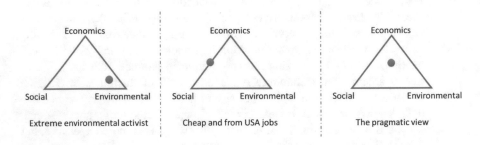

shows an example of three possible choices from three different users. This system will record the choice of each user and will display. The overall ranking calculated from all users. This information, "the voice of the average citizen", will be of paramount importance paving the way for research and policy decision-making in the energy sector. At the present time there is very limited understanding about how the public addresses trade-offs between the different 3BL elements and which indicators are more relevant. Moreover, users will be asked to provide demographic information (e.g. zip code, gender, age, education), allowing for more in-depth analyses. This "feedback data" will be released in a public user-friendly way for the benefit of the public, policymakers and the scientific community.

4   Discussion

We expect that the development of a web platform for comparing energy sources through easy–to–relate–to metrics will promote dialogues between experts and the general public, and will enable exploration and visualization of the public's points of interests, so that the policymakers can correctly understand the needs and priorities of their constituents. Unlike typical top-down approaches with predefined recipes and query items, the proposed system lets the end–user prioritize metrics of interest, provide additional metrics not originally included, provide suggestions and evaluate other users' ideas. While such sense of trust is sought providing technically reliable data sources and models, interpretation into a straightforward rather than technical language is essential.

This platform will implement best practices derived from similar existing efforts like "my2050", but it will largely depart from the underlying philosophy of such tools. We strongly believe that a visually attractive platform is necessary to attract users to engage with critical energy issues. Nevertheless, the success of the platform will be determined by the rate at which users return to the platform and make constant use of it. The unique features that we propose allow users to express their opinions and concerns, and to understand the impact of their choices on easy–to–relate–to metrics. We expect that the personalization aspect and the focus on the user's interest, rather than providing a pre-packaged solution, will make the user want to come back and bring other users to the platform. Furthermore, energy policymakers in general will want also to come back to the site and continuously monitor it as data and metadata evolves with time and events. A transparent interface with social media will further facilitate users' participation.

References

1. World Commission on Environment and Development, *Our Common Future* (Oxford University Press, Oxford Paperbacks, 1987)
2. J. Elkington. *Cannibals with Forks: The Triple Bottom Line of 21st Century Business* (New Society Publishers, 1998)
3. T.L. Saaty, *What is the Analytic Hierarchy Process?* (Springer, Berlin Heidelberg, 1988)
4. T.L. Saaty, *Decision Making with Dependence and Feedback: The Analytic Network Process* (RWS Publications, 1996)

Part III
Barriers Against Transition into Resilience

What Cultural Objects Say About Nuclear Accidents and Their Way of Depicting a Controversial Industry

Aurélien Portelli

Abstract Nuclear accidents have prompted the creation of numerous cultural objects such as novels, films, cartoons, or posters. Here we show what these objects can teach us about the social representations of nuclear power. The object is both a product and a representation. It can influence attitudes and partially contributes to the cognitive context of controversy about atomic power. Consequently, it leads to diverse practices defined by the interests and goals of the groups that own it. French documentaries on Fukushima Daiichi constitute a coherent corpus that makes it possible to identify both ruptures and continuity in the story that is told. These films borrow from the symbols, myths, and analogies provoked by Chernobyl to evoke Fukushima. They also show that the accident ends the myth of 'Soviet neglect' and creates a form of social resilience that has changed the way the Japanese population is seen in France.

Keywords Cultural object · Representation · Socio-technical controversy · Nuclear disaster · Citizen mobilization

1 Introduction

A monumental tower occupies the center of the picture. To the left, the viewer can discern a city; to the right, a body of water and a dock with ships. In the foreground, stonemasons pay homage to King Nimrod; behind them, there is a densely-populated city. A multitude of tiny workers are working on the tower, which is still under construction. The people decided to build a city, *"with a tower that reaches to the heavens"* [1]. But God intervenes, He garbles the words of the workers, work stops and the city is named Babel. In this picture, painted by Bruegel the Elder in 1563 [2], the tower is *"the spiral of knowledge, the relentless swarm and the image*

A. Portelli (✉)
Centre for Research on Risks and Crises (CRC),
MINES ParisTech/PSL University Research, Paris, France
e-mail: aurelien.portelli@mines-paristech.fr

© The Author(s) 2017
J. Ahn et al. (eds.), *Resilience: A New Paradigm of Nuclear Safety*,
DOI 10.1007/978-3-319-58768-4_11

137

of pride" [3]. In 1975, the graphic designer Peter Brauchli reproduced this painting as a poster, replacing the top of the building with the cooling tower of a nuclear reactor.[1] The poster was shown in 1979 in the context of an anti-nuclear campaign in Switzerland [4]. It symbolizes the technological excesses and the catastrophes that nuclear power could lead to.

The image that this picture paints reflects societal fears that were materialized by the 1986 Chernobyl accident. It happened again, 25 years later, with the Fukushima Daiichi accident. Commentators evoked a tragedy, for example in the article entitled *"The Japanese Nuclear Industry, or a Punishment for Hubris"* [5]. The media noted that there would be no end to the Japanese disaster, which became a very controversial element in debate about the industry. Some representations, which had been mothballed when the dust of the 'battle of Chernobyl' had settled, were reactivated in the collective consciousness and were reflected in novels, films and comic books. In turn, this material, collectively termed 'cultural objects' [6] interacted with existing images and produced new ones.

This chapter shows what cultural objects can teach us about the representation of accidents and the nuclear industry. The first section defines the terms 'representation', 'socio-technical controversy' and 'cultural object'. The second describes the characteristics of the object, as both a representational product and agent. Finally, the third section examines French documentaries about Fukushima, and discusses the consequences of the nuclear accident on representations.

2 Social Representation, Controversy, and Cultural Objects

The literature on social representations reveals analytical models that are based on several types of data. Representations may relate to a controversial technology, such as nuclear energy. The controversy arises from identities and practices that impact the production and reception of cultural objects. Works related to nuclear accidents therefore have several things in common: they dramatize it; they often confuse civil and military applications; and they therefore have a more-or-less close relationship with reality.

2.1 Social Representation

The analysis of representations in the social sciences has led to many definitions. Nonetheless, a certain degree of consensus emerges, such as Gaffié, "*A social*

---

[1]Bruegel executed at least two versions of The Tower of Babel. The first (1563) was shown in Vienna, the second (1568) in Rotterdam. It is the Viennese version that was used by Brauchli.

representation takes the form of a set of knowledge, beliefs, patterns of under-standing and action about a socially-relevant subject. It is a particular form of common-sense knowledge that defines reality for the society that developed it with the aim of action and communication" [7]. It manifests both as the sum of content and as a process that can modify thoughts and actions.

Moscovici, the pioneer of the theory of social representations, identifies two major processes in their development: objectification and anchoring [8]. The first designates the origin of the formalization of knowledge about an object. It embodies the meaning of things and facilitates discussion [9]. Objectification is divided into three phases: selection (sorting and taking ownership of information depending on the cultural context), schematizing the selected elements (construc-tion of a coherent picture leading to the simplification of the represented object), and naturalization (the constructed image becomes, for the actor, an autonomous and objective entity). Anchoring extends objectification with the ultimate aim of integrating the novelty into the social space [10]. This process corresponds to the instrumentalization of the object, so that actors can actually use it. Anchoring the novelty can then lead to a change in thinking. Objectification and anchoring therefore show how society transforms knowledge into representations, and how this in turn transforms society [9].

Various analytical models subsequently appeared, such as genetics, dynamic or structural perspectives [11]. The latter sees social representation as a socio-cognitive system consisting of a central core and a peripheral system that complement each other [12]. The central core is the fundamental and non-negotiable element of the representation. It functions as both a generator and organizer. The first determines the meaning of the components of the representa-tion; the second defines the nature of the links that connect these elements. The core is characterized by its stability and its ability to resist contextual developments or the introduction of new practices. The peripheral system is more permeable to change. It is subordinate to the core and provides the connection between the central elements and the social reality of the object. The representation of the peripheral system can be adapted to the context; this ensures the integrity of the central core, which remains protected. Although the structural approach has evolved since its initial definition [13], it remains relevant for understanding social representations.

Their study is based on several types of data. Research has focused on various discursive elements, in the form of (undirected or semi-directed) interviews or questionnaires. A second source is conversations and other forms of verbal inter-action in a group or public place. The third is a textual and iconic corpus (archives, novels, newspaper articles, cartoons, paintings, comic books, photographs, and films), while word association tests form a fourth set [10]. All of these content types highlight statements, analogies, and meaning from which social representations can be developed. They can also prove to be a major societal issue, as the controversy over nuclear technology demonstrates.

2.2 Socio-Technical Controversy

Socio-technical controversy refers to a public debate in which opposing arguments are used to interpret an object or a particular technical system. This debate is not limited to a community of specialists—unlike, for example, scientific controversy [14]—and involves heterogeneous actors who introduce technical and non-technical (economic, social, health, moral, etc.) arguments.[2] The development of nuclear energy in France was uncontroversial until the late 1960s, when the anti-nuclear movement emerged. The initial phase (1970–1974) established the founding ideas [16]. Environmental activists, who emerged from the events of May 1968, accused the nuclear sector of shifting society towards a technical and authoritarian model. They proposed the establishment of an alternative, decentralized libertarian and autonomous society and carried out a diverse range of actions: demonstrations, boycotts, site occupations, petitions, etc.

The movement expanded to a national level in its second phase. This began in 1974 with the announcement of the Messmer Plan.[3] Protestors rejected this massive nuclearization of France. The nuclear program, in addition to threatening the environment and human health, was accused of strengthening capitalism, the technocracy and state authoritarianism. The attitudes of militants hardened: there were bomb attacks, clashes with police (for example, Creys-Malville in 1977), and acts of sabotage. Faced with mounting opposition, pro-nuclear supporters took action and the state-owned power company EDF launched an information campaign to reassure the public, who remained largely in favor of nuclear energy. The fact that the accident at Three Mile Island (in March 1979) had little effect on public opinion confirmed the relative acceptance of nuclear power in France, and the suffocation of the protest movement.

The Chernobyl accident, however, marked a break and initiated a period of controversy. Although there were very few antinuclear demonstrations in France, the reassurances issued by the authorities, coupled with a lack of information and preventive measures, led to a change in how things were done. Research into alternative solutions expanded and there was a massive increase in the number of alerts. In return, industry actors changed their message and attempted to make nuclear power less opaque and more democratic.

New groups appeared in the 1990s (although not at the same level as in the 1970s), while the debate was revived by the Fukushima accident in 2011. The media questioned the safety of French installations, waste management, the profitability of the sector, the working conditions of subcontractors, and the feasibility

---

[2]For some authors, controversy is different to institutional crisis. The first involves an audience made up solely of peers, while the second involves a wider, non-specialized public. However, in most cases, controversies "go beyond the circle of peers and engage social forces and individuals who are located beyond the institutional scope in which they arose" [15].

[3]Prime Minister Pierre Messmer announced, in March 1974, the expansion of the nuclear power program, together with plans to build 13 nuclear plants in 2 years.

of decommissioning. The reaction of the European Union was ambivalent: Germany brought forward the date of the closure of its plants, while France refused to abandon nuclear power. Several factors explain the failure to radically reexamine the French nuclear sector. Politicians simply relegated the Japanese accident to an exceptional event: *"Twenty-five years after the Chernobyl accident, the exception has again served as the main element of arguments that aim to provide reassurance: the Japanese geographic exception has now replaced the Soviet technological exception (victims of their lack of technical expertise)"* [16]. Experts issued reassurances that they would take into account feedback from the Fukushima accident to "imagine the unimaginable", and strengthen plant safety. Political and industrial actors, profiting from the Chernobyl experience, promoted transparency.[4] Politicians managed to make nuclear energy acceptable, and effectively prevented any further escalation of the dispute.

The French nuclear energy controversy therefore has several particular characteristics. It is both technological ('the industrial war' between supporters of the French and American technology in the late 1960s) and post-technological (site selection, rejection of large-scale nuclearization).[5] The controversy over the industry, and more generally the issue of radioactivity [19], has developed slowly, with periods where it has received little attention, followed by periods of intensification or renewal. This periodicity implies changes in the composition of social groups that are involved, and a change in their strategic action plan. Going beyond issues of energy policy, accident risk, or environmental and health impacts, French controversy is intrinsically linked to power struggles in society, underlined by the concept of technopolitics.[6] It also questions other ideas, such as the relationship between Man and Nature. Controversy emerges from the construction of identities and strongly-held collective practices (the pride of nuclear energy pioneers, the collective memory of militants at Plogoff or Creys-Malville), which plays a part in the production and reception of cultural objects.

---

[4]The contribution of the anti-nuclear movement to the history of the industry illustrates the thesis of Lascoume on the social production of controversy [17]: "Criticism has helped to modernize nuclear institutions, forcing them to improve their communication strategies and crisis management systems, requiring them to improve their communication with the public, bypassing the dominant idea in 1986 that all means—first and foremost extreme secrecy—were useful in order to preserve the future of nuclear energy" [16].

[5]This distinction mirrors that of Callon in his article about the sociology of technological controversies [18], where he suggests that controversies emerge during, but not after technological invention.

[6]This concept corresponds to strategic practices that consist of designing or using technology to implement policy objectives [20]. Technopolitical regimes designate groups of individuals, engineering and industrial practices, technical objects, political programs and institutional ideologies. These elements are connected to each other and interact to govern the development of technology and in turn, enable technopolitics to develop.

2.3 Cultural Objects

Taking actual objects as our starting point (rather than the social groups who appropriate them) offers a new perspective on representations. In this case, objects are not considered as independent and decontextualized entities, but are observed through the prism of the history of aesthetic representations, which *"makes it possible to think about knowledge, ideologies and the techniques that are implemented—which should certainly avoid a linear story form—the relationships between the object of the representation and its production or reception"* [21].

The cultural object is a reference to *"all concrete objects (books, writings, paintings, photographs, films, architecture, sculpture, etc.) resulting from a formal production and intended to produce in those who receive it a symbolic 'effect' (aesthetic contemplation, subscription to its values, producing a belief, etc.)"* [6]. This type of object *"is part of a civilizational and historical context and participates in the definition of the worldview of which it is a part"* [22]. It is disseminated in communities of varying homogeneity and scope. Its influence on social reality is therefore difficult to assess and the researcher must collect external data. They can then compare the cultural object with other sources to determine its function in the public space and its contextual significance, or examine its reception by critics. Comparisons, which depend on the availability of sources, should not however be carried out at the expense of the internal analysis of the object. In practice, the object is involved in the formation of social processes, but remains a product whose constituent elements must be carefully studied, as it is the starting point for research.

Cultural objects relating to nuclear accidents are very diverse in terms of both form and content. However, they all approach the subject from a dramatized angle, as is shown in the confusion between a nuclear reactor and a nuclear weapon. Objects draw upon radioactive imagery marked by Hiroshima and Nagasaki, the anxiety generated during the Cold War, and the proliferation of weapons of mass destruction. The image of the mushroom cloud is particularly evoked as a representation of the explosion of a nuclear reactor. As an example, the city of Springfield is destroyed in an episode of *The Simpsons* [23] as a result of the explosion of a nuclear plant, in the same way that Hiroshima was razed to the ground by Little Boy. The similarity with reality highlights the discursive potential of a cultural object, which maintains a dual relationship with social representations.[7]

---

[7]The position of the author, as a representational actor, must be taken into account in this process. Their role is easily identifiable when it concerns a novel or a comic book. The notion of authorship becomes more ambiguous with respect to a film. The director cannot by themselves embody the author, and is seen more as a "virtual home", a "speaker", or the "subject of filmic discourse" [24]. The discourse is therefore not always individualized and can result from a collective effort.

3 The Cultural Object in Representations of the Nuclear Sector

First and foremost, the cultural object can be seen as a product of social representations. The Chernobyl accident caused a representational crisis. What resources and symbolic systems were available to represent the radioactive 'evil'? The nuclear crisis therefore led cultural actors to reflect on how to represent an 'invisible evil'[8] and its impact on perceptions of reality. But the cultural object is more than a product of social reality. It is also an agent involved in the creation of representations and in the structuring of socio-technical controversies.

3.1 The Object as a Product

One approach suggests a cultural history of nuclear accidents, consisting of "*all forms of collective representations, how societies see and represent what is external to them, symbolically through values, spiritually through belief systems, intellectually through the construction of ideas, and pragmatically through image- or text-based techniques (discursive practices) or other means (non-discursive practices)*" [4]. The researcher must therefore explain how the nuclear accident is a break with (or a continuation of) representations of risk perception and its relationship with social reality.

The nuclear accident is represented as a catastrophe, which can be defined as an "*event causing a disaster of major magnitude, whose social and symbolic consequences are on a historical scale*" [26]. The unending consequences of Chernobyl marked a radical change: they created an area that was out of bounds to the world, manifested by the abandonment of vast areas where the invisible threat of radiation lurked [27]. Beck notes, in the preface to his book *Risk Society* [28] that mankind has always responded to suffering, misery and violence by resorting to the 'Other'. Chernobyl put an end to this ability to put events at a distance. Since the Soviet accident, it has not longer been possible to ignore the dangers of the nuclear age. From that point, fear became the product of the most advanced levels of progress in the modern world.

Using an apocalyptic scenario, cultural objects represent the anguish that the Soviet accident triggered. The novel *Silence, we're irradiating* (*Silence, on irradie*) [29] suggests that even the normal operation of a nuclear installation foreshadows disaster. In this novel, workers suffer from all kinds of illnesses: headaches, loss of teeth, dark spots on the skin. Children suffer from physical defects, mental

---

[8]The term radioactive 'evil' comes from the work of Dupuy on "the empowerment of evil with respect to the intentions of those who commit it" [25]. In the case of Hiroshima, evil comes from the intention to commit evil, while in the case of Fukushima, evil comes from the intention on the part of industrial actors to do good.

handicap, and lose their hair. It is rumored that the lake close to the plant is populated with monster fish. The novel paints a picture of degeneration that reflects concerns about the sector. The description of the explosion of the plant reflects both the destruction of Hiroshima and the Chernobyl accident, "*Over more than a kilometer in circumference, an irradiated graphite storm hit, corroding vegetable, mineral, and animal alike. The building that housed the families of engineers and workers had been pulverized. It left the distressing spectacle of a crater containing rubble magma, covered with a layer of tilled soil*" [29]. Sven, a teenager, falls into a ditch and survives. He returns to his village but finds it has been replaced by a post-apocalyptic scene: "*It was a ghost town that he saw before him*" [29].

This image of the apocalypse demonstrates the violence of the technique, which some authors interpret as a form of transcendence. Viewers of the sarcophagus of Chernobyl are seized by a "*holy terror*" [30]. This sacredness is represented as a nightmarish vision in *Mount Fuji in Red* [31]. This short film begins with an outline of Mount Fuji, which has begun to erupt. Below, people flee in panic. Balls of flames shooting into a scarlet sky, rise behind the volcano. Nuclear power plants explode in succession, triggering huge fireballs. The awakening of Mount Fuji, the sacred mountain of the archipelago, causes a chain reaction that is a punishment for the hubris of mankind. The sequence, linking the cataclysm to spirituality, is a representation of divine punishment that is both specific (the volcano is revered by the Japanese) and universal (the myth of the end of the world). The use of religious symbolism and variations on the myth of the Apocalypse reflect an extreme situation in the sense that Eliade understands as "*what mankind discovers by understanding its place in the Universe*" [32]. The representation highlights the appetite for technology in modern societies and the powerlessness of mankind in the face of the forces it has unleashed.

3.2   A Representational Crisis

The Chernobyl accident can be compared to the Holocaust or Hiroshima in the sense that it marked the start of a new world, "*The world of Chernobyl, the product of technoscience, appears to have made all cultural resources mobilized in such situations obsolete: it becomes impossible to use such a system of representation, analogy or experience to understand this world that has become so alien to mankind, denatured and unrecognizable, while at the same time, familiar*" [26]. The use of color quickly proved decisive in circumventing the problem of the 'impossible representation'. In *Mount Fuji in Red*, the explosion of plants produces red and purple clouds of radioactive particles. A survivor explains this strange phenomenon, "*Human stupidity is boundless. Radioactivity is invisible, while technology has been developed that colors it when it spreads in the air*" [31]. In this case, the discourse links the nuclear industry to an industry that is as dangerous as it is counterproductive. The use of color is seen again in the film *The Land of Hope* [33]. At the same time as an accident occurs in the Nagashima plant (a contraction of Nagasaki, Hiroshima and Fukushima), a resident goes to a clinic and learns that she is pregnant. She heads for the exit and sees behind the glass door red smoke that obscures the street. A reverse shot shows a close-up of the character. The color red, symbol of radioactive hell, is reflected in her eyes. A blinking eyelid breaks the illusion, but the young woman is now possessed—by the fear of contamination.

The sensory impact of the accident is approached differently in *A Springtime at Chernobyl* (*Un printemps à Tchernobyl*) [34]. In this comic book, Lepage recounts the story of his stay in the prohibited area in 2008. The graphics use black and white and sepia tones. A two-page spread presents a twilight landscape, where electricity pylons fill the foreground, symbol of the 'electricity fairy' and the modernity of which it is the corollary. The antinomic and threatening shadow of the plant fills the background, represented as a haunted mansion. The use of half-light and violent contrasts add to the gloomy gravity of the scene. Drawings of the area, with its twisted trees and buildings in ruins, follow one after the other. An overview of Pripiat shows a petrified town that has been overtaken by vegetation, like cites abandoned by the Maya. The town, designed to be a showcase of Soviet modernism, "*takes us back to the light of an extinct empire [...] like a dead star*" [34]. A sudden patch of green forest marks a break with the gloom of previous images. Lepage changes the iconic registry and shows a world without humans, where a transfigured nature has resumed its rightful place. The artist describes his discomfort: "*It's calm all around me. These places suggest pleasure... But I'm at Chernobyl! How can I reflect this improbable situation? Only through scientific artifice. The number of micro Sieverts shown under each drawing. What's in front of me, what I'm drawing is not the truth! I don't see the disaster but an explosion of magnificent colors. (...) How can I draw the invisible?*" [34]. This inner monologue expresses the paradox posed by the area, which is as horrible as it is radiant. The indecent beauty of nature, created by nature itself exerts a fascination reminiscent of the romantic landscapes of the nineteenth century. However, the voluptuous forms and colors suggest a betrayal of reality. This is a defining characteristic of the representation of radioactive areas, whose construction depends on the contradictory phenomenological elements that compose it.

3.3   The Object as Agent

The cultural object, in addition to being a product, is also a representational agent. The process of integrating the object into social reality involves the selection and crystallization of certain elements of the representation of the nuclear accident. This can influence thinking and contribute to the controversy over the industry. The object is associated with diverse practices, defined by the interests and objectives of the groups that own it. Crystallization may therefore require the instrumental use of the cultural object, especially when groups that engage in the controversy are ideologically motivated. Its use therefore becomes strategic, as it allows group members to mobilize, act and to rally other individuals who are less concerned about the cause.

The anti-nuclear movement contains many examples that illustrate this function of the object. The online shop belonging to the French Nuclear Exit Network (*Réseau Sortir du Nucléaire*, RSN)[9] is full of merchandise: the *Nuclear Exit* journal, books, DVDs, games, leaflets, posters, stickers, flags, T-shirts, banners, etc. A first category of objects concerns products that are sold commercially, and in the Network's webshop, such as the novel *Silence, we're irradiating*. The second category consists of articles specially designed by the RSN, such as the DVD entitled *Short Films* (*Films courts*). This DVD contains twenty short films screened at a festival organized by the Network on April 26, 2008, for the 22nd commemoration of Chernobyl. The liner note points out that "*cinema is an important tool for raising awareness, it is often more widely accessible than the written word*". The group invites its militant members to organize public screenings of the DVD, in the context of an information campaign about nuclear power. A further example is the poster *Gaul under nuclear occupation* (*La Gaule sous occupation nucléaire*) [35], which uses humor to denounce the sector. The poster is a parody of the map found in the comic book *Asterix*,[10] which shows the occupation of Gaul by the Romans following Caesar's conquest. In this case, it is nuclear power plants that occupy France, 70 years after the bombing of Hiroshima. Facing the nucleocrats entrenched in their garrisons stand the indomitable Gauls, who lead the resistance. The illustration Latinizes the names of nuclear sites, which become *Marculus* or *Golfechus*. The European Pressurized Reactor (EPR) at Flamanville becomes the *Extra Problemus Reactus* and the plant at Fessenheim is renamed *Fissurnhum*. Corsica is renamed *Postum Tchernobylum* in reference to the (controversial) radiological consequences of the Soviet accident. The eagles and standards of Rome are replaced by the names of industrial actors in the sector. The poster is a prophecy of a nuclear disaster in France. A place called *Tsunamus* appears on the Atlantic coast, while the Chinon plant is renamed *Fukushinum*. The image therefore makes reference to an object that is very representative of French popular culture to synthesize the main arguments of the public debate and denounce the country's widespread nuclearization. Cultural objects sold on the RSN website allow the Network to disseminate its ideas, encourage the user to take direct action, and strengthen its militant identity, by creating content that can be easily used by individuals. Objects therefore help to build a shared culture and create a permanent construction for the collective.

---

[9]The aim of the Network, founded in 1997, is to secure the exit of France from nuclear energy and promote alternative energies. It currently brings together 931 associations, while 60,290 individuals have signed its charter. Its actions take many forms: information campaigns, raising awareness with elected officials or unions, petitions, lawsuits, etc.

[10]Asterix is a series of French comic books, created in 1959 by Goscinny and Uderzo. It narrates the adventures of a group of French Gauls who resist the Roman invasion thanks to the power of a magic potion.

3.4 The Value of the Cultural Object

For the researcher, the value of the cultural object lies in its symbolic specificity. By highlighting what is 'built', the object creates a synergy between cultural references. Its analysis makes it possible to access social representations in ways that do not involve interviews or surveys. The object has a dual role, notably in terms of structuring day-to-day conversation—objects feed discursive dynamics and stimulate the production of new representations.

Regarding the nuclear industry, the production and circulation of content is a battleground that the groups involved in the controversy must conquer. While the use of military terminology may seem excessive, it is not possible to orient decisions and actions without first changing the representations that are associated with them. Therefore, how the sector is represented can, depending on the societal context, become a fundamental issue in drawing up energy policy.

Content analysis, however, should not be at the expense of a more aesthetic approach to representations. The iconic and narrative dimension of the object (through the formalization of content), is a major contributor to the constitution of knowledge and understanding the world. Furthermore, the study of forms (in the broadest sense), is useful in identifying which of the elements created by an event lead to rupture or continuity in the representation of reality. In this respect, French documentaries about Fukushima constitute a coherent corpus that can be used to analyze the impact of the disaster on representations.

4 The Representation of Fukushima in French Documentaries

French documentaries use the Japanese accident to illustrate the globalization of risk and the repetition of history. The 'never-ending' accident is dramatized: the actors involved in the management of the crisis are incompetent, workers involved in decontamination are sacrificed on the altar of atomic power, and populations are abandoned to their fate. Faced with disaster, civil society reorganizes itself and the various initiatives taken by citizens lead to some sort of social resilience.

4.1 The Space-Time of the Disaster

The documentary *Nuclear Disasters: Secret Stories (Catastrophes nucléaires: histoires secrets)* [36] begins with a close-up of the Fukushima site, filmed on Saturday, March 12, 2011. Reactor 1 explodes in the next shot. Extracts from televised news programs emphasize the scale of the disaster, while the damaged reactor appears once again. Images of the explosion are shown around the world, reminiscent of those of the planes that crashed, 10 years earlier, into the World

Trade Center. This repetition reflects the extent of the trauma, which is permanently engraved in the collective memory.

These representations superimpose the natural and technological dimensions of the disaster. In the documentary *The World after Fukushima* [37], a lateral tracking shot reveals a landscape devastated by the earthquake and tsunami. Shots of abandoned buildings and heaps of debris follow. The commentary indicates that the accident was caused by the earthquake and tsunami of March 11, but that the real causes lie in the weaknesses of a system *"whose arrogance was matched by its blindness"* [37]. These words modify the images by associating the damage caused by the tsunami with the consequences of the accident. It is the infinite extension of a war zone, says the narrator. In the documentary *Fukushima, Particles and Mankind* (*Fukushima, des particules et des hommes*) [38], color drawings show familiar places: a neighborhood, houses, cultivated fields. A whitish vapor, full of radioactive particles, spreads over the scene. This represents the 'invisible evil'. It emphasizes the deviousness of the threat, and calls into question the reliability of our senses and reality.

The Fukushima accident is open-ended and illustrates the theory of the risk society [28]. The documentary *Fukushima, a step towards global contamination* (*Fukushima, vers une contamination planétaire*) [39] reveals that tuna caught off San Diego (United States) were contaminated with cesium waste from the damaged plant. The same observation is made in frozen fish sold in Switzerland. The globalization of risks is such that all consumers are potential victims of Fukushima. Furthermore, the disaster is shown to be a repeat of history. As the lateral tracking shot in *The World after Fukushima* fades out, it is followed by a new shot. This sequence consists of archive black and white images, recorded on August 6, 1945 at Hiroshima. The landscape scrolls across the screen and ends with a field of ruins, which appears to be the continuation of the previous shot. Hiroshima and Fukushima are therefore merged into one, single atomic energy catastrophe.[11] The analogy is highlighted by a vertical panorama from the Genbaku dome,[12] where the sequence ends. The camera rests on the reflection of the building on the surface of the River Ota, highlighting the parental relationship between the two events.[13]

---

[11]The Fukushima crisis accentuates the media conflation between the civilian and military nuclear sectors, "The nuclear power plant is the twin sister of the atomic bomb: it uses the same, extremely dangerous substances. Its civilian character does not remove that fact that is the twin of nuclear weapons. The nuclear power plant is deadly technology that has been tamed" [40].

[12]This building, built in 1915, is the former Prefectural Industrial Promotion Hall at Hiroshima. It was the only building left standing following the explosion of Little Boy. It has been listed as a UNESCO World Heritage Site since 1996.

[13]The media has advanced many reasons for the development of nuclear power in Japan. According to T. Tanaka, Professor at Hiroshima University, "It is precisely because we were victims of the atomic bomb that the arguments of civilian nuclear supporters appealed to us. For them, the technology that killed our loved ones could not only treat cancer, but also bring us comfort". On the other hand, for the essayist M. Katayama, nuclear power was developed in response to a desire for revenge, "Based on the fact that Japan lost the war because of its scientific backwardness, we concluded that we had to take our revenge by triumphing in exactly that domain" [41].

Fukushima is not only represented as a new Hiroshima, but also as yet another nuclear disaster. The warnings provided by Three Mile Island and Chernobyl were not heeded, and it was therefore no surprise that another accident occurred in Japan. The event is all the more unacceptable as it had been foreseen. In the documentary *Japan: Nuclear energy, the sector of silence* (*Japon: Nucléaire, la filière du silence*) [42] the former Prefect of Fukushima presents a series of damning documents. Workers had highlighted, well before the accident, technical problems and safety breaches at the site. "*The failings revealed by whistleblowers were never addressed. This is what led to the accident we see today*" [42]. The narrator states that these failures and the accumulation of errors did not prevent the operator receiving, in February 2011, authority for Reactor 1 to operate for another 10 years, although the facility was already 40 years old.

The commentary implicitly refers to the debate taking place at the same time in France about the Fessenheim reactor in Alsace. In April 2011, thousands of protesters demanded that this *grande dame* of French nuclear power be decommissioned, "*A month after the earthquake and tsunami in Japan, which caused a serious accident at the nuclear plant of Fukushima, antinuclear protestors focus on the age of the Fessenheim plant, built in 1977, arguing that it is also located in a seismic zone and is subject to the possible flooding of the Rhine*" [43]. The documentary, *Nuclear Energy: the Human Bomb* [44] shows images of anti-nuclear demonstrations. A sign held up by a young boy says "*No Fukushima in Alsace*". This prophetic message illustrates the idea of "enlightened catastrophism" [45]: the message is that the warning provided by Fukushima should lead to the closure of the Fessenheim plant, if history is not to repeat itself.

4.2 The Story of an Accident that Has no End

These images denounce the lack of responsiveness and the incompetence of the actors involved in the crisis. In *Nuclear Disasters*, connecting car batteries to control panels with alligator clips is not seen as an ingenious solution. It demonstrates a lack of resources and becomes proof of the amateurism of the operator.[14] Similarly, the documentary discredits Prime Minister Naoto Kan. Government actions are labored and inefficient. For example, a helicopter is sent to spray water on the plant. The Japanese population saw these images projected on a giant screen and "*noted with dismay the inability of their country, the third largest economy in the world, to cool the Fukushima reactors*" [36]. The sequence shows the end of the myth of "Soviet carelessness" [47]: faced with a nuclear accident, even a great technological power like Japan is nothing more than a house of cards.

---

[14]On the other hand, in the testimony of Masao Yoshida, director of Fukushima Daiichi it highlights the ingenuity of teams in the field during the management of the accident [46].

This negative representation of Naoto Kan should be seen in the context of the criticism he was subject to in Japan. Much of the population believed that the Prime Minister was not up to the situation. His resignation was therefore seen as a milestone in the management of the crisis, "*The opposition, but also most of the media, never stopped saying that with the departure of Kan it would at last be possible to begin to resolve the crisis caused by the March 11 earthquake. The psychological and emotional processes at work here seem to have been typical of that of the scapegoat*" [48].

The timeframe for the nuclear crisis is not limited to the management of the accident. It seems more relevant to think of Fukushima as a chain reaction, which continues to cause regular crises [49]. Documentaries take particular note of this 'never-ending accident', which is viewed as an unfolding drama. The final scenes of the documentary *Welcome to Fukushima* [50] evoke a terrifying scenario: the evacuation of more than 50 million Japanese, should a major earthquake occur before the operator can empty the plant's fuel ponds. Such apocalyptic scenarios only prolong the current disaster. Meanwhile, those responsible for remediation and decontamination at the site are subject to extreme working conditions. The exploitation of the human body represents the dark side of post-accident operations, and reminds us of the martyrdom of Soviet operators, convicts of atomic power, "*(they) take incredible risks, with just a mask and two gloves, they spend the day in an environment than can exceed three times the amount of radiation tolerated for nuclear workers in France*" [51].

The social dimension of the crisis is very apparent. In the documentary *Japan: the Silent Nuclear Sector* [42] a gymnasium has been transformed into a center for displaced people. A horizontal pan shows the living conditions of 'nuclear refugees', who live in total uncertainty while waiting to be rehoused. Authorities seem to underestimate the extent of the contamination. In *Fukushima, the Sacrifice of a Population* (*Fukushima, une population sacrifiée*) [51], a CRIIRAD technician[15] takes measurements in a primary school in Fukushima City. Off-screen, he reads radiation levels, while students prepare their sports equipment in the background. The depth of field reflects the resignation of the government, which is allowing radioactive pollution to take root in the lives of children.

Documentaries like those made by the Japanese media [52], emphasize that social bonds were destroyed by the accident. Radioactivity isolates and fragments communities, says the narrator of *Fukushima, Particles and Mankind* [38]. It prevents any return to normality in the contaminated area. The consequences of the disaster disrupt not only lifestyles, but also the relationship between people and things. Decontamination destroys landscapes. In *Welcome to Fukushima*, a Japanese resident turns this point into an ontological caesura, "*If a tree is a 100*

---

[15]The Commission for Independent Research and Information on Radioactivity (CRIIRAD) was created in May 1986 after the Chernobyl accident. The role of this independent body is to monitor radioactivity in the environment and materials, evaluate the impact of radioactive releases from nuclear facilities, provide information about radioactivity and its civil and military applications, and protect populations against the risks created by ionizing radiation.

years old, it means that a man planted it two generations earlier. And the tree connects us with that person. [...] But by cutting down trees to clean up, no-one bothers to communicate with them. The tree is nothing more than an object. It's cut down, that's all" [50]. The disaster depopulated territories, ripped society apart and made any dialogue about what is and what is no longer, impossible.

4.3   The Resilience of Civil Society: A New Representation of the Japanese Population

The consequences of Fukushima suggest a sacrificial system; a Japanese replica of Chernobyl's 'nuclear Gulag' [53, 54].[16] This extreme situation has led to rising levels of criticism. A refugee publicly challenges the CEO of TEPCO, *"It might not look like much, but challenging a leader in this way is very rare in Japan"* [36]. The disciplined Japanese culture has cracked and the anti-nuclear movement has made significant progress *"in a country where demonstrations are extremely rare"* [42]. The literature confirms that opposition to nuclear power, which appeared in Japan in the 1970s, has strengthened since the Fukushima accident [56]. But documentaries do not only evoke a change in attitudes and behaviors. They also reflect the surprise of seeing certain archetypes disintegrate. The French media has a particular image of the Japanese population, which is recognized for its spirit of resignation in the face of disaster [57]. The images shot after Fukushima paint a very different picture of Japanese society. A new representation, highlighting the insubordination of a part of the population, appeared on screens.

Documentaries also show the impact of the nuclear accident in France. *"Since Fukushima, the debate, which until then French society was only moderately interested in, has taken on a new dimension"* [58]. *Nuclear Energy: the Human Bomb* shows images of anti-nuclear protests, *"In France, for the first time, we see that a serious accident is possible"* [44]. In an interview, Alain de Halleux talks about the reception of his documentary, *NTR Nuclear: Nothing to Report (R.A.S Nucléaire—Rien à signaler)* [59]. His film, which received little attention in 2009, was re-broadcast on the French television channel Arte on March 25, 2011 and appears to have found its audience, *"Now that a plant has blown up at Fukushima, everybody's asking questions about safety and suddenly, we're finally paying attention to warnings and emergencies"* [60]. For Greenpeace, local officials realized that their town, even though it was located 30 km from a plant, was not immune to accidents, *"It's this grass-roots change that we believe will lead to the death of nuclear power in France"* [58].

---

[16]Japanese history is often viewed through this prism: the sacrifice of soldiers during the WW2, the sacrifice of Okinawa for the installation of United States' bases, the sacrifice of populations that are allowed to reside in contaminated zones [55].

The question of site safety has restarted the debate on the French 'nucleocracy'. In *The World after Fukushima* the Japanese nuclear sector is associated with a form of autocratic rule. In *Nuclear Power: the French Exception (Nucléaire: exception française)* [61] the sector is said to be dangerous, inevitably leading to a militarized and centralized police state. The narrator reminds us of the post-Fukushima context: Germany and Switzerland are planning their exit from nuclear power; Italians declare their opposition to the revival of the sector; and Japan stops the operation of its facilities. France, however, refuses to give up atomic power, *"Our governments have made nuclear power a state religion, protected by an all-powerful technocracy"* [61]. French dogmatism is presented as an obstacle to common sense: despite the Fukushima accident, the state wants to preserve the industry, which it sees as the guarantor of energy independence and French influence.

Critics of the Japanese and French nuclear sector have united to denounce authoritarian abuses of the system and promote a more separatist society. Documentaries reflect this aspiration, which has emerged from the anti-nuclear movement, by showing the effects of the accident on Japanese civil society. In these representations, the crisis provokes an unprecedented mobilization of the population. New solidarity networks emerged spontaneously. The clearance of contaminated zones has united volunteers from throughout the archipelago, *"Me and my parents are from Hiroshima. So we feel particularly affected by radioactivity. So I came here to help"* [51]. The victims of military and civilian atomic power appear to belong to a community with a shared destiny.

Such documentaries highlight collective resilience and the re-conquest of social action. The organization 'The Renaissance of Fukushima' brings together farmers, doctors, psychologists, geologists, physicists and retirees. Their mission is to make *"the invisible visible"* [38]. Members measure soil contamination, exchange data and expertise, and try to find solutions to restart agriculture. The accident interrupted the ancestral practice of working the land. Recovery symbolizes their hope of a renaissance in the contaminated areas.

These images reflect a constellation of positive initiatives and interactions: sharing knowledge and skills, establishing a support system, strengthening links between mankind and nature, and thinking about the meaning of community. The social balance that was destroyed by the disaster must be reconstituted. According to one farmer, *"This nuclear accident, it's not only about Fukushima. We all need to take responsibility because we let ourselves be overwhelmed by nuclear power. We must be united and think together"* [37]. Local action celebrates the struggle for life, while radioactivity causes necrosis. This neo-activism contrasts with the vertical management of the disaster. We are witnessing the emergence of an alternative society, driven by a deep desire for autonomy, which could regenerate the foundations of democracy.[17]

---

[17]The sociopolitical implications of these initiatives should however be carefully noted, "Since March 11, we have seen the growing importance of local action. But we cannot say if this is a positive development, a move to greater autonomy and better cooperation" [62].

5   Conclusion

Representations contribute to social reality by formalizing content that can change ideas and behaviors. They appear during the development of a technical system but also further downstream, especially if the development leads to debate in civil society. The issues that are at stake with respect to nuclear energy have caused deep ideological divisions and sparked renewed controversy, coinciding with the various accidents that have occurred in the sector. Nuclear accidents occupy a special place in the story, in the sense that their effects concern not only the industrial context but also living beings as a whole.

Cultural objects address this immeasurable dimension of the disaster. While they have multiple uses and purposes, what they have in common is that they manifest mourning for technical power and the dramatic consequences of nuclear accidents. Their value for the researcher is that they make it possible to examine the (more or less standardized) formalization of the representation. An analysis of the aestheticization process helps towards a better understanding of how representational elements are realized in a particular cultural system. It is this real-world realization that forms the perimeter of the study of objects.

Although research based on documentaries cannot aim for completeness, it provides food for thought about the impact of Fukushima on representations. The event does not seem to have upset the representational system of nuclear accidents. Films describe a world that could be a duplication of Chernobyl. A radioactive world, which is the demonstration of mankind's technological pride: abandoned territories, sacrificed populations, destinies that have been changed in a Faustian pact with the atom. Behind these images and words, the unwavering shadow of the sarcophagus of the Chernobyl plant looms. The Soviet accident remains the great caesura which changed representations. It created new sensitivities, new ways to show the invisible and say the unsayable. Chernobyl created a mythology, symbolic references, practices, a whole range of signs, almost a language that subsequent documentaries borrowed from to tell the story of Fukushima.

But the representation of Fukushima is not just the transposition of a Japanese Chernobyl. If the central core appears to be intact, new elements appear in the peripheral system.[18] Fukushima shows that the major technological powers are not safe from a nuclear accident. The idea is not new: the Three Mile Island accident demonstrated this in 1979. But low environmental emissions, skillful government communication,[19] and the limited means of the anti-nuclear movement have reduced the impact of the accident in France. The scope of Fukushima is very different. The event reactivated the public debate, including elements such as the myth of safety, the nightmare of waste, the exploitation of workers, the nucleocracy, and the desire for social autonomy. While Fukushima has not buried nuclear

---

[18]Caution should be exercised and only broader research will test these hypotheses.

[19]After Three Mile Island, French Prime Minister Raymond Barre announced that "The same scenario that took place in the United States could happen in France" (Le Monde, April 3, 1979).

power in France (or even created a national protest movement), Japan was shaken to its core and is experiencing a social renaissance. The telegenic virtue of the disaster is that it has pedagogical value: it reveals our carelessness and leads us to correct our position [63]. The accident has destabilized technological certainties and anathematized actors in the crisis. It has given renewed meaning to collective action, while the representation of the Japanese population has changed archetypes, showing how civil society can become resilient.

References

1. Ecole Biblique de Jerusalem (dir.), *La Bible de Jérusalem* (Editions du Cerf, Paris, 1998), 2117 p.
2. P. Bruegel, *La tour de Babel*, Kunsthistorisches Museum, Vienne, huile sur panneau de bois, 114 × 155 cm. (1563)
3. P. Robert-Jones, F. Robert-Jones, *Bruegel* (Flammarion, Paris, 1997), 351 p.
4. F. Walter, *Catastrophes, une histoire culturelle XVe—XXIe s* (Editions du Seuil, Paris, 2008), 380 p.
5. G. Mc CORMACK, Le Japon nucléaire ou l'hubris puni, *Le Monde Diplomatique* (avril 2011)
6. J. Davallon, Réflexions sur l'efficacité symbolique des productions culturelles. Langage et société, no **24**, 37–52 (1983)
7. B. Gaffie, Confrontations des représentations sociales et construction de la réalité. Journal International sur les Représentations Sociales **2**(1), 6–19 (2004)
8. S. Moscovici, *La psychanalyse, son image et son public* (PUF, Paris, 1961, 1976), 506 p.
9. Valence, *Les représentations sociales* (Editions de Boeck, Bruxelles, 2010), 174 p.
10. J.-M. Seca, *Les représentations sociales* (Armand Colin, Paris, 2010), 217 p.
11. D. Jodelet, Représentation sociale, in *Le dictionnaire des sciences humaines*, ed. by S. Mesure, P. Savidan (PUF, Paris, 2006), 1277, pp. 1003–1005
12. J.-C. Abric, L'organisation interne des représentations sociales: système central et système périphérique, in *Structures et transformations des représentations sociales*, ed. by C. Guimelli (Delachaux et Nestlé, Paris, 1994), 277 p.
13. P. Moliner, A. Martos, La fonction génératrice de sens du noyau des représentations sociales: une remise en cause? Papers on Social Representations. Peer Rev. Int. J. (2005)
14. D. Raynaud, *Sociologie des controverses scientifiques* (PUF, Paris, 2003), 222 p.
15. C. Lemieux, A quoi sert l'analyse des controverses? *Mil neuf cent*, n 25 (2007), pp. 191–212
16. S. Topçu, *La France nucléaire. L'art de gouverner une technologie contestée* (Seuil, Paris, 2013), 349 p.
17. P. Lascoumes, La productivité sociale des controverses, intervention au séminaire *Penser les sciences, les techniques et l'expertise aujourd'hui*, EHESS - CNRS (25 janvier 2001)
18. M. Callon, Pour une sociologie des controverses technologiques. Fundamenta Scientiae **2**(3/4), 381–399 (1981)
19. S. Boudia, Naissance et rebonds d'une controverses scientifique: les dangers de la radioactivité pendant la guerre froide. Mil Neuf Cent. Revue d'histoire intellectuelle **25**, 157–170 (2007)
20. G. Hecht, *Le rayonnement de la France, énergie nucléaire et identité nationale après la Seconde Guerre mondiale* (Editions de La Découverte, Paris, 2004), 385 p.
21. R. Chartier, P.-A. Fabre, Histoire des représentations, in *Le dictionnaire des sciences humaines*, ed. by S. Mesure, P. Savidan (PUF, Paris, 2006), 1277, pp. 1005–1007
22. E. Diet, L'objet culturel et ses fonctions médiatrices. *Connexions*, n° 93, pp. 39–59 (2010)

23. M. Groening, *Treehouse of Horror XV*. The Simpsons Season 16, 20th Century Fox TV & Gracie Films, 30 min. (2004)
24. J. Aumont, M. Marie, *Dictionnaire théorique et critique du cinéma* (Armand Colin, Paris, 2008), 300 p.
25. J.-P. Dupuy, Un paradis habité par des meurtriers sans méchanceté et des victimes sans haine: Hiroshima, ernobyl, Fukushima. Ebisu **47**, 49–57 (2012)
26. Y. Dupont, (dir.), *Dictionnaire des risques* (Armand Colin, Paris, 2003), 421 p.
27. D. Le Breton, *Sociologie du risque* (PUF, Paris, 2012), 127 p.
28. U. Beck, *La société du risque. Sur la voie d'une autre modernité* (Flammarion, Paris, 1986, 2001), 521 p.
29. C. Leon, *Silence, on irradie*. Editions Thierry Magnier, (2009), 111 p.
30. J.-P. Dupuy, *Retour de Tchernobyl, journal d'un homme en colère* (Editions du Seuil, Paris, 2006), 179 p.
31. Kurosawa, *Rêves - Le Mont Fuji en rouge*, Warner Bros, 119 min. (1990)
32. M. Eliade, *Images et symboles* (Gallimard, Coll. Tel, Paris, 1952), 252 p.
33. S. Sion, *The Land of Hope*, Rapid Eye Movies & Third Window Films, 133 min. (2012)
34. E. Lepage, *Un printemps à Tchernobyl* (Futuropolis, Paris, 2012), 163 p.
35. B. Aflallo, *La Gaule sous occupation nucléaire*, affiche sur papier recyclé, 29, 7 × 42 cm
36. C. Le Pomellec, *Catastrophes nucléaires: histoires secrètes*. Tac Presse & Canal +, 93 min. (2012)
37. K. Watanabe, *Le monde après Fukushima*, ARTE France & Kami Productions, 117 min. (2013)
38. G. Rabier, C.-J. Parisot, *Fukushima, des particules et des hommes*, Kami productions & France Télévisions, 52 min. (2014)
39. L. Coninck, *Fukushima, vers une contamination planétaire*, Code 5, 52 min. (2014)
40. J.-J. Delfour, Nucléaire et jouissance technologique. *Le Monde* (11 avril 2011)
41. Y. Shiokura, Comment un pays irradié est devenu pronucléaire, *Courrier Internationale* (18 août 2011)
42. S. Lebrun, C. Barreyre, *Japon: Nucléaire, la filière du silence*, Envoyé Spécial, 30 min. (2011)
43. AFP, Des milliers de manifestants demandent l'arrêt de la centrale de Fessenheim, *Le Monde* (10 avril 2011)
44. Fayner, *Nucléaire: la bombe humaine*, France Télévisions & Chasseurs d'Etoiles, 52 min. (2012)
45. J.-P. Dupuy, *Pour un catastrophisme éclairé* (Editions du Seuil, Paris, 2002), 214 p.
46. F. Guarnieri, S. Travadel, C. Martin, A. Portelli, A. Afrouss, *L'accident de Fukushima Daiichi, le récit du directeur de la centrale, Volume I – L'anéantissement* (Presses des Mines, Paris, 2015), 341 p.
47. G. Geal, Il faut sortir de la religion de l'atome, *Le Monde*. (17 mars 2011)
48. T. Guthmann, Gestion de crise et culture politique du Japon contemporain: le Premier ministre Kan Naoto a-t-il réellement failli? Ebisu **47**, 81–87 (2012)
49. F. Guarneri, S. Travadel, Engineering thinking in emergency situations: A new nuclear safety concept. Bulletin of Atomic Scientists **70**(6), 79–86 (2014)
50. A. De Halleux, *Welcome to Fukushima*, Simple Production, Crescendo films, L'Indien Productions, 59 min. (2013)
51. D. Zavaglia, *Fukushima, une population sacrifiée*, LCPAn & Scientifilms, 52 min. (2012)
52. D. Boilley, Initiatives citoyennes au Japon suite à la catastrophe de Fukushima, 30 p. (février 2012)
53. G. Ackerman, *Tchernobyl, retour sur un désastre* (Gallimard, Paris, 2006), 162 p.
54. W. Tchertkoff, *Le crime de Tchernobyl. Le goulag nucléaire* (Actes Sud, Arles, 2006), 717 p.
55. P.-F. Souyri, Les gens de Fukushima ne se sentent pas comme de victimes mais comme des sacrifiées, *Le Monde* (10 mars 2012)
56. M. Gaulene, Le mouvement antinucléaire japonais depuis Fukushima, CERI - Sciences Po. org, (2012), 11 p.

57. J. Lagane, Catastrophe environnementale au Japon: apport des savoirs profanes et mouvements citoyens. Ebisu **47**, 143–150 (2012)
58. J.-C. Deniau, *Nucléaire: la grande explication*, JEM productions, 75 min. (2012)
59. A. De Halleux, *RAS: Nucléaire rien à signaler*, Arte & Crescendo Films, 52 min. (2009)
60. J. Chandoutis, *Entretien avec Alain de Halleux*, 15 min. (2011)
61. F. Biamonti, *Nucléaire: exception française*, Morgane et Kami Productions, 70 min. (2013)
62. P. Marmignon, Communautés de quartier et associations: le retour du local après le 11 mars 2011. Ebisu **47**, 215–221 (2012)
63. R. Debray, *Du bon usage des catastrophes* (Gallimard, Paris, 2011), 112 p.

Why Is It so Difficult to Learn from Accidents?

Kohta Juraku

Abstract After the Fukushima nuclear accident, the whole Japanese society swiftly achieved a consensus to have comprehensive accident investigations to identify the root cause of the disaster. The Government and other major actors established several accident investigation commissions to meet this public will. However, the author has to say the lessons have not been learned and absorbed well so far, with deep regret. Because the issues centering on responsibility and social justice have not been dealt with well, the outputs of the investigations transformed into alternative sanction on nuclear industry and poorly articulated regulatory reformation, for example. This trajectory has been considered as a result of the particular and common culture of East Asian societies, but the author would argue that it should become more and more important global problem in the future world with high-reliability and complicated technological systems and their failures. The integration of the concept of risk governance to build prescribed consensus of responsibility distribution is strongly suggested as a key idea of remedy to this problem.

Keywords Accident investigation · Responsibility · Social justice · Risk governance

1 Introduction

Why is it so difficult to learn from accidents? This is the given question of this chapter. In fact, it seems that the learning process from the Fukushima nuclear accident has not been satisfying so far, while the major investigation reports attracted very strong public attention both domestically and internationally when they were published. For example, the deficits in risk governance of Japanese nuclear program has not been getting better, but has even become worse in some aspects [1]. Difficulty in post-accident (or "disaster") social learning process often

K. Juraku (✉)
Tokyo Denki University, Tokyo, Japan
e-mail: juraku@mail.dendai.ac.jp

© The Author(s) 2017
J. Ahn et al. (eds.), *Resilience: A New Paradigm of Nuclear Safety*,
DOI 10.1007/978-3-319-58768-4_12

shows similar symptom to it, although many people point out the importance of comprehensive accident investigation and reflection from its result, and actually try to carry out those processes. People often notice that we are taking wrong trajectory again, but it is always difficult to breakthrough it. Why do we have to have such frustrating, regrettable and disappointing experiences again and again in many fields of modern technical enterprise? The author would try to explore the background of this aporia from the point of view of sociology of science and technology. This is not based on strict empirical analysis, but would rather be something discursive illustration. The author, however, believes that it should still be suggestive to stimulate the interdisciplinary discussions to elaborate the concept of "resilience." This chapter will also touch upon the ethical issues and their strong relation to the post-accident social learning process.

2   Post-Fukushima Accident Investigation and After

It is common understanding that deliberate, comprehensive and careful investigation for terrible technological accidents is essentially important and must be done officially. This tradition had evolved especially well in Anglo-Saxon countries such as the United Kigdom and the USA since nineteenth century. In the mid twentieth century, the modern accident investigation paradigm was established in these countries. These are well-known early cases that the Tacoma Narrows Bridge collapse in 1940 and the Comet disasters in 1954. In the former case, the US Government established the Board of Engineers consisted of the three experienced engineers under the Federal Works Agency. They conducted a comprehensive scientific survey on the cause of devastating shake that destroyed the just four-months-old suspension bridge and provided the findings on important aerodynamic phenomenon–self-induced vibration. It opened our eyes on the importance of aerodynamic considerations to design and construct safer suspension bridges and the lesson encouraged the progress of research and development in the relevant engineering fields.

Also, in the later case, a couple of in-flight disintegration accidents of the world-first jet airliner called for the world-first modern, systematic and uncompromised airplane accident investigation by the RAE (Royal Aircraft Establishment) with the strong commitment of then UK Prime Minister W. Churchill. They let us know many things about both of the mechanism of metallic fatigue and the methods of aviation accident investigation. All of those contributed a conspicuous progress of aviation safety.

Such a great success in the early stage of modern technology strongly imprinted us the effectiveness and necessity of post-accident investigation to have technological improvement (especially in terms of engineering safety) based on the lessons of tragic disaster. It meant the progress of our technology, and the society. It meant the prevention of the next similar disaster. It did make amends for the victims and their families of the accident. Especially, it is the most successful that accident investigation tradition in aviation field. Almost all countries now have their permanent accident investigation institution for aviation (and sometimes for public transportation in general).

It was also quite natural that people took over and extended the paradigm from aviation to space vehicle accident because these two fields are closely related each other and have many similarities. The Presidential Commission on the Space Shuttle *Challenger* Accident (Rodgers Commission) was chartered after space shuttle Challenger crash in 1986. Also, the NASA (National Aeronautics and Space Administration) convened the *Columbia* Accident Investigation Board responding to the space shuttle *Columbia* accident as the official accident investigation body, although it was not presidential commission. Other than these symbolic cases, many official accident investigations have been conducted after serious accidents in space development in many countries.

This was also the same story in the field of nuclear utilization. The President's Commission on the Accident at Three Mile Island (so-called Kemeny Commission) have often been acknowledged as the milestone in the history of post-accident investigation and safety improvement in the nuclear field. Now many people believe that we learnt and could learn from disaster through accident investigation process.

This belief has seemed to be well-shared in Japan, too. Soon after the Fukushima accident happened, public discussion about formal and comprehensive accident investigation was begun. This belief was acknowledged and adopted by the Japanese Government, the Japanese Diet, TEPCO (Tokyo Electric Power Co.) themselves and a non-profit organization (NPO) established for an independent investigation. All of these four major committees (or commissions) published their final reports by the mid of 2012. We can now have them on the web and/or as printed matter, and some of them have already been translated into English, for non-Japanese readers. Post-Fukushima formal accident investigation seems to be concluded.

This fact creates an expectation that many lessons have already been learned well, next Fukushima will be prevented by the measures responding to those lessons and the society as a whole should have become more resilient to similar (and even other) type of disaster. Also, some people have believed and even still believe that such changes should have positive effects on public opinion/sentiment about Japanese nuclear program.

However, the reality in Japan now is pretty far from these expectations. It has taken different trajectory than people's belief of "learning from disaster through investigation" theory.

As mentioned above, some deficits in Japanese nuclear governance have remained, or even become worse than before the Fukushima accident. The majority of public opinion is still negative on nuclear program as a whole, for relevant organizations and on restart of safety upgraded nuclear power plants, while the Abe Administration officially decided to maintain Japan's nuclear power utilization

[2, 3]. Nuclear advocates continue the discussion to restore public trust, to build the consensus and to promote their program again. Critics persist in their counterarguments on the efficiency, risk, transparency and feasibility of "nuclear village's" theory. This landscape is almost the same to the scene BEFORE the Fukushima nuclear disaster happened. Sandwiched in between those polarized discourses, the rate of pro- and con-nuclear poll has been stabilized—at the point of a little bit negative against nuclear—for these six years. General public gradually lost their interest on nuclear dispute as well as trust towards the people relevant to nuclear activities. On the other hand, nuclear power station restart program is still walking randomly, not articulated well and the experts in nuclear field are pretty demoralized.

It is hard to say the learning process through accident investigation was successfully finished and we overcome the accident. It is really far from the oracle of "learning from disaster through investigation" theory.

3 Untaken Responsibility: Unsuccessful Prosecution and Alternative Sanction by Tightened Regulation

Then, a question comes up: what has the Japanese society been doing after the accident investigation? The author's answer is "unsettled discussion about the locus of responsibility." Not only the direct stakeholders, such as the Government, TEPCO, and the investigation commissions, but also the whole society, of course including general public, have experienced the difficulty in coping with the separation of the issues centering on social justice and practical improvement based on the lessons learned from the accident since it happened.

Severe nuclear accident could be interpreted as one of the most extreme and typical cases of organizational accident with serious consequences [4]. Needless to say, the Fukushima case was the first experience of this kind for Japanese society. It is well known that Japanese society (and perhaps other East Asian societies as well) have relatively strong retributivism and martinetism (severe punishment policy) for individuals involved in the cause of disaster even when the nature of accident is organizational [5, 6]. There is a long history of controversy about separation of criminal prosecution process and accident investigation activities in Japan, and it has never been settled down. People, as well as the victims and their familiy members, have pretty strong feeling of unjust without strict punishment for individual's fault that cause and/or worsen the damages caused by accident.

In this respect, no one has been officially punished through criminal prosecution process on the Fukushima accident so far. This fact should be very uncomfortable for accident victims as well as for many members of Japanese society, thinking about the strong tradition of socially embedded retributivism and martinetism. Of course, damage compensation and life recovery assistance for suffering people have been carried out by incrementally improved official schemes. But it does not work instead of the punishment of responsible person. The author would like to argue, the major accident investigations and their reports did play another role in society than practical learning of lessons from the accident, to fulfill this unfocused public outrage.

It was the most authoritative one that the National Diet of Japan Fukushima Nuclear Accident Independent Investigation Commission (NAIIC) among four major accident investigation committees (or commissions) established after the Fukushima accident. NAIIC was established on December 8, 2011, with the legal basis by a special act. Their report was published on July 5, 2012 and some of its statements attracted the strongest public attention and even encouraged public anger. The author would take two examples from their provocative theories here: (1) "manmade" and "Made in Japan" disaster theories on the root cause of the accident and (2) "regulatory capture" criticism against the corruption of the past nuclear regulation [3].

The first case, "manmade" and "Made in Japan" disaster theory was suggested in the "Message from the Chairman" page of the Executive Summery written by Chairman Kurokawa, not in the body text of full report [7]. That page was exclusively for its English version and no counterpart in the original Japanese report. However, this expression was broadly cited in its media coverage. The word "manmade" attracted rapid and positive attention mainly in Japanese domestic public opinion because this interpretation was consistent with the tradition described above: it legitimated the lay theory accusing the relevant persons and their faults. It seemed to even encourage the legal criminal prosecution process for the relevant officials in the Government, TEPCO and other institutions.[1] The process was virtually started on August 2, 2012, after the NAIIC's report published, although no one was finally indicted after prosecutor's investigation.

Kurokawa also suggested another message at the same time in his letter—the theory of "Made in Japan" disaster with "Japanese Culture" explanation. It was spread all over the world very quickly, as wells as in Japan. This could obscure our analytical understanding on the root cause of the accident and there were negative responses on this point from foreign major journalism [8, 9]. It also seemed to be odd because this was something contradictive to individual prosecution approach supported by his own "manmade" theory (because everyone could be dismissed their responsibility if the root cause was the "culture"). But, these keywords are often cited simultaneously without any inconvenience, and considered those as the most important messages of the NAIIC report.

---

[1]In Japan, the result of accident investigation could be used as the evidences in criminal case. For example, the Aichi Prefectural Police, the Nagoya District Public Prosecutors Office adopted the materials and final reports of then Aircraft Accident Investigation Committee (AAIC, transformed into Japan Transportation Safety Board (JTSB) in 2008) as their evidences to prosecute the captain in the case of the JL706 accident happened in 1997. In this case, the Nagoya District Court and Nagoya High Court formally admitted those materials as evidence. Then AAIC commissioner was summoned by the courts as the witness. These practices fallen foul of the Convention of International Civil Aviation (Chicago Convention), but the Japanese Government formally issues the difference notification on the separation of accident investigation and criminal prosecution to the council of International Civil Aviation Organization (ICAO).

The second eye-catching narrative suggested by the NAIIC report was "regulatory capture" criticism against the corruption of the nuclear regulation in Japan. It strictly pointed out the deficits of past nuclear regulatory system, then proposed the fascinating keyword—"regulatory capture." Shuya Nomura, a member of the NAIIC, a jurist and the proponent of this concept described its outline precisely: "Regulatory capture is a theory posited by George Stigler in *The Theory of Economic Regulation*. It refers to a condition in which regulators are "taken over" by the operators due to their lack of expertise and information, which results in the regulations becoming ineffective" [10].

However, this Nobel Prize awarded concept was never used as an analytical framework in the report. It just exemplified the historical process of collusive regulatory practices as a case of "regulatory capture." This was interpreted as just a strict criticism against the corruption and became very popular. But, causal relationship between any particular factors and the result (= corruption) has never been demonstrated by using this concept in the NAIIC report.

These NAIIC's narratives inspire us an approach to punish victimizers: sanction through regulation.

People's unsettled outrage has seemed to result in unlimited and never-ending efforts to reduce the risk from hazard created by any nuclear activities. New regulatory authority (NRA, Nuclear Regulatory Authority) adopted decisively strict approach that calls for further measures to increase and to demonstrate plant safety in bit-by-bit manner (i.e. additional safety measures against similar scenario to the Fukushima case, safety review with "new regulatory standards," earthquake resistance retrofit, on-site active fault survey, and so on). This sequential regulatory actions has made operators and manufacturers impoverished by never-ending review process while public trust has not been effectively recovered in proportion to their efforts. It could be interpreted that regulation fulfills the public will to punish "evil" nuclear industry instead of legal prosecution process.

Additionally, it should be noted that the final conclusion of the investigation reports and the actual design of nuclear regulatory reform did not have causal relationships as a matter of fact. The discussion about the reform of regulatory system was carried out at the Government and the National Diet before NAIIC and other major final reports were published. Japanese Government established the NRA in September 19, 2012, three months after the Act was approved on June 20, 2012. The sessions about the change in law was held during spring of the year. At that moment, only the report of so-called "Independent Commission" (established by an NPO) and the interim report of the Governmental Commission (ICANPS, Investigation Committee on the Accident at the Fukushima Nuclear Power Stations) were released. It was chronologically impossible to reflect the final recommendation of NAIIC report on this institutional reformation in formal and traceable manner.

Actually, NRA themselves do not admit that the recommendations of NAIIC report was a part of the background of their establishment, according to their website [11]. It seems to be quite unreasonable that the National Diet did not wait for their own commission's conclusion and recommendation, as well as other major committees', though their final reports had almost been finished.

4 Implications for "Resilience": Beyond Cultural Essentialism

This situation should cause a serious contradiction—random-walking of policy and practice of nuclear power utilization, while more and more rigorous regulation without strategic and effective safety upgrade program. Past discussions on this issue, trying to sort out the incoherence, have focused on the cultural background of Japanese society that was described above, and tend to suggest ways to "redress" it to comply with "global standard" of separation of prosecution and investigation [6, 12]. This tendency shows interesting consistency with Kurokawa's "Made in Japan" theory. However, the author would like to discuss further implication of this responsibility issue to deepen and broaden the discussion centering on the concept of "resilience."

Indeed, retributivism and martinetism could be interpreted as particularly prominent characteristics of East Asian societies. It should be admitted that these "cultural" differences are observed and need to be considered. Separation of prosecution and investigation is still essentially important to make accident investigation effective, reliable and just, in principle. We should call for careful arrangement when we think about institutional and legal harmonization with international standards such as multilateral treaties on utilization of nuclear and other advanced technologies.

Nonetheless, the author would argue that this issue should become more and more global, and more and more difficult to cope with. It can no longer be trivialized as a local issue caused by particular "cultural" context. It should become more and more unsolvable even in other societies that have been considered as not so "retributivistic" or "martinetistic" so far.

As a series of studies including the ones on the cases of nuclear accident have shown, the contribution of so-called human factors has increased both in causes of accidents and amplification of damages by them, inversely proportional to the improvement of reliability of advanced technological system [4]. This trend is perhaps an irreversible and historically inevitable tendency. It must become one of the most central questions in many fields of contemporary society to cope with the problems that are relevant to human factors to prevent or to improve the "resilience" to possible disasters. Every leading nuclear engineer knows that the most dominant and variant factor in the PRA (probabilistic risk assessment) is human factor. For this very reason, it is still under discussion how to appropriately include it to make the PRA method reliable and suggestive. After the long history of engineering efforts to improve technologies in "technical" sense, human factors are coming back to the core of the discussion about the success and failure of our artifacts.

On the other hand, accident always creates its victims in some sense, regardless of the nature of the cause of accident. This proposition has been unchanged since ancient times. Rather, the more society become advanced and deeply interconnected with technological systems, the more diversified types of "damage" and "victim" to be compensated and cared. Some of those must trigger big public outrage that could never been soothed so easily. We always have to face the issue of social justice: to remedy damaged and violated rights of them.

These two contexts would make more and more difficult to separate the issue of responsibility from the other parts of discussion to learn lessons from accident. Separation of prosecution and investigation might become much more practically difficult even in the societies that adopt this principle and have long experience of practice of it.

In our conventional idea, especially in engineering field, we use an analyticl approach of accidents. We break down an accident as a whole into "factors" and find causal relationships among them. Finally, we identify the root causes of each accident. In this approach, "factors" caused by or relevant to individual or organization also need to be dealt as "human factors" or "organizational factors." Experts consider those "factors" as manipulable (operational) elements. Thus, many researches have invented various ways to prevent undesirable human behaviors or to encourage desirable ones (so-called human engineering).

But, this conventional approach of engineering effort has little impact on the post-accident human-related issues—responsibility issue and its ethical consequences. As described above, the increasing weight of human factors in causes and amplification of damages of future accident is inevitable. It should lead people's attention and even anger to responsibility issue. Without to invent and implement the way to deal with this aspect appropriately, every society would experience the similar social deadlock that has been observed in post-Fukushima Japan centering on nuclear issue. This is no longer a local requirement but the universal condition to realize more resilient governance on advanced technologies, including nuclear technology, of course.

5 An Idea of Remedy: Revisit the Origin of "Resilience" Concept

It is obvious that this aporia very badly spoils our post-disaster resilience. It prevents our recovery process from the damages caused by the accident, spoils improvement of technology based on the lessons of it, delays advancement of robustness of the society and obstracts any other proactive efforts responding to the disaster. Public outrage can never have positive effect for society to exert its resilience thoroughly, if it is just neglected or poorly cared.

Some (engineering) experts may still argue that this problem is solvable by "education" of public (that encourage them to accept experts' notion): we should let them know that we need to keep those two things separated to make our society resilient and to prevent the next similar failure. If this is a discussion about just "failure," not "accident," it might be possible to maintain this strategy. However, it is an indisputable principle of modern democratic society that we must take care of the issue of damage and responsibility with deliberate ethical considerations and the deep sense of social justice. That is the nature of "disaster." Therefore, it is inevitable to cope with those aspects integrated with technical and practical activities to understand and overcome accident.

At this point, the author would argue that an idea to remedy this problem might be found in the original context of "resilience" concept. Needless to say, "resilience" is the word originally used in the field of psychology and psychiatry to describe, analyze and encourage the human beings' ability to adopt and cope with stress and adversity. In those fields, resilience is considered as an inherent ability of us, but, it is studied that it could be encouraged by appropriate support by relevant people and society, at the same time. Thus, if we extract some implications from this original context of "resilience" concept, we should recognize the importance of social healing process to remedy post-disaster emotional trauma of society and individuals. The process must be clinical and call for very intensive and grass-root but sometimes low-profile efforts beside victims. Their damages need to be cured by psychological, bedside and ethical practices, while compensation, assistance funding, town reconstruction program and any other institutional and pragmatic supports are still important and effective in many cases. We need this kind of careful emotional treatments *before* we talk and do something about engineering resilience under the actual post-disaster situation like post "3.11" Japan.

Furthermore, it is even required to apply the similar healing process to encourage the resilience of expert community, as the victimizer's side. Japanese STS (Science and Technology Studies) researcher Ekou Yagi introduced her own experiences of "just be there with victimizers" after serious technological disasters, JR Fukuchiyama line train derail accident in 2005 and Fukushima Daiichi nuclear accident in 2011 [13]. She points out that the members of victimizing companies/organizations of serious accidents also require emotional support to be proactive and to build sound and respectful relationship with victims and their families. She tried to be a supporter, not by doing something actively with them, but by just being beside them and hearing their voices calmly. She reported that the existence of such an escort person with the knowledge on their business (it was the advantage of her as a STS researcher) seemed to be supportive for "victimizers" and did encourage the positive signs of their own changes towards the rebuilt of public trust.

6 Concluding Remarks: At the Heart of Risk Governance

To enhance post-disaster social resilience, however, we should arrange some appropriate arrangement to reduce the burden of such a psychological healing process after the actual occurrence of serious accident. As the concluding remarks of this chapter, the author would suggest an idea to help to make the arrangement more properly: to build a consensus about framing, characterization and evaluation of risks and distribution of mandates, roles and responsibilities among stakeholders. At this point, it should be noted that this is the very substantial goal of risk governance, and its core activity is risk communication.

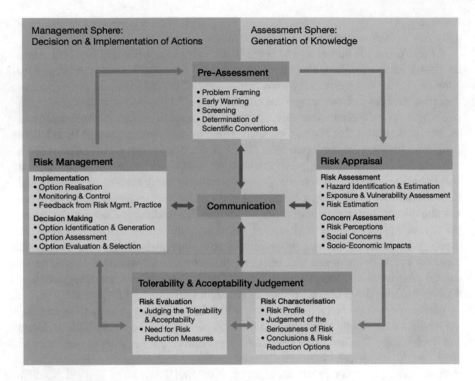

Fig. 1 IRGC Risk Governance Framework

It is one of the contributing things to prevent settlement of public anger that the locus and distribution of responsibility becomes vague, is intentionally changed or makes trivialized afterward. These injustices often trigger public outrage and emotional trauma of society and individuals more serious. It also heavily destroys public trust towards major actors and makes the recovery process slower and ineffective significantly. Once it is damaged, the trust would never be restored in the short term, as classical social psychological work demonstrated [14]. Loss of public trust should cause deadlocked situation that spoils public interests in the same way to Japanese case described in the earlier sections of this chapter.

To avoid such a fault and to realize effective risk governance, the figure illustrated by IRGC (International Risk Governance Council) would be suggestive even for "resilience" discussion in engineering fields (Fig. 1) [15]. It shows four phases of activities to cope with activities and the "communication" element connects these four factors. The important implication of this figure is that the concept of "risk" is not so evident, clear-cut, quantitative and easily operational thing. It should be discussed that the framing of risk issue in the "pre-assessment" phase. It should be included the "concern assessment" process in "risk appraisal" phase in parallel with so-called "risk assessment." To make any judgment on risk issue, we need to "characterize" the profile of risk. It is not an automatic output from the result of assessment, but a proactive and qualitative process to make the discussion concrete.

Finally, "managing" risk, but it is not the end of once-through cycle but the beginnings of next cycle. Communication is the key element for all of these four phases and not on-directional flow of the result of evaluation or decision made by limited number of experts.

In this multi-dimensional process of risk governance, all of stakeholders is encouraged to participate in the discussion about the distribution of mandate, role and responsibility to keep risks smaller than tolerable level. This discussion should be done on a daily basis to build consensus before something wrong happens. Prescribed (formal and informal) agreement would help post-accident remedy process by encouraging proactive collaboration among them to learn from and overcome the disaster.

Considerations on ethical implications of accident and integration of the concept of risk governance with risk communication will cultivate constructive and collaborative pathway towards more "resilient" engineering practices in the reality of our societies.

Acknowledgements Part of this chapter is based on the author's recently published and forthcoming book chapters [3, 16] and supported by the JSPS (Japan Society for Promotion of Science) academic funding program "Higashi-Nihon Dai-shinsai Gakujutsu Chousa" (Academic Survey Program for Great East Japan Disaster).

References

1. T. Taniguchi, Lessons learned from deficit analysis of nuclear governance, *Proceedings of the International Symposium on Earthquake, Tsunami and Nuclear Risks after the accident of TEPCO's Fukushima Daiichi Nuclear Power Stations*, Kyoto, Japan, 30 October 2014, Research Reactor Institute, Kyoto University (2014)
2. T. Sata, Oi-hanketsu ga Toikakeru-mono (Implications of the judgment of the court on Oi Nuclear Power Plant Case), J. At. Energ. Soc. Jpn., **57**(2), 119–122 (2015) (in Japanese)
3. K. Juraku, S.G. Knowles, S. Schmid, in *After Fukushima: Legacies of 3.11*, ed. by S.G. Knowles, K. Cleveland, R. Shineha. Learning from disaster: experts and the contested meanings of 3.11 (University of Pennsylvania Press, Philadelphia, Pennsylvania, *Forthcoming*)
4. J. Reason, *Managing the Risks of Organizational Accidents*, (Ashgate Publishing Ltd, 1997)
5. Y. Ikeda, Problems on criminal negligence and aircraft accidents, Bull. Sch. High-Technol. Hum. Welf. Tokai University, **4**, 81–91 (Tokai University, 1995)
6. Science Council of Japan, Ningen to Kougaku Kenkyu Renraku Iinkai Anzen Kougaku Senmon Iinkai Houkoku: Jiko-Chousa no Arikata ni kansuru Teigen (Report of the Safety Engineering Expert Committee, Committee of Human and Engineering: Recommendations on Practice of Accident Investigation). (Science Council of Japan, June 23, 2005) (in Japanese)
7. NAIIC, Executive Summery of The Official Report of Fukushima Nuclear Accident Independent Investigation Commission, July 5, 2012, National Diet of Japan Fukushima Nuclear Accident Independent Investigation Commission, Tokyo, Japan (2012)
8. Bloomberg, *Japan's Unsatisfying Nuclear Report, Bloomberg*, (July 9, 2012)
9. M. Dickie, *Beware post-crisis 'Made in Japan' labels*, (The Financial Times, July 8, 2012)

10. NAIIC, The Official Report of Fukushima Nuclear Accident Independent Investigation Commission, July 5, 2012, National Diet of Japan Fukushima Nuclear Accident Independent Investigation Commission, Tokyo, Japan (2012)
11. Nuclear Regulatory Authority, Background of the Reform of an Organization in charge of Nuclear Safety Regulation, Nuclear Regulatory Authority, http://www.nsr.go.jp/english/e_nra/outline/03.html (2015)
12. Y. Hatamura, *Shippai-gaku no Susume [An Encouragement of Learning from Failure]* (Kodansha, Tokyo, Japan, 2000) (in Japanese)
13. E. Yagi, Staying beside persons identified as responsible for preventing accidents—case studies on the Fukushima Nuclear Power Plant accident and the JR Fukuchiyama line train derailment, J. Sci. Technol. Stud. **12**, 106–113 (Tamagawa University Press, Tokyo, Japan, 2016) (in Japanese)
14. P. Slovic, Perceived risk, trust, and democracy, Risk. Anal. **13**, 675–682 (1993)
15. International Risk Governance Council, IRGC's *White Paper Risk Governance—Towards an Integrative Framework, International Risk Governance Council* (2005)
16. K. Juraku, Deficits of Japanese nuclear risk governance remaining after the Fukushima accident: case of contaminated water management, in K. Kamae (ed.), *Earthquakes, Tsunamis and Nuclear Risks: Prediction and Assessment Beyond the Fukushima Accident*, (Springer, 2016)

Decision-Making in Extreme Situations Following the Fukushima Daiichi Accident

Sébastien Travadel

Abstract The Fukushima Daiichi accident raises questions about current decision-making models. Faced with an overwhelming situation, which threatened both their own lives and that of the entire population, the plant's operators were obliged to take action, despite the lack of resources. In these conditions, decision making cannot be reduced to an optimization exercise based on a range of possibilities, or the application of planned operational responses to an emergency situation. The inevitable catastrophe, the social pressure it generates, the moral dilemmas it creates and the psychological drivers for action are characteristic of an extreme situation. The action plan must therefore be reinvented and individuals mobilised to these ends. It is therefore in a broader context of 'action' that decision making takes shape, and finds its logical foundations, meaning and temporality. Understanding decision making in extreme situations first requires a grasp of the development of a specific value system (that is mediated by the physical experience of the situation) in which the individual and social representations play a central role.

Keywords Decision-making · Extreme situations · Uncertainty · Ambiguity · Rationality · Temporality

1 Introduction

When Circe warned Odysseus about the dilemma he would face when passing through the Strait of Messina, she reminded him that it would be a decision of the heart. However, the priestess suggested that he choose the peril of Scylla: better to lose a few companions than to see them all engulfed by Charybdis. Odysseus followed this advice. Therefore, how can we say he 'decided'?

S. Travadel (✉)
Centre for Research on Risks and Crises (CRC), MINES ParisTech,
PSL – Research University, Sophia Antipolis Cedex, France
e-mail: sebastien.travadel@mines-paristech.fr

J. Ahn et al. (eds.), *Resilience: A New Paradigm of Nuclear Safety*,
DOI 10.1007/978-3-319-58768-4_13

The word 'decision' typically refers to the "*end of the deliberation in a voluntary act that results in the choice of an action*".[1] The meaning given to each of these terms varies according to different schools of thought. In Aristotle's *Nicomachean Ethics*, the decision relates to the available resources needed to reach a given, desirable, conclusion, while Cartesians (e.g. René Descartes in his *Metaphysical Meditations*, 1641) would argue that without Circe's advice, Odysseus was taking a decision that was beyond his ability to understand and he therefore did not choose. Leibniz argues in his *Theodicy* (1710) that understanding is not a necessity—it only provides a guide—and that the decision exists only through the effort of action. In terms of expected utility theory [1], Odysseus rationally opted for the path that minimized the maximum damage.

Beyond these considerations, if the sacrifice of companions can be made acceptable, it must be integrated into a social and symbolic universe that gives it meaning. It is the will of the gods that allows Odysseus to return to his kingdom, which places him above other men. Of course, nowadays Man tends to be emancipated from the gods and can think for himself. He decides after careful reflection on causes, which "*must always be mixed with chance in order to form a basis for reasoning*".[2] However, despite this distancing that is at the heart of Technology, the human being must still find meaning in their actions.

In his testimony [2], the Director of the Fukushima Daiichi power plant shows how operators, who were obliged to decide between the survival of some and the sacrifice of others, gave meaning to such decisions. Some of their critical decisions are set out below (Sect. 2). The inability of current theories to account for the magnitude of such decisions (Sect. 3) leads us to introduce the concept of 'projected time' (Sect. 4), and to explore mechanisms for the development of meaning in extreme situations (Sect. 5).

2   Fukushima Daiichi: Faulty Decision-Making?

It has become normal to describe the Fukushima accident as the result of poor decisions or a failure to act.[3] In particular, the nuclear community has recognised that the accident could have been avoided through appropriate prevention measures. This would have entailed raising the height of the dykes that surrounded the site and protected it against waves following the results of numerical simulations, conducted in 2008, which indicated the potential for flooding. However, a consensus needs to be reached on the usefulness of such information: TEPCO indicated that it found

---

[1]Source: Dictionnaire culturel, A. Rey, Paris, Le Robert (2005).

[2]D. Hume, Of the Understanding. A Treatise of Human Nature. Book I, 1739.

[3]The Japanese Parliamentary Commission investigating the accident concluded that it was a 'man-made disaster' [3].

the scenario unlikely given additional geological studies that invalidated the predictions [4].

Furthermore, how the accident was managed in the first days following the earthquake illustrates how decisions were taken in a context of crisis. The Japanese government's Investigation Committee was particularly interested in the circumstances in which Reactor 1 exploded on the afternoon of 12 March, 2011, when on-site teams thought they had vented it and cooling via the injection of seawater was about to begin [5]. Investigators asked the superintendent of the plant about his decision-making process given the information available to him [2]. To understand their approach, it is useful to recall certain facts.

On 11 March, the superintendent, Masao Yoshida, ordered preparations to begin for the injection of water into Reactor 1 using fire conduits and fire engines, to ensure cooling when the isolation condenser (IC)[4] stopped working because of electrical failure. The water level in the tank was checked by reading an indicator. The level indication was normal, which supported the belief of operators that the reactor was correctly cooled. The level, however, could not be constantly monitored. At around 02:00 on 12 March, the indicator showed a stable or slightly rising water level, while at the same time the pressure inside the tank fell. Communication between the control room and the crisis room was difficult and the superintendent had not been informed of problems in the IC conduit. Perplexed by the rise in radioactivity, he concluded that the water level indicator was malfunctioning, that the IC had potentially been non-operational for several hours and that the core was probably exposed [2]. This was confirmed by measuring the increase in pressure in the containment vessel, which had exceeded its structural limits.

The plant's superintendent said that he regretted having placed too much confidence in the water level indicator and not having asked the control room about the IC conduit. For their part, investigators expressed their surprise at the apparent passivity of operators, given that an IC valve had been closed for no apparent reason and that the injection of water was impossible, with the result that the reactor was not cooled between 18:25 and 21:30. It appeared that none of the shift leaders had alerted the crisis cell.

Yoshida also stated that, *"anyway, in terms of solutions, we did not have anything much better than the diesel pump, injecting with the fire pump and using the fire engine, which we finally did. Could we have reacted more quickly if we had known? I think that physically, we could not have gone faster"* [2] (our translation).

Faced with pressure that exceeded the structural limits of the containment vessel, the superintendent asked for Reactor 1 to be vented. This operation required the activation of a valve, which proved particularly difficult and dangerous because of the high levels of radioactivity, the lack of electricity or pneumatic equipment, and the lack of indicators or light in the control room. As none of the operators had

---

[4]A backup system in some boiling water reactors. It cools the core when power cannot be evacuated by the main condenser. The system condenses the steam produced in a heat exchanger, and then re-injects it into the tank using gravity.

taken the initiative and studied the plans and diagrams of the network in anticipation of venting, investigators asked Yoshida about when he had proposed this
solution. He said that he had only considered venting when he had sufficient data to
confirm that there was excess pressure in the chamber. The priority was therefore to
obtain information on the key parameters indicating the state of the reactor [2].

The Commission was also surprised that operators had not considered the possibility of a hydrogen leak from the tank to the containment vessel into damaged pipes,
although it was known that core fusion can produce large amounts of hydrogen. The
Director said he had been aware that, if the core was damaged, hydrogen was produced; notwithstanding, he felt that it would remain confined within the vessel and he
had focused on the threat of a container explosion, given the high pressure that had
been observed. He stated, "*the top of the reactor building is covered and ventilation
panels are arranged on the side. We had not even imagined that these panels were
closed and that hydrogen and oxygen had accumulated. We focused on the containment vessel. [...] We were prisoners of our* a priori *assumptions*" [2] (our translation). Moreover, the entire international nuclear community was unaware of this
scenario (*Ibid.*). It was only several hours after the explosion, following an investigation of the destroyed buildings, that operators concluded that the accumulation of
hydrogen was probably the cause. They then studied the measures that needed to be
taken in order to prevent a similar scenario at the plant's other reactors.

At dawn on 15 March, although Reactors 1 and 3 had already exploded, operators felt a strong jolt and heard a loud noise, which they could not immediately
identify the source of; at the same time they noted damage to the building of
Reactor 4 and that the pressure in the containment vessel of Reactor 2 had fallen to
zero [5]. Although they gave little credibility to the reading from the pressure
indicator and, on a scientific level, the hypothesis that Reactor 2 had exploded was
not consistent with the available information, Yoshida considered the noise to be
the most important factor and ordered an evacuation.

These examples suggest that the criteria for decision making, the relevance of
the decision and the resources available to the decision maker were deficient, to the
extent that they hindered the management of the accident. According to Yoshida
himself, "*it was total confusion. And that was in this atmosphere that it was
necessary to give orders. So I recognize that it was not done in a logical and
considered order*" [2] (our translation). However, identifying potential derivations
from logical reasoning and understanding the circumstances presupposes that the
processes at work can be formalized.

3 Testing Decision Models Using the Fukushima Daiichi Accident

A classical approach in management science is to model decision-making in four
phases. After collecting the information necessary to diagnose the problem, the
decision maker formulates potential ways to resolve it, based on a necessarily

limited rationality; they then select a particular mode of action, and evaluate a provisional satisfactory solution before iterating the process if necessary [6]. The comparison of potential scenarios results in mathematical formulations, developed from an economic perspective. The choice is equivalent to the optimization of a function (the utility). A set of axioms expresses postulates about the properties of preferences (for a discussion of the various theories see [7]). If an objective assessment of the probabilities of the consequences of an act is not possible, the decision maker can use 'subjective' probabilities. The probability distribution is therefore measured according to their knowledge of the possible states of the object they are interested in and upon which they wish to act.

Nevertheless, the assessment of subjective probabilities can be arbitrary and, in many practical situations, the behaviour of agents does not reflect preferences that are consistent with these axioms, without being necessarily called an 'error' [1]. To account for the assumed ignorance of the decision maker of certain states and their aversion to uncertainty, Gilboa and Schmeidler [8] showed that decisions can sometimes be seen as the maximization of a form of expected utility that takes into account the worst case scenario, which is consistent with the 'maxmin' model. It is also possible to broaden the spectrum of reactions in uncertain situations [9], in order to model less paranoid attitudes [1].

From this perspective, the behaviour of the decision maker is interpreted according to a concept of ambiguity relative to their knowledge of the world—he could not be sure of the meaning of the few information he got regarding the state of the facilities—, which is how the questions of the Japanese investigators should be understood (see Sect. 2). Faced with ambiguity about the state of Reactor 1, it could be argued that Yoshida violated expected utility theory by not deciding to immediately cool the core; or, alternatively, he demonstrated an aversion to uncertainty, by deciding to rely on information about the water level, while simultaneously preparing cooling mechanisms.[5] The fact that he did not foresee hydrogen leaks can be interpreted in two ways using the 'maxmin' model: either he failed (cognitively) in his assessment based on all of the objective information available; or his attitude or imperfect knowledge led him to limit his choices to a subset of possible states. As for his decision to send staff to gather information on the state of the reactors before considering operations such as venting, this can be interpreted as an example of incomplete preference, where the status quo is maintained until such time as a conclusively better, new alternative appears.

Can we conclude that Yoshida acted irrationally, or do the models provide an incomplete description of such decisions? Gilboa [10] argues that we qualify behaviour as 'irrational' if whoever violates its precepts regrets their actions. The regret expressed by Yoshida about the confidence he placed in Reactor 1's water level indicators (see Sect. 2) would suggest that the decision was irrational.

[5]The superintendent had to manage limited resources and set priorities. Therefore, in this case it is not possible to apply a principle of dominance and conclude that the "ignore the water level" option was better.

However, the superintendent immediately qualifies his comments by stating the impossibility of conceiving an alternative solution. Through this remark, Yoshida integrates his decision into a wider context of action.

This is to be compared with observations from current research in Natural Decision Making (NDM) that aims to account for decision making in the presence of changing conditions, ill-defined tasks, time pressure and significant personal risks in the case of error [11]. In these conditions, the decision maker's accounts of their decision making "*do not fit into a decision-tree framework*"; they are not "*making choices*", "*considering alternatives*", or "*assessing probabilities*", but they see themselves as acting and reacting on the basis of prior experience, generating and modifying plans to meet the needs of the situation [12]. The decision maker acts on the basis of heuristics, then develops a mental simulation to assess the feasibility of the proposed response. These studies are consistent with those of Gilboa and Schmeidler [8], who axiomatized a *Case-Based Decision Theory*, which postulates that the decision maker acts by comparing the current situation to one already experienced.[6] Coordination and leadership modalities also change when tasks are unpredictable and interdependent, as is the case in an emergency context [13].

However, if the context for the intervention of firefighters or emergency room surgeons is sometimes called an 'extreme situation' [14], in practice these 'dynamic' situations constitute the predictable working environment of the decision maker. The problem relates more to the definition of the case in question, than the solution once the diagnosis has been made. Ultimately, the 'extreme' nature of a situation is assessed differently by different researchers and does not necessarily imply that the decision maker is completely overwhelmed or out of their depth [15]. Such individuals have substantial resources at their disposal, a well-established set of procedures and the impact of their actions is limited at the scale of society.

In addition, whether they focus on decision making processes based on scenarios or on more empirical approaches, investigations of the influence of stress [16], a hostile physical or social environment [17], or the formal organization [18] on the performance of the decision maker are simplistic. They lead the analysis to be focused on the physical or emotional factors that could have led Yoshida to make errors (for example his decision to evacuate the site). This cognitive approach is indicative of the common sense meaning of 'emotion', i.e. a complex state of consciousness, usually sudden and momentary, accompanied by physiological disorders.[7]

However, this perspective largely ignores the role played by emotions in decision making [19]. Ellis [20] considers that emotions and values are necessary components of a decision, which does not mean that the decision becomes 'irrational'. This assertion is illustrated by the way in which the plant's staff decided to return to the field following the explosion at Reactor 3 on 15 March. According to

---

[6]However, this theory is not specific to an emergency situation.

[7]Dictionnaire culturel, op. cit.

Yoshida, *"everyone was in shock, frozen, unable to think. So I got them all together to talk to them. [...] I also told them [...] that if we did not respond, the situation would become even more catastrophic [...] It was at that point that I experienced one of the most emotional moments of my life. They all wanted to go back, they even pushed each other out of the way to get there"* [2] (our translation).

The Fukushima accident demonstrates how difficult it can be to make decisions when the realities of the situation are not conducive and the unfolding scenario cannot be stopped (due to a serious lack of resources); at the same time, the physical integrity of individuals cannot be guaranteed and the consequences of taking action —or not—have societal significance. The inevitable catastrophe, the social pressure it generates, the moral dilemmas it creates and the psychological drivers for action are characteristic of an extreme situation, as described by Travadel and Guarnieri [21]. In extreme situations, the action plan must be reinvented and individuals mobilised to this end. Yoshida therefore stated that he initially had no solutions, no idea how to react, and fell back on 'administrative' procedures in an attempt to regain self-confidence [2]. Similarly, organizational theory tends to regard a crisis as a situation where, *"not only are there insufficient resources, but it is a situation where the rules were not thought of yet"* [22]. The new order must be acceptable, when life itself—that of workers or of an entire population—is threatened.

Current models do not make it possible to deepen our understanding of the practical management of such situations. To progress, the analysis must be based on suitable metaphysics and integrate the world of the decision maker, in which their decisions make sense. The following sections consider these two dimensions.

4 Decision Making and Catastrophe: Back from the Future and Return

The concept of decision making inevitably refers to concepts of causation and rationality. Through the introduction of probabilistic links between states of the world and actions, expected utility theories opened up the debate on the causal link. Savage's axioms (for a discussion, see [7]) apply to actions that do not have a causal influence on the state of the world in which their consequences are experienced.[8] Consequently, the final state of the system is often associated with a fixed point,[9] which rules out many decision-making scenarios. To overcome this problem, (unconditional) utility was replaced by a concept of utility which conditions the probabilities of states of the world (those leading to the expected consequences) to the execution of the act. These probabilities are interpreted either in terms of classical conditional probabilities, or causal probability [23]. In the first case,

---

[8]The problem can be reformulated to make it the case [23].

[9]This is the case in economic models such as the 'perfect competition' model in which the actions of one agent do not change the overall balance.

evidential decision theory computes an act's expected utility using the probability of a state given the act $P(S|A)$[10]; in the second case, causal decision theory replaces $P(S|A)$ with $P(A \rightarrow S)$ or a similar causal probability [24]. This choice defines the decision making perspective. Thus, according to Jeffrey, "*in decision-making it is deliberation, not observation, that changes your probabilities. To think you face a decision problem rather than a question of fact about the rest of nature is to expect whatever changes arise in your probabilities for those states of nature during your deliberation to stem from changes in your probabilities of choosing options*" [25].

Rationality is therefore at least dual, and Lewis's counterfactual decision theory can account for both aspects [26]. The fundamental idea of this analysis is that the counterfactual "*If A were the case, C would be the case*" is true just in case it takes less of a departure from actuality to make the antecedent true along with the consequent than to make the antecedent true without the consequent [27]. The causal dependence is then stated as follows: Where c and e are two distinct actual events, e causally depends on c if and only if, if c were not to occur, e would not occur.

A dual rationality goes hand-in-hand with the concept of temporality. Either the decision maker operates by projecting a set of possible futures and seeks to maximize the consequences of their actions, or they try to make their actions as consistent as possible with a desirable state of the world. Dupuy [28] thus states that: either, at every moment in 'occurring time', regardless of the predictions of an infallible Predictor, "*agents have the power to act in such a way that, if they were so to act, they would render inaccurate the predictions of the supposed Predictor*", which means that causal links are probable ones; or, at all times in 'projected time', causal links are fixed and the agent has the power to do something such that, if he were to do it, the 'script' of his life would have been different. Dupuy suggested merging these two concepts of temporality form a loop, in which the past and the future determine each other. In particular, a future state, when represented by a favourable probability and another that is disastrous (with a very low probability) can serve as an anchor point for ongoing action in an approach that is constantly under review [26].

We argue that Yoshida was guided by these two representations of temporality. He applied a causal type of reasoning in order to deduce from the information at his disposal that the core of Reactor 1 was undergoing fusion. He expected to find proof of an 'event e', for which it was then necessary to find the cause. Using the same reasoning (this time in anticipation) he decided to start the venting manoeuvre and avoid an explosion, based on the objective data available to him. Given the information at his disposal, he assessed the plausibility of a causal link to unwanted consequences, based on an appreciation of the law of physics. At the same time, he organized actions to be taken based on information that he did not yet have in a measurable form, but which he had nevertheless convinced himself was true.

---

[10]This theoretical orientation has led to debate about the ability of an agent to assess the likelihood of their actions—a problem that is resolved by invoking the predisposition of the agent to act [23].

Although it had not yet happened, the future catastrophe seemed real enough to him to guide his actions. The destruction of an entire region—not simply an official accident—constituted, at least a partial anchor point for the decisions that he took. He repeated this point many times during the hearing, "*it was clear that we were heading for a major accident and we had to prepare for it*" [2] (our translation). He stated that he had always had such a situation at the back of his mind, beginning with his initial instructions to prepare alternative cooling methods. Yoshida's decisions therefore took place in a sort of 'projected time' temporality. It would therefore be wrong to say that he 'expected' the loss of the cooling systems, as this would place his actions in a causal type of rationality and an 'occurring time' temporality. This idea is similar to that of Dupuy [28], who argues that temporality should not be seen as a container in which human activities take place, but as a result of human activities.

This observation may seem trivial. Naturally, where the cooling systems of a nuclear reactor are damaged, the operator must consider the potential for disaster. However, limiting the investigation to the critical bifurcations of the decisions as the situation unfolded, and trying to understand the rationality of choices based on available information or resources—an exercise that is at the heart of traditional investigation processes—means that the disaster is only looked at in terms of its potential, which is likely to significantly weaken its power of determination. A contrario, when rationality is viewed in terms of 'projected time' it "*is a fundamental existential problem that rears its head every time we are confronted with absolute uncertainty concerning a variable on which our 'salvation' depends*" [28]. By comparison, an approach in terms of scenarios sees the future as a distant objective reality for the individual. In this context, the paradox highlighted by Dupuy [26] is as follows: prospective methods make it possible to socially create an image of the future; at the same time, they empty it of its physical dimension, they do not acknowledge reality of any kind.

Difficulties arise when the decision is counterfactually examined using a probabilistic-causal approach, with a view to arriving at a moral judgment. The expression of a causal link is inherently relative. Core fusion, overpressure or leaks in the containment vessel of Reactor 1, like the accumulation of hydrogen and the failure of ventilation equipment, may be considered as the 'causes' of the explosion of Reactor 1. Emphasising one over the other is relevant depending on the class of situation in question, the context or the contrast to be established between a situation and an event. For this reason some authors have suggested the reformulation of the counterfactual causal dependence as follows: If c^* had occurred instead of c, then e^* would have occurred instead of e [27]. Whichever is the case, this observation shows that causality has no transcendent reality, except in the narrow field of science (i.e. excluding human affairs). An assertion of causality requires adopting a point of view, the mark of subjectivity. The a posteriori allocation of probabilities in causal reasoning leads to short-circuiting the infinity of potential future bifurcations, and the retention of only a few of them. Reasoning is thereby biased because, unlike moral judgments, "*the foundation for probabilistic judgment cannot include any information that is only available after action has been taken*" [26] (our translation). This may explain the discrepancies between the questions of the investigators and the action taken by Yoshida (see Sect. 2). The Commission's investigators mixed two interpretations: their causal analysis in an 'occurring time' frame (reconstructed a posteriori) support a value judgments of the actions taken by the plant's superintendent. This approach does not take full account of the decisions made at Fukushima Daiichi. To make sense, Yoshida's actions must be understood in their entirety; it is impossible to separate decisions from the close connection that the individual has with the realities they confront. When it comes to give it meaning, action in extreme situation emerges at least in part within a 'projected time' temporality.

5 From Decision Making to Taking Action in Extreme Situations

The plant's superintendent was unambiguous in his description of the relationship between his staff and the production facility, which he characterised as a fight, dominated by fear and suffering. His lexicon and register provide further evidence: he speaks of "*three monsters*", "*three nuclear units that were unleashed*", and tries to "*achieve the impossible with very few staff*" to "*tame this thing*" [21] (our translation). In this context of sensory stimulation, impressions and perceptions had a strong influence on his decisions. His order to evacuate the site is an example of this (see Sect. 2).

However, emotions do not always disrupt the ambient order and well-regulated planning. Damasio [29] showed that they are a key component in the development of rational thinking. Several experiments have since confirmed the need to reintroduce emotion into the process of conceptualization. Similarly, it has been demonstrated that individuals produce concepts according to their perceptual experience [30]. Such studies show that the embodiment of emotion can ground concepts. An illustration of this is found in Yoshida's testimony. He stated that following the explosion of Reactor 1 (when the pressure was about 500 kPa), the number '500' left him "ill at ease". He went on to add, "*I know this is totally irrational, it was just a feeling*" [2] (our translation); this feeling would influence his decisions concerning the other reactors.

Leontyev [31] put forward similar arguments. He claimed that human activity forms the foundation for consciousness—a back-and-forth process that operates between an individual and an object, guided by a pattern and determined by sensual contact with the outside world. Leontyev goes on to say that beyond this circular process, which influences interactions between the organism and the environment, mental representations of the objective world are governed by processes in which the individual is in physical contact with it and thereby obeys its intrinsic properties and its own relationships. It is the object that initially determines how actions unfold and, secondly, it is how it appears as a subjective product of the action that records and stabilizes the objective content of the activity. The resistance of the object breaks the cycle of internal mental processes and provides an opening to the outside world. Moreover, Vygotsky [32] showed that actions are social by nature; they always take place in the presence of others and are mediated by signs. Recent studies characterize the moderating role of social relations in the relationship between the individual and their body as a conceptualization tool: the *"current (social or other) context influences the way in which a concept is represented in a conceptual task and the extent people recruit embodied information to solve it"* [30].

According to these theoretical results, meaning is grounded in bodily processes of perception and action. The organism's bodily interaction with the environment is of crucial importance to its cognitive processes [33]. What is meant here by the 'body' is not the body as a functional system with input and output, but rather, as the enactive approach defines it, *"an adaptive autonomous and therefore sense-making system"* [34]. From an anthropological perspective, Mauss had already defined the body as the *"primary technical object"* [35], which appears therefore as a support and provides meaning. *"It is impossible for a man not to be permanently changed and transformed by the sensory flow that runs through him. The world is the product of a body which translates it into perceptions and meaning, one cannot exist without the other. The body is a semantic filter"* [36] (our translation).

From this perspective, phenomenology, which is specific to the experiential aspect of emotion, can be linked to values. Here, 'values' represents what gives meaning to action, although not exclusively as an abstract object to consciously work towards. At a more primitive level, 'values' are more properties of a particular type that are exemplified by contexts, objects or behaviours: emotions—such as fear—link us with exemplifications of these evaluative properties—for example danger [37]. In an emotional state, the body becomes prepared to potentially take action and *"the specific way it is prepared is interpreted, very naturally, as an apprehension of some of the evaluative aspects of the environment. Here, then, bodily sensations are not understood as simply the effect of thinking about the environment, they play a direct role, by virtue of their phenomenology, as an explicit presentation of their own objects, i.e. values. Emotions are what is felt by a body that is prepared for action: it is therefore in this narrow sense that one can say that emotion is an experience of value"* (*ibid.*, our translation).

Plunged into an unprecedented sensory universe, the operators at Fukushima Daiichi had to redefine the meanings and values in their world. The scene was apocalyptic: high levels of radioactivity, extreme temperatures, piles of debris, aftershocks, floods, darkness and exploding reactors formed the context in which they were required to take action. Decision making was shaped by their contact with this material and social reality, physical challenges, and the way they behaved and saw others behaving.

Individual commitment was influenced by the need to take action and the resources required, the rules and shared representations of the action to be performed. Group behaviour is indeed determined by the image it has of its task [38]. In extreme situations, the construction of this image integrates current social representations from the public sphere. In this case, public opinion of the Fukushima Daiichi accident hinged on interference from the Prime Minister in the management of the crisis and the incompetence of TEPCO [39]. The validity of the actions taken by on-site operators was not acknowledged until long after the events of 11 March, 2011. Three months after the accident, workers suffered from an unusually high level of psychological problems, linked to the social discrimination they were subject to [40].

The decisions of the plant's superintendent were therefore influenced by the need to protect the physical and mental wellbeing of his colleagues [2]. Nevertheless, when it became clear that the only way to vent Reactor 1 and prevent its explosion was to manually open a valve located in a highly radioactive zone, technicians reported for work. The superintendent stated, "*We decided to do the operation by hand, as a last resort. We decided to do this because we thought that it could be done, if all it took was to accept being irradiated*" (*ibid.*, our translation). It could have been the case that economic considerations dictated this decision: the loss of a few employees could have brought a solution to the crisis as long as it would not have jeopardized the remaining resources. However, such a decision can only be seen as acceptable as a result of a personal and interpersonal journeys of the decision maker and his colleagues, through an action process leading to a singular system of values. In order to understand decision making in extreme situations, one must first understand the development (mediated by physical experience of the situation) of this value system in which individual and social representations play a key role.

6 Conclusion

Investigations into the accident at Fukushima Daiichi highlighted failures in communication, and a lack of foresight and anticipation on the part of operators in some of the decisions that were taken. These analyses implicitly suggest that there was a range of options based on a known state, and they reduce decision making to an optimization exercise. Consequently, the feedback from experience becomes focused on the lack of coordination between the operator's headquarters, the Japanese government and the plant, obsolete instrumentation, or the effects of stress on behaviour.[11] Of course it is clear that our understanding of the impact of stress or emotions on behaviour or decision making is still preliminary [41], and merits further examination. However, paradoxically, this type of 'safety science' approach seeks to place behaviour in a theoretical context made up of bloodless social mechanisms that take no account of the humanity of those who must act. Moreover, in the nuclear context, executives are sometimes tempted to resort to formalisms to demonstrate a high degree of control, even if it means negating the difficulties faced

---

[11]See for example the conclusions of the Japanese Parliamentary Investigation Commission [3].

by operators [42]. In extreme conditions, the risk is that the factors that determine the 'entry into resilience' are ignored [43].

In their current state, the lessons that have been learned from the accident may hide some key drivers for the planning and development of actions taken in the face of devastation. The concept of the extreme situation invites us to supplement these lessons and reintroduce factors related to the human body, emotions, and kinaesthetic, which create our initial relationship with the world, and constitute a socio-sensual structure for behaviour. In order to understand behaviour in extreme situations we must first understand the experience, which is marked by radical changes that cannot be easily aggregated into logical arguments. Moreover, the integration of a more sensitive approach to the behaviour of others into an otherwise rational approach to decision making would appear to be a promising avenue for better management [44].

On-site management problems were compounded by the injunctions of remote decision makers. The conflict between the plant's staff and the Japanese government cannot simply be reduced to a failure to share information or a lack of awareness of each other's problems. It is the result of different relationships with danger, through social pressure, to moral issues. More generally, the management of the accident demonstrates the intrinsic limit of an "optimization" type decision making process. At the political level, there is a tendency to expect from rational decision maker to base their decision on technical consideration, whereas it has some necessary social and ethical implications which cannot be avoided. In any event, the people rely on their own criteria in order to assess the rationality of an evacuation for instance, based on the social meaning of their decision, personal feelings, etc. A 'resilient community' is the result of these multiple nucleus of action/decision which have to be coordinated with a full acknowledgement of their specificities.

It is therefore necessary to establish an ethical framework that is appropriate to extreme situations, which articulates different temporal and rational registers. The intervention by the residents of Olympus offers a universal interpretation of Odysseus's decision. A contrario, due to the fact that all parties sought to make their own sense of the situation, the Japanese population, its government and authorities did not understand the magnitude of Yoshida's actions or those of his colleagues.

References

1. I. Gilboa, M. Marinacci, Ambiguity and the bayesian paradigm, in *Advances in Economics and Econometrics*. Tenth World Congress, ed. by D. Acemoglu, M. Arellano and E. Dekel, vol. 1 (Cambridge University Press, 2013), p. 179–242
2. F. Guarnieri, S. Travadel, C. Martin, A. Portelli, A. Afrouss, L'accident de Fukushima Daiichi: le récit du directeur de la centrale. vol. I, L'anéantissement, (Presses des Mines, Paris, 2015), p. 341
3. NAIIC, The Official Report of The Fukushima Nuclear Accident Independent Investigation Commission. Executive Summary, The National Diet of Japan (2012)

4. H. Drumhiller, *International Experts' Meeting on How did Individual and Organizational Use of Probability and Risk Assessment at TEPCO Contribute to the Fukushima Accident?* presented at Human and Organizational Factors in Nuclear Safety in the Light of the Accident at the Fukushima Daiichi Nuclear Power Plant, Vienna, 21-mai-2013 2000, International Atomic Energy Agency (2000)

5. A. Portelli, A. Afrous, L'accident nucléaire de Fukushima Daiichi: rappel des faits, in *L'accident de Fukushima Daiichi: le récit du directeur de la centrale* ed. by F. Guarnieri, S. Travadel, C. Martin, A. Portelli, A. Afrouss, vol. I, L'anéantissement, p. 53–68

6. J.G. March, H. Simon, *Organizations*, 2nd edn. (Wiley-Blackwell, Cambridge, 1993)

7. P.C. Fishburn, *Utility Theory for Decision Making* (Wiley, New York, 1970)

8. I. Gilboa, D. Schmeidler, Case-based decision theory. Q. J. Econ. **110**(3), 605 (1995)

9. P. Ghirardato, F. Maccheroni, M. Marinacci, Differentiating ambiguity and ambiguity attitude. J. Econ. Theory **118**(2), 133 (2004)

10. I. Gilboa, Rationality and the Bayesian Paradigm: An Integrative Note (2014), http://itzhakgilboa. weebly.com/uploads/8/3/6/3/8363317/gilboa_rationality_and_bayesian_paradigm.pdf. Accessed May 2015

11. G. Kelin, K. Klinger, Natural decision making, Hum. Syst. IAC Gatew, **XI**(3), 16 (2000)

12. G.A. Klein, A recognition-primed decision (RPD) model of rapid decision making, in *Decision Making in Action: Models and Methods*, ed. by G.A. Klein, J. Orasanu, R. Calderwood (Ablex Publishing Corporation, Norwood, 1993), p. 138–147

13. K.J. Klein, J.C. Ziegert, A.P. Knight, Y. Xiao, Dynamic delegation: Shared, hierarchical, and deindividualized leadership in extreme action teams. Adm. Sci. Q. **51**(4), 590 (2006)

14. J.-F. Lebraty, P. Lievre, M. Recope, G. Rix-Lievre, *Stress and Decision*. The role of experience, in Proceedings of 10th International Conference on Naturalistic Decision Making (University of Central Florida, Orlando, Florida, 22 May–3rd June 2011), p. 266–270

15. C. Gode-Sanchez, V. Hauch, M. Lasou, J.-F. Lebraty, Une singularité dans l'aide à la décision: le cas de la Liaison 16. Systèmes Inf. Manag. **17**(2), 9 (2012)

16. M.A. Staal, *Stress, Cognition, and Human Performance: A Literature Review and Conceptual Framework*, NASA/TM-2004-212824, NASA Ames Research Center (2004)

17. A. Strommen-Bakhtiar, E. Mathisen, Sense-giving systems for crisis situations in extreme environments, in *Proceedings of InSITE 2012*, (Informing Science Press, Montréal, 22–27 June 2012), p. 91–112

18. G.A. Bigley, K.H. Roberts, The incident command system: High reliability organizing for complex and volatile task environments. Acad. Manage. J. **44**(6), 1287 (2001)

19. G. Loewenstein, J.S. Lerner, The role of affect in decision making, in *Handbook of Affective Science* ed. by R.J. Davidson, K.R. Scherer, H.H. Goldsmith, (Oxford University Press, New York, 2003), p. 619–642

20. G. Ellis, The myth of a purely rational life. Theol. Sci. **5**(1), 87 (2007)

21. S. Travadel, F. Guarnieri, L'agir en situation extrême, in *L'accident de Fukushima Daiichi: le récit du directeur de la centrale* ed. by F. Guarnieri, S. Travadel, C. Martin, A. Portelli, A. Afrouss. vol. I (L'anéantissement, Presses des Mines, Paris, 2015), p. 283–321

22. G. de Terssac, Théorie du travail d'organisation, in *Interpréter l'agir. Un défi théorique* ed. by B. Maggi, (PUF, Paris, 2011), p. 97–121

23. J. Joyce, Causation in decision theory, presented at Causality Study Fortnight 09-sept-2008, University of Kent (2008)

24. P. Weirich, Causal decision theory, The Stanford Encyclopedia of Philosophy. Winter 2012 Edition, 05-oct-2012. [online]. http://plato.stanford.edu/archives/win2012/entries/decision-causal/. Accessed 15 May 2015

25. R. Jeffrey, *Subjective Probability: The Real Thing* (Cambridge University Press, Cambridge, 2004)

26. J.-P. Dupuy, *Pour un catastrophisme éclairé* (Seuil, Paris, 2004)

27. P. Menzies, Counterfactual theories of causation, Stanford Encyclopedia of Philosophy Archive. Spring 2014 Edition, 10-janv-2001. [online]. http://plato.stanford.edu/archives/spr2014/entries/causation-counterfactual/. Accessed 15 May 2015

28. J.-P. Dupuy, Two temporalities, two rationalities: a new look at Newcomb's paradox, in *Economics and Cognitive Science*, P. Bourgine, B. Walliser, ed. by Pergamon (London, 1992), p. 191–220
29. A. Damasio, *The Feeling of What Happens: Body and Emotion in the Making of Consciousness* (Mariner Books, San Diego, 2000)
30. P.M. Niedenthal, P. Winkielman, L. Mondillon, N. Vermeulen, Embodiment of emotion concepts. J. Pers. Soc. Psychol. **96**(6), 1120 (2009)
31. A.N. Leontiev, Activity and consciousness, in *Philosophy in the USSR, Problems of Dialectical Materialism* (Progress Publishers, Moscow, 1977), p. 180–202
32. L. Vygotsky, *Mind in Society* (Harvard University Press, Cambride, 1978)
33. R. Kerkhofs, W.F. Haselager, The embodiment of meaning. Manuscrito **29**, 753 (2006)
34. E.A. Di Paolo, E. Thompson, The enactive approach, in *The Routledge Handbook of Embodied Cognition*, ed. by L. Shapiro (Routledge, New York, 2014), p. 68–78
35. M. Mauss, Les techniques du corps. J. Psychol. **32**(3–4), 365 (1936)
36. D. Le Breton, Anthropologie du corps et modernité, (PUF, Paris, 2013)
37. J.A. Deonna, F. Teroni, L'intentionnalité des émotions : du corps aux valeurs, Rev. Eur. Sci. Soc. **XLVII**(144), 25 (2009)
38. S. Moscovici, G. Paichelier, Travail, individu et groupe, in *Introduction à la psychologie sociale*, vol. II (Librairie Larousse, Paris, 1973), p. 9–44
39. K. Cleveland, Significant breaking worse. Crit. Asian Stud. **46**(3), 509 (2014)
40. J. Shigemura, T. Tanigawa, I. Saito, S. Nomura, Psychological distress in workers at the fukushima nuclear power plants. JAMA **308**(7), 667 (2012)
41. K.M. Kowalski-Trakofler, C. Vaught, T. Scharf, Judgment and decision making under stress: an overview for emergency managers. Int. J. Emerg. Manag. **1**(3), 278 (2003)
42. C. Dejours, Pathologie de la communication, situations de travail et espace public : le cas du nucléaire, in Raisons Pratiques, 3, ed. by A. Cottereau, P. Ladriere, (Éditions de l'Ecole des Hautes Etudes en Sciences Sociales, Paris, 1992), p. 177–201
43. F. Guarnieri, S. Travadel, Engineering thinking in emergency situations: A new nuclear safety concept. Bull. At. Sci. **70**(6), 79 (2014)
44. D. Van Hoorebeke, L'émotion et la prise de décision, Rev. Fr. Gest. **34**(182), 33 (2008)

An Ethical Perspective on Extreme Situations and Nuclear Safety Preservation

Hortense Blazsin

Abstract Extreme situations lead to the collapse of systems together with all existing rules, including symbolic ones. Therefore there are no longer any procedures to comply with, nor any outside guidance to help in making the complex decisions imposed by such situations. The decision-maker has to look elsewhere to find the resources to guide their actions, all the more since they are likely to be held responsible when the situation returns to normal. We argue that ethics, based on practical reason, offer a way out of the dead-end. Practical reason is anchored in individual motivation, as opposed to external rules, and is ultimately guided by solicitude towards other human beings. As it rises from the inner desires, feelings and reasoning of a person it offers a guide for action, even when artefacts collapse. Furthermore ethics could provide common ground on which to build an interdisciplinary approach to resilience in extreme situations, as ethical questions run through all disciplines. Building on Paul Ricoeur's practical philosophy, we describe what an ethical approach to decision-making in extreme situations could look like, as well as its implications for organizations. We show that such an approach requires that organizations allow their members to use their practical reason and act autonomously not only when accidents occur, but also in normal situations. Such a transformation could lead to building "safe institutions", i.e. organizations within which people would preserve safety, rather than organizations that manage safety through people.

Keywords Decision-making · Ricoeur · Ethics · Practical reason · Responsibility

H. Blazsin (✉)
Centre for Research on Risks and Crises, MINES ParisTech/PSL Research University,
Sophia Antipolis, France
e-mail: hortense.blazsin@mines-paristech.fr

© The Author(s) 2017
J. Ahn et al. (eds.), *Resilience: A New Paradigm of Nuclear Safety*,
DOI 10.1007/978-3-319-58768-4_14

185

1 Introduction

This book and the underlying workshop, entitled "Becoming Resilient in Extreme Situations: A New Paradigm of Nuclear Safety", bring together scientists from Europe, Japan and the United States, from disciplines as diverse as anthropology, sociology, history, philosophy, epidemiology, physics, engineering and biology.

Such heterogeneity raises the question of what could possibly unite us, and whether we would be able to find common ground. The topic definitely is not enough, especially one as complex as the collapse of a nuclear power plant following a tsunami. The issues are so complex that each discipline alone would find it difficult to reach scientific certainty, or even consensus. If even this initial level of agreement is not guaranteed, it seems fair to question the ability of a multidisciplinary group to find common ground.

However, there is one thread that seems to have run throughout the workshop. Either explicitly or implicitly, the question of ethics has filled the air, and occasionally fueled debates. At this stage we use the term "ethics" in a broad sense; it refers to the set of moral principles guiding human (individual and collective) action, which determine what is acceptable, i.e. respectful of human nature (that of the actor and that of those potentially touched by their actions). Diverse ethical questions, ranging from the victim's identity to evacuation-related decisions and public engagement with scientific knowledge have surfaced. We must not forget that under sophisticated concepts such as "resilience" or "extreme situations" lie more basic aspects of safety, i.e. respect for the humanity of those impacted by the management of extreme situations, either as victims or as actors. Therefore, it seems reasonable to argue that an ethical framework could provide some common ground for our multidisciplinary efforts, forming a "dictionary" that we can use to translate [2] individual standpoints into a shared language.

Here, we sum up and put into perspective the most pressing ethical issues that were raised during the workshop. We raise many questions, but answer few, if any at all. Nevertheless, hopefully we show that this lack of answers does not render the exercise useless. Rather, we endeavor to show that an ethical perspective on safety and the management of extreme situations not only provides a moral safeguard; it also offers a very practical guide for individual and collective action. To achieve this, we reflect on the management of the Fukushima Daiichi accident through the prism of "practical reason". The concept, which first appeared in the work of Aristotle [9], reappears in the work of Kant [8], and more contemporarily, Paul Ricoeur [11]. We use the work of the latter to illustrate the heuristics of ethics for safe action and the management of extreme situations.

2   The Management of Extreme Situations, Multiple Dilemmas

Broadly speaking, the application of ethics to major accidents and the management of extreme situations leads to a focus on the relationship between people (both individuals and collectives) and these events.

During the workshop, the concept of the victim (which makes the relationship between people and the accident explicit) emerged as a heuristic that united the various ethical questions that were raised. For instance, how should the public (who are all potential victims of major accidents) be taken into account, and involved in decisions that may impact them at some point? As major accidents unfold, how do we determine what truly helps victims, and what is in their best interest? Should such a "best interest" be acted upon, even if victims do not give their consent? Finally after an accident has occurred, who should be considered as a victim? How can victims be recognized as such—and ultimately, be compensated for their loss, assuming that such compensation is possible?

One important issue that was raised is that of a mediator who arbitrates between victims and the accident. Such a mediator is necessary to establish a relationship between people and what has happened to them. This relationship is a prerequisite for the evaluation of the post-accident situation and attempts to restore harmony, from which a new cycle can begin. The question becomes even more difficult when one looks at non-human victims (i.e. nature), that have no voice to express the damage it has suffered and where it may not be possible to restore harmony.

During the workshop, the Sorites paradox[1] was used to illustrate the immense difficulty of giving an identity to victims of major accidents. This is not only a question of the number of victims: tens of thousands could be named. But, for example, who should be considered as a victim of the Chernobyl accident, where millions were exposed to very small doses of radiation? If someone lives in an area that was affected by radiation from Chernobyl and develops cancer, how can we determine whether s/he is a victim of the Chernobyl accident? This philosophical argument is supported by epidemiology, which highlights the difficulty of correlating radiation maps with actual damage to human health. Together, these issues question the ability of traditional models to shed heuristic and instrumental light on phenomena that are as complex as major accidents.

The issue of traditional models was only one way in which the relevance of current scientific approaches to major accidents was questioned. Other issues concerned how to establish a relationship between scientists and the public, and how to make scientific knowledge available to less-expert audiences and include them in decisions that may ultimately disrupt their life. Consequently, public

---

[1]The Sorites paradox, also called the "little-by-little paradox" highlights the difficulty that arises from indeterminate or fuzzy boundaries. Using the concept of the "heap", it shows that if one takes grains out one by one, it is impossible to establish which grain was the limit that turned the heap into a "non-heap".

engagement, disclosure of scientific information and consent emerged as questions to be addressed.

Another series of questions concerned the management of evacuees. For instance, should elderly people be evacuated from the disaster zone? Although they may suffer from nuclear radiation if they stay, uprooting them from their environment can trigger other effects, such as desocialization and loss of reference points. This reminds us that major accidents do not only trigger physical damage, but also psychological and social damage. Furthermore, what should be done if people are unwilling to be evacuated?

The question of public engagement and preparation is not only an ethical matter (i.e. maximizing people's involvement in life-changing decisions). It is also a matter of social resilience. It is an illustration of how ethics and safety can strengthen one another. It supports the idea that—far from being a purely abstract and theoretical perspective—ethics may have very concrete implications, and offer a practical guide for individual and collective action to manage extreme situations and preserve safety.

3   Ethics: Not Only a Way Out, but also a Way up

We use the definition of ethics given by the contemporary French philosopher Paul Ricoeur. According to him, ethics refers to a person's attempt to lead "a good life, with and for others, within just institutions" [11]. Therefore it lies at the heart of human behavior towards oneself as well as others. It offers a guide for both day-to-day actions and "extreme situations". In extreme situations everything collapses, from rules to structures to meaning to values, leaving nothing to guide action but one's inner conviction that a particular decision is the right one. To act when faced with this "tragic dimension of action", where no rule is applicable, calls for a higher purpose. Ricoeur calls it solicitude towards others. Such solicitude is expressed through "practical wisdom" [10], which is the result of a process of "deliberation", i.e. critical analysis and the comparison of existing rules with the higher purpose of solicitude.

In the following section, we show how practical reason can help when reflecting on safety and the management of extreme situations. Based on the testimony of Masao Yoshida, who was director of the Fukushima Daiichi plant at the time of the tsunami, we illustrate the role played by individual intention and voluntary action in the preservation of safety. We then generalize these lessons to show how Ricoeur's practical philosophy can provide inspiration for an approach to safety that is different to traditional, engineering approaches; we term this "practical safety".

3.1 Voluntary Action and Decision-Making, the Ethical Way Out of Extreme Situations

As Masao Yoshida indicates in his testimony, in Japan, the law on nuclear catastrophes is invoked when an exceptional situation develops in a nuclear power plant. Under this law, he became responsible for crisis management [5]. He had the power to make decisions, and take responsibility for any actions supervised by him. Considering the potentially immense consequences of any decisions made in this context, it seems reasonable to reflect on the implications of such a responsibility for an individual, to question whether s/he is in a position to hold it, and the attribution of responsibility by legal and organizational systems. Furthermore, such situations make individuals responsible "not because one is free by nature, but because society judges it 'fair' to place responsibility in a particular social location", displacing the source of responsibility "from the individual onto society" [3].

The imposition of such a responsibility therefore appears to be heteronomous (i.e. obeying an external rule), rather than autonomous (i.e. following a self-imposed rule). This is no surprise, as heteronomy lies at the heart of contemporary organizations. They rest on engineered processes and rules and therefore on a reified rationality, that is external to the individual and imposed upon them [1]. Consequently, actions that are carried out in such a context do not result from individual free will, which implies that they cannot be deemed voluntary. Indeed, according to Ricoeur [10], following Aristotle [8], to qualify as voluntary an action has to have a specific goal and result from free will. From the perspective of these philosophers, individuals can only be held responsible for their voluntary action. As Irwin [7] comments on *Nicomachean Ethics*, "only voluntary actions can be assessed for praise and blame. To find a voluntary action is to find an action it is reasonable to consider for praise and blame. Aristotle (…) assumes that the same conditions make actions candidates for moral and for legal scrutiny and reactions— praise, blame, reward, punishment, and so on." If only voluntary actions can be subject to moral or legal judgment, is it fair to consider the decision-maker in extreme situations as responsible? Given that their responsibility stems from a heteronomous, rather than an autonomous source, and that organizations usually restrict autonomy, can an individual truly be held responsible?

Obviously the question is far from new. However, the fact that the extraordinary collapse of the Fukushima Daiichi plant placed a specific individual, Masao Yoshida, under moral and legal scrutiny, indicates the need to reflect on the relationship between responsibility, voluntary action, and safety when managing extreme situations. Furthermore, if individuals are to be considered responsible for their actions, not only is it fairer to build systems that provide them with the means to act voluntarily and responsibly, it is also a necessity imposed by extreme situations.

Extreme situations lead to the collapse of systems and vitiate all preexisting rules, including symbolic ones [4]. There are no longer any procedures that guide actions and decisions. The individual is no longer heteronomous, and autonomous, voluntary action is the only option left. The decision made by Yoshida and a number of operators to remain at the plant in order to contain the damage exemplifies the role of autonomous action oriented at preserving safety. The fact that they were willing to sacrifice their own lives to save others may be considered as an example of *solicitude*.

As such, their actions can be deemed ethical and considered as an expression of "practical reason", in the meaning developed by Ricoeur. Ricoeur asserts that rationality is only one of the many forms reason can take; another is practical reason. The main characteristics of Ricoeur's practical reason are that: it stems from individual desires, and therefore free will; these desires are "reasonable", i.e. understood by others as possible motivators for action, meaning that they are made explicit to others who can confirm such reasonableness; it is strategic, i.e. it articulates a means to an end, and leads to the development of complex reasoning, triggering a dynamic teleology; and finally, it is ethical [10]. The concept seems to be an appropriate heuristic for the management of extreme situations. It is anchored in free will, with an explicit connection to ethical aims. Furthermore, the need to make motives explicit (to ensure reasonableness) encourages reflexive thinking about motives and action, which increases the relevance of action. It also emphasizes the importance of complex reasoning and its ability to trigger teleological causation,[2] in a context where it may be crucial to the preservation of safety (cf. Weick's theory of enactment [12]).

According to Ricoeur, all human beings have the capacity for practical reason, while the collective environment may be more or less favorable to its expression. He calls the most favorable collective environment the "just institution". However, as mentioned above, contemporary organizations are built around rational, engineered logic and tools, which estrange people from practical reason and therefore from voluntary, responsible action. Organizations therefore need to transform if they are to enable their members to use practical reason.

3.2 *"Safe Institutions" for the Management of Extreme Situations*

Managing extreme situations requires people to be able to act voluntarily and responsibly, both when there is no other option, and in "normal" situations. This approach strengthens their ability to act appropriately should an extreme situation arise. Building systems that favor voluntary and responsible action therefore becomes a matter of both individual and systemic resilience.

---

[2]Teleological causation posits that creating a specific goal contributes to the creation of the conditions that ensure that it is reached, by "disposing" the individual to reach it. The logic is similar to that of Weick's theory of enactment according to which picking up certain cues leads the situation to develop in a certain way.

Ricoeur's concept of the "just institution" offers another lever for the construction of systems that are better able to manage extreme situations and preserve safety. The "just institution" rests on the fact that it responds to the individual's sense of justice, and achieves an appropriate balance between the obligations that are imposed on them and the privileges that are granted [11]. Its primary aim is to ensure that individuals are free to act. The concept of the "institution" refers to a structure with shared interests, or belonging to a specific community: according to Ricoeur, "the idea of the institution rests fundamentally on shared customs, not on binding rules" [11]. It argues that the individual is free to act and pursue ethical aims. It relates to the way people interact, and how they are brought together and live with one another, as these interactions (reified in the form of shared customs) constitute the institutional environment. Finally, it rests on a principle of plurality, which in turn implies diversity and the potential for (potentially contentious) dialogue.

This perspective appears to be at odds with current organizations, which rest on engineered methods and designs and therefore primarily seek uniformity and abstraction from individuality. However, it may be possible to translate it into concrete actions that organizations can implement. Such organizations would be called "safe institutions". At a general level, this means refocusing organizations on people, their sense of justice, their ability to exercise practical reason, and their goal of leading "a good life for oneself as well as others", rather than the design or procedures of organizations themselves. On a more practical level, it suggests actions related to training (i.e. ensuring that people are in a position to maintain safety), and redefining how careers are built and what success means.

Organizations that apply the "safe institutions" philosophy and favor practical reason could be said to enable "practical safety". We define practical safety as "the ability of individuals to appropriate safety as an internal value, which enables them to decide on a course of action that preserves the safety of others as well as their own, when a situation requires them to do so". Such an approach asserts the idea that safety can only be managed *by* people *in* organizations, rather than *by* organizations *through* people. It also reminds us that safety can only be maintained through concrete actions, rather than predesigned, abstract rules and procedures.

More broadly, the concept of the just (safe) institution resonates directly with a number of questions raised during the workshop. For example, the question of consent and self-sacrifice: if an individual's obligations include the potential need to sacrifice oneself, how is it possible to establish a sense of justice? What should the status of voluntary actions be, from both a moral and legal standpoint, when things "go back to normal"? What criteria should be used to judge, both morally and legally, an action that could be nothing but ad hoc and voluntary?

Once again, this raises the question of the moral and legal status attributed to an action and the actor by the collectivity once the situation goes back to normal. And, once again, Ricoeur's philosophy provides a heuristic. He argues that action can be compared to texts, in the sense that once they are achieved, they are out in the world, available for interpretation by third parties who may not share the frame of "reference" for the action, i.e. the situation from which it stems. As action is primarily a way to bring about a change in the world, it leaves a trace. This "mark" inscribes the action in the world, making it an "archive" of the initial act. Therefore, in addition to the intention that provided the motivation, an action can also be evaluated against the "persisting configurations" that it brought about. This evaluation can only be carried out by third parties at a later date. This difficult task involves retracing the path back to the initial action, which due to the complexity of the world, may be extremely distant from its consequences. Such an approach resonates with some of the problems inherent in actions carried out in extreme situations, and offers a fruitful avenue for further research.

4   To Conclude: Ethics, a Way up and Out of Extreme Situations, not a Set of Solutions

As promised, we raise many more questions than answers. Most of these questions emerged during the workshop, and they range from the decision-making process in high-risk technologies, to the question of how to determine who victims are. This highlights the transverse nature of such questions and the role ethics may play in the emergence of an interdisciplinary approach to major accidents and the management of extreme situations.

The workshop examined the question of potential or actual victims, and this appears to be a possible avenue for an ethical approach to major accidents. It leads us to ask *who* the victims are, in the strongest sense. How—if at all—can they be identified as victims of a specific accident raises a philosophical and social question? How should they be helped, either during the management of the event or afterwards, through medical treatment or compensation? Finally, how should the concept of the victim be defined, both in scientific terms and for decision-making purposes? It is deeply anchored in the concept of solicitude and could offer a heuristic approach to the examination of major accidents and the management of extreme situations from an ethical perspective.

The question of individual action and responsibility opens a second door to ethical reflection on extreme situations. Yoshida's testimony on the management of the Fukushima Daiichi accident shows that individuals were held both legally and morally accountable for actions taken during extreme situations. As we have seen, such judgments require that voluntary action is possible. This in turn requires a radical shift in organizations and how individual actions and their contribution to safety are considered. Ricoeur's practical philosophy shows how ethics not only help to develop a better understanding of what underlies the management of extreme situation, but also open the way for actions that are more favorable to the preservation of safety.

References

1. H. Blazsin, *De l'ingénierie de la raison à la raison pratique: vers une nouvelle approche de la sécurité*. Doctoral thesis, Ecole des Mines de Paris, 2014
2. M. Callon, B. Latour, Unscrewing the big Leviathan: how actors macro-structure reality and how sociologists help them to do so, in *Advances in social theory and methodology: Towards an integration of micro- and macro-sociologies*, ed. by K. Knorr-Cetina, A.V. Cicourel (Routledge & Kegan Paul, Boston, 1981), pp. 277–303
3. F. Ewald, The return of Descartes's malicious demon: an outline of a philosophy of precaution, ed. by Baker & Simon, *Embracing risk: The changing culture of insurance and responsibility*, The University of Chicago Press., pp. 273–301 (2002)
4. F. Guarneri, S. Travadel, Engineering thinking in emergency situations: A new nuclear safety concept. Bulletin of Atomic Scientists **70**(6), 79–86 (2014)
5. F. Guarnieri, S. Travadel, C. Martin, A. Portelli, A. Afrouss, *L'accident de Fukushima Daiichi, Le récit du directeur de la centrale*, vol. 1 (Presses des MINES, L'anéantissement, 2015)
6. E. Hollnagel, D. Woods, N. Leveson, *Resilience Engineering. Concepts and precepts*, Ashgate, Aldershot, UK (2006)
7. T.H. Irwin, Reason and Responsibility in Aristotle, in *Essays on Aristotle's ethics*, ed. by A. O. Rorty (University of California Press, Berkeley, Los Angeles, London, 1980), pp. 117–155
8. E. Kant, *Critique de la raison pratique*, Paris: PUF, 2012 (1788)
9. C. Taylor, *Nicomachean Ethics, Books II–IV, Translated with an introduction and commentary* (Oxford University Press, Oxford, 2006)
10. P. Ricoeur, *From Text to Action: Essays in Hermeneutics II*, trans. Kathleen Blamey and John B. Thompson, Evanston: Northwestern University Press, 1991 (1986)
11. P. Ricoeur, *Oneself as Another*, trans. Kathleen Blamey, Chicago: University of Chicago Press, 1992 (1990)
12. K.E. Weick, *Making Sense of the Organization* (Blackwell Publishing, Oxford, 2001)

Japan's Nuclear Imaginaries Before and After Fukushima: Visions of Science, Technology, and Society

Kyoko Sato

Abstract Two recent insights regarding social imaginaries are of particular relevance in thinking about the Fukushima disaster and its aftermath. First, social imaginaries are consequential for social resilience. Second, imaginaries play a significant role in the way a society addresses science and technology. In light of these insights, the chapter explores nuclear imaginaries in Japan before and after Fukushima, and presents several key historical factors that shaped such imaginaries in the lasting manner. It presents how Japan's nuclear imaginaries have persistently embraced certain ideals of science and technology, and excluded people subject to radiation risks. The chapter concludes by calling for explicit engagement with our nuclear imaginaries, in terms of social resilience, and also as an arena where we can explore more democratic approaches to science and technology. Such engagement is also consequential to larger visions of society.

Keywords Social imaginaries · Sociotechnical imaginaries · Resilience · Public engagement with science and technology · Fukushima nuclear disaster · Science, Technology, and Democracy

1 Introduction

For decades, scholars in humanities and social sciences have explored the role of imagination in social life. After influential works by Anderson [1],[1] Castoriadis [6], Appadurai [2] and Taylor [36], the concept of social imaginaries—imagined collectivities, together with shared assumptions about social relations and practices, as

---

[1]In his seminal account of the emergence of nation-states, Anderson defined the nation as "an imagined political community"—the product of the shared imaginations of those who perceive themselves as members.

K. Sato (✉)
Program in Science, Technology, and Society, Stanford University,
Stanford, CA 94305, USA
e-mail: kyokos@stanford.edu

© The Author(s) 2017
J. Ahn et al. (eds.), *Resilience: A New Paradigm of Nuclear Safety*,
DOI 10.1007/978-3-319-58768-4_15

195

well as collective representations and narratives about a society's past, present, and future—has become a common analytical tool in such fields as anthropology and sociology.

Two recent insights regarding collective imaginaries are of particular relevance in thinking about the Fukushima disaster and its aftermath. First, social imaginaries are consequential for a group's social resilience. Reviewing empirical works (including their own) on groups facing various adversities (e.g., poverty, health disparities, discrimination, marginalization), Hall and Lamont [11, 12] argue that social imaginaries are constitutive of the collective capabilities of a community or society, as they not only bind its members with narratives of its past accomplishments and a vision of what it means to belong to it, but also indicate how its members understand what they are capable of doing together. Defining social resilience as "the capacity of groups of people bound together in organizations, classes, racial groups, communities, or nations to sustain and advance their well-being in the face of challenges" [11], they argue that imaginaries can provide significant resources for such a capacity. For instance, social recognition and status of a group—i.e., where it stands in collective imaginaries—shape capabilities of its members by influencing how they can secure support from others, as well as how they perceive their own efficacy [11]. Put another way, how collective imaginaries specify and support group identities and define who qualify as valued members of the collectivity can be direct sources of resilience for group members [12]. Conversely, when these imaginaries stigmatize a group, they constrain capabilities of its members. At the same time, groups might turn to collective imaginaries for tool to cope with difficulties: In coping with discrimination, members of marginalized groups in the United States often rely on such cultural resources as principles of equality as a key American creed (e.g., [26]), whereas a strong group identity has been found to alleviate the impact of adverse experiences. In sum, imaginaries serve as important resources—or constraints—for social resilience, independently of *and* consequentially for a group's access to material resources.

Second, imaginaries play a significant role in the way a society addresses science and technology. Jasanoff ([17]; also see Jasanoff and Kim [18]) argues that visions of future developments in science and technology are inevitably and intricately connected to collective visions of good and attainable futures, positing a concept of "sociotechnical imaginaries." Defined as "collectively held, institutionally stabilized, and publicly performed visions of desirable futures, animated by shared understandings of forms of social life and social order attainable through, and supportive of, advances in science and technology" [17], this concept allows us to explore and better understand the complex interplay between developments in science, technology, and society. For instance, examining the development and regulation of nuclear power in the United States and South Korea, Jasanoff and Kim [18] identify different ways in which nationhood and nuclear power have been imagined together in the two countries, as well as how such different imaginaries illuminate their different responses to nuclear disasters and the spread of anti-nuclear movement. Importantly, such imaginaries are variable across groups within society; they are durable yet changeable; and they are not only descriptive, but also prescriptive—engaging with what kinds of future should be pursued and how they should be achieved through science and technology. They also encompass shared fears of harms that might result from invention and innovation.

These recent insights both suggest that social imaginaries are consequential to the future of the collectivity, whether by affecting resilience of a social group and society as a whole, or by positing what futures are desirable and how to attain them using science and technology. This calls for an explicit examination of Japan's nuclear imaginaries, both before and after the 2011 disaster. What visions of society were pursued through nuclear technology? Who were imagined as relevant actors in the development and regulation of nuclear power? How did nuclear imaginaries enable and constrain capabilities of different actors? What visions do current politics and policy approaches embody? Do they facilitate resilience of communities affected by the disaster? In the following, I present a few findings from a larger project that systematically traces such imaginaries.[2] To clarify, different groups in society might harbor and advocate different, competing imaginaries, but some can be more dominant than others. Policy is a particularly important site that presents and institutionalizes certain imaginaries as authoritative and representative.

2 Nuclear Imaginaries in Japan: At the Time of the 2011 Disaster in Fukushima

Among the most striking aspects of the way nuclear technology was imagined by the general public in Japan right before the 2011 disaster are: (1) how decoupled and dissociated nuclear energy production was from nuclear weapons; and (2) how rarely the former was imagined at all. The public was overwhelmingly indifferent, and took it for granted that there were 17 nuclear power plants (NPPs), supplying 30 percent of the country's electricity. As former-nuclear-engineer-turned-opponent Tanaka [43] argued, nuclear energy solicited little public attention: "what supports the national policy to promote nuclear power more than anything is the unrecognized indifference of people in big cities" (my translation). Pointing out that none of Tokyo Electric Power Company (TEPCO)'s 17 reactors existed within the areas (including Tokyo) to which the company supplied electricity—all were located in Fukushima and Niigata Prefectures—he said, "We have 55 reactors, but most of us live our daily life as if they don't exist" (my translation). This was the climate in which any critiques or even reservations about nuclear power would have one

---

[2]Together with work with Jasanoff and Lamont, I am part of a multi-year project, "The Fukushima Disaster and the Cultural Politics of Nuclear Power in the United States and Japan," which traces the development of dominant nuclear imaginaries in Japan and the United States. Our data include media coverage, policy documents, organizational documents, interviews, and ethnographic data.

labeled as "unrealistic dreamer," as Murakami [28] described after the disaster.[3] Nuclear phase-out was unthinkable to many Japanese.

This paradigm was to continue: A month before the March 11 disaster, the Japanese government decided to extend the operation of existing nuclear power plants, partly as a measure to reduce greenhouse gas emission. At the time of the disaster, TEPCO had plans to start constructing two additional reactors at the Daiichi (or 1F, or *ichiefu*, as it has been locally called) site the following year. In general, the predominant political discourse about nuclear energy centered on the following ideas: (a) it is a source of stable supply; (b) it is economically efficient; (c) it produces zero carbon emission; and (d) by ensuring safety and gaining public understanding, we need to expand it. For instance, in the 2010 Basic Energy Plan, long-term energy plans announced every several years, nuclear energy was categorized as "non-fossil" and "zero-emission sources" together with "renewable energy," and it was proposed that the ratio of these two types of energy be raised from the then 34% to more than 50% by 2020, and about 70% by 2030.[4] The Plan stipulated that more nuclear power stations ("at least 14 reactors") be built and the operating rate of the facilities be increased (to "about 90%") by 2030, while gaining public understanding and trust, especially of local residents, and on the condition of ensuring safety.

At the time, for the government and other proponents of nuclear power production, it was linked to the nationhood both as an indispensable technology that allowed the country with scarce natural resources to prosper and as a technological domain in which Japan excelled. This is evident in the 2006 "Nuclear Power Nation-Building Plan," a report submitted by the Nuclear Energy Subcommittee in the Advisory Committee for Natural Resources and Energy.[5] The plan urges Japan to increase the share of nuclear power, spread its nuclear technology globally, and contribute to nuclear non-proliferation. Similarly, in his ill-timed book, writer-critic Toyota [37] argued that Japan's nuclear technology was the safest in the world and that the country should promote it both for the economic gains and for the good of humanity, such as a solution to climate challenges.

Organized opponents existed throughout Japan, though marginalized as Luddites, hippies, or "unrealistic dreamers." They had called attention to various issues of concern, such as unfounded "safety myth," the insularity of the nuclear community as "village," and the industry "capture" of policy processes (e.g., [10, 23])—many of these have become accepted as shared understandings after the 2011 accident, even presented by major investigative reports put out in 2012 by the Diet, the Cabinet, and a private foundation.

---

[3]Murakami's speech in Barcelona on June 9, 2011, upon receiving the International Catalunya Prize.

[4]http://www.enecho.meti.go.jp/topics/kihonkeikaku/100618honbun.pdf (Last accessed on May 28, 2015).

[5]http://www.enecho.meti.go.jp/topics/images/060901-keikaku.pdf (Last accessed on May 28, 2015) Based on the Framework for Nuclear Energy Policy, approved by the Cabinet in October 2005.

Notably, not only the general public, but also many activists against nuclear power and weapons did not necessarily consider the two applications tightly connected. Even a great number of *hibakusha* and anti-nuclear weapons activists were uncritical of nuclear energy production.[6] In a July 2011 interview, Terumi Tanaka, Secretary General of *Nihon Hidankyo* (Japan Confederation of A- and H-Bomb Sufferers Organizations), said: "I have been thinking since the nuclear accident, perhaps we *hibakusha* may not have thought very much about nuclear power. These days I think that we need to revisit and more thoroughly study the background of the technology, the system of management, how the industry and government addressed it, etc., and continue to debate about what we can say and do as *hibakusha*" (my translation).[7]

In sum, in the pre-Fukushima dominant imaginaries, future Japan is ecological, efficient, economically prosperous, and equipped with clean energy and strong science and technology, and these objectives are facilitated by nuclear technology. In this vision, technological prowess is an important part of the country's national identity (see Hecht [13] for the French case), and major social and economic problems are solved by advances in science and technology, which are supported by both the government and market forces.

Also implied in these imaginaries is the so-called "deficit model," in which the public's skepticism toward and/or rejection of a specific scientific or technological development is attributed to its ignorance and incomprehension. In this model, knowledge is monopolized by experts, and solutions to the public objection consist of educating them with more and better information about science and engineering and raising their "literacy." Scholars in the field of science and technology studies (STS) have presented various critiques to this model, showing how sophisticated and productive "lay" knowledge can be [9, 40, 41], how "local" and parochial—as opposed to "universal"—expert knowledge can be [41], and how increased scientific "literacy" does not always lead to acceptance and appreciation of science and technology [4]. These insights have problematized how decision-making about science and technology—enormously consequential to the whole society—is left to a small group of experts and the political and corporate elite, and challenged us to explore more democratic approaches to science and technology. For instance, critics of the deficit model have called for public engagement in various aspects of science and engineering: not only final assessment of a given option for policymaking objectives (e.g., public hearings, consensus conferences, deliberative polls), but also

---

[6]One of them, Gensuikin (Japan Congress against A- and H-Bombs), has long been calling for nuclear phase-out in energy production; it held its annual meeting in Fukushima in July 2013, clearly signaling its opposition to the two related applications of nuclear technology. Another group, Kakkin (National Council for Peace and against Nuclear Weapons), has long supported "peaceful use" of nuclear technology, even after the 2011 disaster. The other group, Gensuikyo (Japan Council against Atomic and Hydrogen Bombs), has been cautiously against certain aspects of nuclear power production, but has not opposed to the idea itself completely.

[7]"Hoshasen to mukiai nagara ikiru: Fukushima Genpatsu jiko—Hibakusha to shite dou uketomeruka" Hidankyo Shimbun, August 2011 issue (No. 391).

early stages of scientific research and technological development (e.g., [31, 39]; see Delgado et al. [7] for a review of current issues regarding public engagement).

Furthermore, lacking conspicuously in these imaginaries surrounding nuclear technology are certain actors, practices, and phenomena: workers at NPPs, local residents, day-to-day operations at NPPs, and risks of radiation for workers and residents. After the 2011 accident, it came as a great shock for many Japanese to learn about the precarious conditions of labor at the plants (as depicted in Higuchi [14], Asakawa [3], Jobin [19]), as well as how "manual" and low-tech some of the workings and physical realities of NPPs are—as opposed to the images of a clean control room, which was a typical representation of an NPP—and how much uncertainty surrounded a control of, and effects of, radiation. Urban Japanese were also largely oblivious to the risks that local residents bore as NPPs supplied energy to their cities. Moreover, decoupling from bombs prevented *hibakusha*'s postwar social, political, and physical struggles from being relevant to discussion of life with NPPs.

3 Historical Factors Behind the Pre-Fukushima Nuclear Imaginaries

While nuclear imaginaries described above are obviously a product of long-term, complex historical processes involving numerous actors, events, and cultural, political, and economic resources, below I highlight several key factors that have significantly shaped them in early postwar years in the way that have persisted since then.

First, in postwar Japan, the public discussion of nuclear technology—and the war experience in general—was significantly shaped by the systematic censorship carried out by the Allied Occupation. Under the censorship apparatus laid out by the United States, discussion or expression of the experience of the bombings was severely restricted, as well as criticism of the US or other Allied nations. Notably, the public had little awareness of censorship or the press code [5, 16, 24]. The insidious nature of this censorship had a profound impact on the way the Japanese talked about and thought about the atomic bombings. Kawamura [24] argues that this kind of manipulation contributed to the way the issue of atomic bombings became meaningless, hidden, and invisible to most Japanese in plain sight during these postwar years.

After the end of the Allied Occupation in 1952, many Japanese saw visual representation of victims of atomic bombs for the first time when *Asahi Graph*—a *Life*-like general interest photo magazine—published a series of photos in its August 6th 1952 issue. While the photos certainly were shocking by most standards, with charred bodies and badly injured children, strikingly missing was any critiques of the acts of bombings themselves. The brutality and inhumanity of the bombs were emphasized without an agent, and also portrayed as a deterrent of another war, or even a purveyor of peace. Remarkably enough, an organized movement against nuclear bombs did not emerge until after the Bikini Atoll incident in March 1954, when the crew of a Japanese fishing boat was exposed to nuclear fallout from the American testing of thermonuclear bomb.

Second, in the Cold War context, Japan came to thoroughly embrace the concept of "peaceful" use of nuclear technology, which was aggressively promoted by the United States. With the 1953 "Atoms for Peace" speech, Eisenhower sought to recast nuclear technology for world redemption and incorporate it into the emerging Cold War order by promising to share it with non-communists countries. Japan's nuclear energy industry came about in this context, simultaneously with the rise of anti-nuclear weapons movement. Here, the rejection of weapons not only did *not* contradict the excitement about "peaceful" use, but also served as a driving force of the latter [42]. In the name of turning a tragedy into inspiration, the US government even launched a campaign to build an NPP in Hiroshima in 1955 [35]. The US found powerful Japanese allies including young politician Yasuhiro Nakasone (later Prime Minister) and media tycoon and politician Matsutaro Shoriki, who ran the *Yomiuri Shimbun*, helped launch the Japanese professional baseball, and later came to be known as the father of nuclear power in Japan. For instance, Shoriki worked with the US government to organize the traveling exhibition on "the peaceful use of atomic power." The exhibition started in Tokyo and visited nine other cities including Hiroshima, where it was co-sponsored by local municipalities, university, and newspaper, and received enthusiastically in spring 1956. While many *hibakusha* were initially cautious about this "peaceful" application of the technology, arguing that no solution had been found to the problem of managing radioactive materials, by summer 1956, even Moritaki Ichiro, an intellectual leader of survivors and nuclear weapons abolitionist, came to embrace the idea of "peaceful" use [35]. Importantly, this dichotomous view in which the tragedy of military use is contrasted to the prosperity of "peaceful" use, as well as eventual decoupling of the two, resulted from concerted efforts by the US government and Japanese supporters of nuclear energy. In the late 1950s, very little opposition existed to the ideas of nuclear power production or plans of building NPPs. Narratives of nation-building through nuclear energy were not hindered by the memories of the bombs or the growing anti-nuclear weapons movement; rather, they were supported by the exceptionalist idea that, as the "sole victim" of the bombs, Japan should lead the world in this technology.

Third, as some scholars argue, behind the de-politicized nature of nuclear energy production was the long-standing history of "internal colonialism" in Japan, whereby Tokyo and the power that be there have exploited and colonized the periphery such as the Tohoku region, of which Fukushima Pref. is part. As Hopson [15] points out, Tohoku-born intellectuals have long described the region a domestic colony of the center, whose subjugated and "backward" status resulted from official policy decisions during the Meiji period (1868–1912) of rapid modernization. These intellectuals were aware—some as early as in the 1890s—that the region's often essentialized "backwardness" was a product of the exploitation of its resources and domination [15]. As the region turned into a significant provider of rice and labor for the growing Tokyo Metropolitan area, the narrative that the backward region needed to be developed also became common, and local support for projects such as NPPs became strong. In this context, Tohoku became the primary provider of electric power for Tokyo, and the constructions of NPPs in Fukushima was an extension of this historical trend.[8] (For more studies of exploitation of the periphery by the center, see Kainuma [22], Takahashi [32], Kawanishi [25]). With this unequal relationship as a backdrop, a sociotechnical system and imaginary that isolate NPPs, their workers, and local residents from urban areas came about, corroborated by narratives of nation-building as a noble goal.

Fourth, although Japan's science and technology nationalism preceded World War II [27], after the defeat marked by the atomic bombs, narratives of nation-(re) building through science and technology became an even more prominent constitutive element of government policy in various areas. Despite a number of debates on the relationships between science, technology and society in the 1950s and 1960s, often carried out by preeminent scientists and engineers such as Hideki Yukawa and Mitsuo Taketani (e.g., [8]), the ideas that science and technology belong to the elite and experts prevailed and survived multiple challenges, including pollution diseases and various NPP accidents both at home and abroad. While this deficit model has been prevalent globally, in Japan it had a particular elective affinity with the country's tradition of powerful bureaucratic elite and enduring scientific nationalism, contributing to the rise of safety myth and nuclear village and the systematic exclusion of lay voices.

4   Nuclear Imaginaries in Japan: After Fukushima

The 2011 nuclear disaster was a colossal event in Japanese history that has prompted unprecedented efforts to review and discuss what happened, how and why, as well as where we should go as society. Issues of nuclear power, long marginalized and depoliticized, have come into the spotlight in the Japanese public discourse. Numerous TV programs, magazine and newspaper articles, blogs, films, and books have explored a variety of issues, from historical backgrounds of Japan's NPPs to causes of the disaster to the effects of radiation on human health to the energy future of Japan. Furthermore, there have been multiple large-scale efforts to investigate the accident, while new regulatory framework was introduced (see Juraku in this volume on post-disaster investigative efforts and their impact).

However, despite these efforts at reflection and momentary openness to change that followed, much of the older imaginaries remain today, dictating policy and political debates, as well as the way the public can engage with decision processes.

---

[8]As an effort to unearth the region's rich culture and history and understand what its reality says about Japan's past and present, noted folklorist Norio Akasaka has been advocating "Tohoku-gaku," or Tohoku Studies.

In particular, the deficit model, the way the polity is imagined as centralized, and the way radiation risk bearers are concealed all persist.

For instance, despite the consistent majority opposition to restarting of reactors in opinion polls[9] (all the commercial reactors in Japan were offline between September 2013 and August 2015), the 2014 Strategic Energy Plan has paved ways to restarting of NPPs whose safety has been confirmed by the Nuclear Regulation Authority (NRA) under "the new regulatory requirements, which are of the most stringent level in the world".[10] The Plan presented nuclear power still as a primary, "base-load" source for the country's energy supply, emphasizing the same rationales as earlier (e.g., efficiency, stability, Japan's scarce natural resources, less greenhouse gas emission than fossil fuel-based energy). In case of restarting a reactor, the government will "make best efforts to obtain the understanding and cooperation of the host municipalities and other relevant parties." With this new framework, several reactors at three NPPs have been restarted since August 2015 (as of July 2017) regardless of fierce local and urban-area protests.[11]

The new regulatory framework epitomizes the challenge of bringing about a fundamental change. On the one hand, it incorporates some new openness, emphasizing the significance of opening up the regulatory processes, increasing transparency over the energy policy planning process, and obtaining public trust. The 2014 Plan even calls for an end to the national government's monopoly over many decision-making processes, as well as more open engagement with various stakeholders. On the other hand, the idea that the issue is communication with the public, rather than the public's democratic participation, still prevails in the Plan. In this line of thinking, nuclear safety is presented as a domain exclusively for elite efforts, whether scientific or managerial, and the public's concerns and anxiety as something to be resolved with explanation and communication. Such ideas of unproblematized expertise and the deficit model still predominate, despite much soul-searching that took place. Furthermore, as Juraku (this volume) points out, extensive post-disaster efforts at reflection and investigation failed to address such vital societal issues as ethics, responsibility, and social justice head-on.

---

[9]See opinion polls published by major newspapers: e.g., Asashi Shimbun (March 18, 2014); Tokyo Shimbun (September 20, 2015); Nikkei Shimbun (August 24, 2014; February 29, 2016); Mainichi Shimbun (August 9, 2015; March 8, 2016); and Yomiuri Shimbun (March 9, 2016).

[10]http://www.enecho.meti.go.jp/en/category/others/basic_plan/pdf/4th_strategic_energy_plan.pdf (Last accessed on May 29, 2015).

[11]The case of Takahama NPP illustrates the tensions between opposition and support of NPP restarting. In April 2015, a district court issued a provisional injunction against the restarting of its two reactors on the basis of safety concerns filed by local residents. However, this was overturned by the same court in December 2015, and the two reactors were restarted in early 2016. In March 2016, however, another district court issued an injunction against their operation, siding with residents who lived within 70 km from Takahama and raised safety concerns. This unprecedented ruling, which led to the shutdown of an operating reactor, was followed by the June 2016 approval by the NRA to operate Takahama's two other reactors (one 40-year-old and the other 39-year-old) for additional 20 years.

In this context, mothers who express concerns about the effects of low-dose radiation exposure on their young children are portrayed as irrational and pressured to be silent; uninformed workers are mobilized to participate in decontamination efforts in a precarious, exploitative manner (e.g., [20, 30]); official discourses continue to deny or undermine the harmful effects of radiation (e.g., [29, 38]); evacuees from some areas with decreased radiation are nudged to return, with financial support about to be reduced or cut; and municipalities within 30 km of a NPP, although now part of emergency evacuation plans, still do not have a formal say in its restarting. These are consistent with the earlier visions of nuclear technology, even though to some extent Japan's nuclear imaginaries have forever been changed by the accident.

5   Conclusions: Toward Democratic Imaginaries

If new nuclear imaginaries are to serve as resources for social resilience, they need to allow those affected by the negative consequences of the disaster to feel that their experience matters, that they are part of this social enterprise that explores new relationships to nuclear technology, and that they have a say. Japan's dominant nuclear imaginaries have consistently excluded their voice, before and after Fukushima, but new imaginaries need to include their voice, bolster their status, and support their identities. The current situation can also be used as an opportunity to reflect further on our general relationship to science and technology. While many countries have incorporated public engagement in their science and technology policy processes, Japan has generally lagged behind in this, although some provisionary attempts have been made since the disaster [34]. Nonetheless, the deficit model and technocratic approaches still prevail, and what kinds of public engagement would be productive in Japan needs to be explored. And if, in addressing nuclear technology, we are also signaling and performing where we are going as society, we should reexamine our approaches to nuclear governance more carefully and explicitly. After all, the key issues are not simply whether we want nuclear energy or not and how to proceed with the decision we make; it is also about whether we want a society that exploits and neglects the vulnerable [33], as well as about how we make decisions as a democratic society. At stake is what kind of future we are creating.

Acknowledgements   This chapter is based on the author's research as part of a multi-year project with Sheila Jasanoff and Michele Lamont, "The Fukushima Disaster and the Cultural Politics of Nuclear Power in the United States and Japan," funded with an STS grant from the National Science Foundation (Award No. SES 1257117).

References

1. B. Anderson, *Imagined Communities* (Verso, London, 1983)
2. A. Appadurai, Disjuncture and difference in the global cultural economy. Public Cult. **2**(2), 1–24 (1990)
3. R. Asakawa, *Fukushima Genpatsu de Ima Okiteiru Honto no Koto* (Takarajima-sha, Tokyo, 2011)
4. M. Bucchi, F. Neresini, Science and public participation, in *The Handbook of Science and Technology Studies*, ed. by E.J. Hackett et al. (MIT Press, Cambridge, MA, 2008)
5. M. Braw, *The Atomic Bomb Suppressed: American Censorship in Occupied Japan* (M.E. Sharpe, Armonk, NY, 1991)
6. C. Castoriadis, *The Imaginary Institution of Society* (MIT Press, Cambridge, MA, 1975/1998)
7. A. Delgano, Public engagement coming of age: from theory to practice in sts encounters with nanotechnology. Public Understand. Sci. **20**(6), 826–845 (2011)
8. Y. Doi, *Genpatsu to Goyo Gakusha: Yukawa Hideki kara Yoshimoto Takaaki made* (San-ichi Shobo, Tokyo, 2012)
9. S. Epstein, *Impure Science: AIDS, Activism, and the Politics of Knowledge* (University of California Press, Berkeley, CA, 1996)
10. Genpatsu Rokyuka Mondai Kenkyu-Kai, *Marude Genpatsu nado Naika no Yoni – Jishin Retto, Genpatsu no Shinjitsu* (Gendai Shokan, Tokyo, 2008)
11. P. Hall, M. Lamont, Why social relations matter for politics and successful societies. Annu. Rev. Polit. Sci. **16**, 23.1–23.23 (2013a)
12. P. Hall, M. Lamont, Introduction: social resilience in the neoliberal era, in *Social Resilience in the Neo-Liberal Era*, ed. by P. Hall and M. Lamont (Cambridge University Press, New York, 2013b)
13. G. Hecht, *The Radiance of France: Nuclear Power and National Identity after World War II* (MIT Press, Cambridge, MA, 1998)
14. K. Higuchi, *Yami ni Kesareru Genpatsu Hibakusha* (Sanichi Shobo, Tokyo, 1981)
15. N. Hopson, Systems of irresponsibility and Japan's internal colony. Asia-Pac. J. **11**(52), 2 (2013)
16. K. Horiba, *Genbaku Hyogen to Ken-etsu: Nihonjin wa Do Taio Shitaka* (Asahi Sensho, Tokyo, 1995)
17. S. Jasanoff, Future imperfect: science, technology, and the imaginations of modernity, in *Dreamscapes of Modernity: Sociotechnical Imaginaries and the Fabrication of Power*, ed. by S. Jasanoff and S.-H. Kim (University of Chicago Press, Chicago, 2015)
18. S. Jasanoff, S.-H. Kim, Containing the atom: sociotechnical imaginaries and nuclear power in the United States and South Korea. Minerva **47**, 46–119 (2009)
19. P. Jobin, Dying for TEPCO? Fukushima's nuclear contract workers. Asia-Pac. J. **9**(18) (2011), (http://www.japanfocus.org/-Paul-Jobin/3523. Last Accessed 29 May 2015
20. P. Jobin, 3.11 jiko ikou no housha sen hogo. *Ohara Shakai Mondai Kenkyujo Zasshi* **658** (2013)
21. K. Juraku, Why is it so difficult to learn from accidents? (this volume)
22. H. Kainuma, *Fukushima ron: genshiryoku-mura wa naze umareta noka (Fukushima Village: How did the Nuclear Village Come into Being?)* (Seido-sha, Tokyo, 2011)
23. S. Kamata, *Genpatsu Retto wo Iku* (Shueisha, Tokyo, 2001)
24. M. Kawamura, *Genpatsu to Genbaku: "Kaku" no Sengo Seishin-shi* (Kawade Shobo, Tokyo, 2011)
25. H. Kawanishi, *Tōhoku o yomu* (Mumyosōha Shuppan, Akita, 2011)
26. M. Lamont, J.S. Welburn, C. Fleming, Responses to discrimination and social resilience under neoliberalism: the case of Brazil, Israel, and the United States, in *Social Resilience in the Neo-Liberal Era*, ed. by P. Hall, M. Lamont (Cambridge University Press, New York, 2013)

27. H. Mizuno, *Science for the Empire: Scientific Nationalism in Modern Japan* (Stanford University Press, Palo Alto, CA, 2009)
28. H. Murakami, Speaking as an unrealistic dreamer, translated by Emanuel Pastreich. Asia-Pac. J. **9**(29), 7 (2011), http://www.japanfocus.org/-Murakami-Haruki/3571/article.html. Last Accessed 29 May 2015
29. C. Perrow, Nuclear denial: from Hiroshima to Fukushima. Bull. Atom. Sci. **69**(5), 56–67 (2013)
30. Reuters, Tokubetsu Ripoto: Fukushima Josen ni Sukuu 'Homuresu Torihiki' to Hansha Seiryoku (2014). http://jp.reuters.com/article/topNews/idJPTYEA0705O20140108. Last Accessed 29 May 2015
31. G. Rowe, T. Horlick-Jones, J. Walls, N. Pidgeon, Difficulties in evaluating public engagement initiatives: reflections on an evaluation of the UK GM nation public debate about transgenic crops. Public Underst. Sci. **14**, 52–331 (2005)
32. T. Takahashi, *Gisei no shisutemu: Fukushima, Okinawa* (Shūeisha, Tokyo, 2012)
33. T. Takeda, *Watashi-tachi ha koushite Genpatsu taikoku wo eranda* (Chuo Koron, Tokyo, 2011)
34. M. Tanaka, Kagaku gijutsu wo meguru komyunikeshon no iso to giron, in *Posuto 3·11 no kagaku to seiji*, ed. by M. Nakamura (Nakanishiya Shuppan, Kyoto, Japan, 2013)
35. Y. Tanaka, P. Kuznick, Japan, the atomic bomb, and the 'peaceful uses of nuclear power. Asia-Pac. J. **9**(18), 1 (2011)
36. C. Taylor, *Modern Social Imaginaries* (Duke University Press, 2003)
37. A. Toyota, *Nihon no genpatsu gijutsu ga sekai wo kaeru* (Shodensha, Tokyo, 2010)
38. P. Williamson, Demystifying the official discourse on childhood thyroid cancer in Fukushima. Asia-Pac. J. **12**(49), 2 (2014)
39. J. Wilsdon, Paddling upstream: new currents in European technology assessment, in *The Future of Technology Assessment*, ed. by M. Rodemeyer et al. (Woodrow Wilson International Center for Scholars, Washington DC, 2009)
40. B. Wynne, *Rationality and Ritual: The Windscale Inquiry and Nuclear Decisions in Britain* (The British Society for the History of Science, Bucks, England, 1982)
41. B. Wynne, May the sheep safely graze? A reflexive view of expert-lay knowledge, in *Risk, Environment and Modernity*, ed. by S. Lash et al. (Sage, London, 1996)
42. A. Yamamoto, *Kaku Enerugi Gensetsu no Sengo-shi 1945–1960: "Hibaku no Kioku" to "Genshiryoku no Yume"* (Jinbun Shoin, Kyoto, Japan, 2012)
43 M. Tanaka, Habikori hajimeta `anzen yoyu' toiu kiken shinwa, in *Marude Genpatsu nado Naika no Yoni: Jishin Retto, Genpatsu no Shinjitsu*, ed. by G. R. M. Kenkyukai (Gendai Shokan, Tokyo, 2008)

The Institute of Resilient Communities

Kai Vetter

Abstract Resilience is the key to a prosperous global and modern society; Efforts to mitigate physical damage, economic loss, and to protect social and political infrastructures in response to catastrophic events, such as a nuclear accident or a natural disaster, are essential for communities to survive and thrive in the aftermath of such incidents. The 2011 Fukushima Dai-ichi nuclear power plant accident serves as an example of the risks associated with advanced technologies and the need to minimize physical as well as psychological effects on local and global communities. Other examples can be found reflecting the misperception of risks including concerns associated with vaccination or genetically modified organisms. While we have to recognize the risks associated with the development, implementation, and utilization of advanced technologies, we also have to recognize that the impact of not adopting them can have much more detrimental effects to individuals, communities, and even societies. We have established the Institute for Resilient Communities in Berkeley, CA in collaboration with Japanese partners to address the needs for better scientific and technological capabilities to assess, predict, and minimize the impact of disruptive events in the future and to enhance the understanding of associated risks to the public. While the initial focus resides in radiological resilience and is closely related to the events in Fukushima more than 5 years ago, the goal is to establish a broader framework for researchers, educators, and communities to enhance resilience locally and globally together.

Keywords Resilience · Communities · Radiological and nuclear accidents · Real and perceived risks of nuclear radiation and advanced technologies · Radiological resilience

K. Vetter (✉)
Applied Nuclear Physics, Lawrence Berkeley National Laboratory,
and Department of Nuclear Engineering, University of California, Berkeley, CA, USA
e-mail: kvetter@lbl.gov

© The Author(s) 2017
J. Ahn et al. (eds.), *Resilience: A New Paradigm of Nuclear Safety*,
DOI 10.1007/978-3-319-58768-4_16

207

1 Introduction

The Institute for Resilient Communities (IRC) is dedicated to providing tools that can be deployed to enhance resilience in communities locally and globally. The goal is to minimize the impact associated with sudden or long-term changes induced by human actions or natural disasters or a combination of both. To achieve this goal, the IRC combines science, technology, education, and outreach and involves academic and educational institutions as well as communities in an international, multi-disciplinary, and multi-cultural context. It offers a framework for research, education, and community involvement to minimize the physical and psychological impact of future disruptive events and development and a forum for dialogue among researchers, educators, decision makers, and communities.

2 The Concept of Resilience

The term resilience has become widely used over the last 15 years, specifically after the terrorist attack on 9/11/2001 with slightly different definitions and interpretations [1, 2]. According to dictionaries, such as Merriam-Webster [3], resilience is defined as the ability to recover *after* a shock or deformation, the latter specifically in a mechanistic context. Sometimes, resilience is not only associated with the ability to recover after an event but also with actions to reduce the probability that a devastating event will occur (e.g. with the goal of avoiding or eliminating the possibility for incidents *before* they happen) [4]. Yet, others associate resilience with robustness or resistance to minimize or absorb the impact of an event, specifically in the context of resilience engineering with the focus on limiting the damage to infrastructure *during* the event [5]. In the following discussion and reflecting the goals of the Institute for Resilient Communities, we focus on the process to enhance capabilities to minimize the impact during and support the recovery after an unexpected event [6, 7]. This is similar to President Obama's statement describing resilience as "the ability to adapt to changing conditions and withstand and rapidly recover from disruption due to emergencies" [8]. While it is paramount to minimize the probability for an event to occur or to enhance the robustness of systems, catastrophic events with potentially devastating consequences—whether induced by nature or by actions or inactions by humans—will occur in the future. Resilience is a complex concept as it does not only include science and engineering to minimize the impact and to accelerate the recovery—ideally to a better state than before—but even more importantly, the concept embodies societal and educational components. Resilience efforts need to actively involve communities ranging from the local to the global scale, as communities need to be the drivers of actions to enhance their resilience and their ability to withstand and to recover more effectively. The goal of enhancing resilience is not to ultimately reduce resources for communities assuming that a state of sufficient resilience has been achieved. To the contrary, resilience implies a dynamic adaption to our ever-changing world, whether driven by environmental, technological, societal, health, or other causes.

The concept of resilience often appears provocative to the public, as it implies the possibility for more disruptions or accidents to happen. However, this is exactly what has to be realized: the fact that there is a finite probability for events with potentially disastrous consequences to happen. We need to enhance our resilience in order to be better prepared so we can minimize the physical and "measurable" impact as well as the psychological and emotional health impact. Only an informed public and educated decision-makers are able to provide an effective response to a disruption. The necessity for a better-informed public on advanced technologies or more broadly for a "technologically literate citizen" is not new [9]. As we can see, for example, from the events associated with Fukushima, in the discussion about vaccinations or genetically modified organisms (GMOs) the public, particularly in the United States but also globally, is susceptible to a perceived risk, rather than a factual risk. Unfortunately, the concept of risk is not well understood in the public and therefore, any discussion about risk is dominated by the potential impact of an event, rather than both the impact and the probability of an event. As a matter of fact, the likelihood for a disastrous event to happen is often neglected. The mis-perception of the concept of risk in the public is compounded by the fact that the potential impact can easily be used in the media to capture attention and increase readership. In addition, social networks are becoming one of the main venues for communications and distribution of information without any review and vetting of information. The fact that social networks operate instantaneously does not help conventional media to convey information that has been vetted and therefore, is always delayed. The delayed release of vetted information by conventional means also increases the potential for early spread of misinformation, which can be difficult to combat after the fact.

Regarding the contentious issue of GMOs, the risk of potential side effects overwhelms the discussion about the benefits in the public. While there is no scientific evidence supporting any claims of detrimental side effects of GMOs in the modification of approved food [10, 11], social media is dominated by claims of disastrous effects on animals and human consumers. Not only is a large portion of the animal feed in the US already genetically modified [12] without any obvious effect, GMOs need to play a critical role in addressing the enormous and expanding malnutrition in many developing countries specifically in Africa or India [13]. The development and utilization of vaccines have prevented devastating epidemics decimating whole communities and regions in the past. While vaccination is associated with a small health risk, this risk is insignificant in comparison to the risk of foregoing vaccinations. Recent increases in measles epidemics in the US reflect the emerging challenges by experts to convince the public about the benefits vs. the risk associated with vaccination [14].

The opposition to GMOs or vaccination is largely driven by the misperception of risk, as the general public often overestimates the detrimental effects and under-estimates or is not aware of the benefits and the effects of not adopting these technologies. GMOs have another challenge in that it is a new and advanced technology that citizens have no exposure to or experience with. Therefore, the general public is not able to rationally anticipate the consequences, thereby adding GMOs as an example of the impacts of the fear of the unknown. The discussion of nuclear radiation, radioactivity, and ultimately nuclear power encounters a similar challenge. Radiation and radioactivity are largely misunderstood concepts in our society, resulting in highly charged discussions and decisions driven by emotions. The Fukushima Dai-ichi Nuclear Power plant accident, which happened as a consequence of the Great-Eastern earthquake in March 2011 and the subsequent tsunami, serves as an important example in which the psychological health effects have been substantial while the actual health effects due to the feared radiation remains minimal. We will discuss the need to enhance the radiological resilience to events like this in more detail in the next section, with the focus on our work in Berkeley and Japan and in an increasing number of other locations around the world. The global expansion reflects the need to enhance resilience—here specifically radiological resilience—globally, as the concept of radiation is not well understood and continues to evoke fear around the world. In addition, any event associated with the actual or potential release of radioactivity will be broadcasted instantly around the world through the social media networks, which may cause mass panic. It is worthwhile to mention that any large release of radioactivity will have an effect globally: First, due to the actual physical transport and second, due to the instantaneous communication through electronic and social media. As discussed above already, this is not only true for radioactive materials but also for viruses or bacteria, although the physical transport of viruses and bacteria will be slower as the "transporter" is not expected to be the atmosphere such as the effective jet streams across the world, but people and goods traveling across the world.

3   The Challenge Associated with Fukushima as Example

The events at the Dai-ichi nuclear power plant in Fukushima, Japan, in March 2011 have highlighted the need to enhance the resilience to radiological and nuclear events. The incidence and the associated large releases of radioactive materials had and still have an enormous societal and economical impact in Japan and globally. Although no casualties and health effects have been and likely will be attributable directly to radiation, these events have manifold and substantial impact on local communities and have provoked ongoing anxiety and concerns globally.

In order to minimize the impact of these and possible future radiological incidents, technologies and scientific understanding have to be enhanced and equally important, the understanding of nuclear and radiological matters in the public. An important reason for concerns and anxiety in local and global communities can be found in the lack of knowledge and accessible information about radiation and the lack of clear and transparent communications. As a result, confusion, misinformation, and conflicting statements from the government or scientists, as was the case after the Fukushima Dai-ichi nuclear power plant, can cause public distrust of the authorities and potentially of the scientific community. In addition, more effective technologies are required in combination with better scientific models to assess and predict the distribution and transport of radioisotopes in the environment and ultimately, to understand the transport into and minimize the impact onto the biosphere.

The events in Fukushima underscore the necessity for advancements in science and technology and improved communication with the public, and provide a unique opportunity to address both. Currently, local communities and global societies are ill-equipped to prepare for and respond to radiological events and the impact of perceived risks associated with nuclear and radiological incidents. Major releases of radioactivity will have a global impact for two reasons: (1) The radioactive materials can be transported globally quite effectively as has been observed after Chernobyl or Fukushima or previously with nuclear weapons testing; (2) Any event related to the releases of radioactivity will end up as headlines in the global and social media as it is seen as a rare, newsworthy incident, and will be—in most cases inaccurately—associated with a significant health impact. Both aspects will cause increased concerns and fear world-wide, which can only be addressed by an enhanced understanding of radiation through improved scientific literacy and public communication, as well as a better understanding of the associated risk for environmental, biological and health effects. Furthermore, the perceived risk and perceived danger of radiation will continue to hamper public acceptance of nuclear power, which has the potential to contribute to the effort to combat climate change by providing another carbon free energy source necessary to meet the increasing future energy demands while reducing the global CO_2 footprint.

4 Combining Research, Education and Outreach, and Communities—The Institute for Resilient Communities

In order to address the above-described need to enhance community resilience, we have established the Institute for Resilient Communities with the initial focus on radiological resilience, reflecting our activities in response to the Fukushima Dai-ichi Nuclear Power Plant accident. Our Institute-related efforts are composed of research, education and outreach, and community involvement. We briefly discuss each component.

4.1 Outreach and Education—The Berkeley RadWatch and DoseNet Projects

After the releases of radioactive materials on and shortly after March 15, 2011 from the badly damaged Fukushima Dai-ichi nuclear power plant, a team of UC Berkeley graduate and undergraduate students set up sample monitoring stations (which were later upgraded with more sophisticated instrumentation) to collect rain water and air samples on the roof of Etcheverry Hall on the UC Berkeley campus. Subsequently, other environmental as well as a wide range of local food samples were measured utilizing state-of-the are radiation detection equipment available at UC Berkeley and LBNL. The goal of this activity was two-fold: (1) To study the characteristics, such as type and quantity, to determine whether measurable amounts of radioactive materials that we could associate with the releases in Fukushima could be detected (including monitoring for the gradual disappearance of measurable materials); (2) To make the results from the monitoring stations easily accessible and digestible to the general public and to engage the public in a dialogue about radiation and to put our findings in the context of the radiation we are exposed to naturally or electively on a daily basis.

The effort described above, as well as ongoing measurements conducted at UC Berkeley, are part of the Berkeley RadWatch project [15]. Within one week of the 2011 accident in Fukushima, we established a website, which hosted results and analysis, as well as a forum for discussion of the results. Responses to claims and results from other sources, in some cases, claims that were scientifically inaccurate or did not have any scientific basis, were also posted on the site. Since the start of this program in 2011, we have performed more than 1000 measurements with all data and results available at the website. Early in 2014, our team installed an automatic and near-real-time air monitor that provides activities of radioactive particulates captured in a filter mounted in front of a high-energy resolution high-purity gamma-ray spectrometer. This system represents a world-wide unique instrument as it continuously measures radioactivity in air with a state-of-the art and expensive gamma-ray spectrometer and provides the raw spectra and isotopic analysis every 15 min.

In parallel with the RadWatch project, we established the Kelpwatch project in collaboration with Steve Manley from California State University in Long Beach, CA [16]. The goal of this activity was to measure radioisotopes in marine kelp that is collected along the Pacific Coast of North America. Just as RadWatch monitored the arrival of radioisotopes from Japan in California via the jet stream within about 70 h after the releases, the main goal of Kelpwatch was to observe the arrival of radioactive materials that were released into the ocean in March 2011 and carried by the Pacific Ocean currents to the coast of California. Oceanographic transport models expected the appearance 2–4 years after the releases [17]. Kelp samples were acquired and analyzed during five collection periods at about 40 sites between February 2013 and May 2016. We were not able to measure Cs-134 in any sample. The Cs-134 radioisotopes can be used as strong evidence to originate from the Dai-ichi Nuclear Power Plant accident. This is in contrast to the observed Cs-137, which also originated from Fukushima but also from the above-ground weapons tests and has not disappeared yet due its 30 year half live. It is interesting to note that the Cs-137 concentration we observed in kelp was in the order of 0.2 Bq/kg for all samples and the limits for Cs-134 were 0.05 Bg/kg, indicating that most of the Cs-137 observed remains from the weapons test more than 50 years ago. Since these numbers do not mean much in general, we need to put them into the context of naturally occurring radioactivity. For example, the concentration of naturally occurring K-40 was about 4000 Bq/kg in these samples, a factor of 20,000 more then the Cs-137 observed. By utilizing a chemical preprocessing step and filtering large amounts of water, the Woods Whole Oceanographic Institution was able to measure Cs-134 off the Northern California Coast in water [18]. The observed concentration of 2 mBq/l is a factor 5000 less than the approximately 10 Bq/l of naturally occurring K-40 observed in the Pacific Ocean [19]. The ongoing releases of contaminated water off the coast of Japan resulted only in fairly small concentrations, even close to the harbor of the power plant. For example, the water concentration of Cs-134 in close proximity to the harbor of the Dai-ichi nuclear power plant is about 20 mBq/l, if detectable at all [19]. While the observation of Cs-134 in the water does confirm the transport of this isotope via ocean currents, it is significantly diluted spatially and temporally and—as expected and as with the other samples mentioned before—never posed a health risk to the public or the environment.

We are still continuing to conduct environmental and food sample measurements to-date, including measurements driven by requests from the public. This even includes materials from Europe with potential contamination from the Chernobyl accident more than 30 years ago. In addition to real-time air-sampling and sample measurements performed utilizing gamma radiation, the team set up alpha spectrometers that enable us to measure alpha particle decay in the same samples. The main goal of this activity is to measure naturally occurring alpha decay, specifically of Po-210. The fact that there is naturally occurring gamma radiation in the environment is not widely known, and the fact there is naturally occurring alpha-particle radiation is even less known. However, naturally occurring background radiation is part of the world we are living in and something that the general public should be aware, so any new information about radioactive contamination can be put into the proper context. It is noteworthy that the Po-210 measurements, particularly in fish, do dominate the radiation exposure, meaning the dose from Po-210 is larger than the dose due to K-40 [20]. This information should not lead to a decision to avoid eating fish, as we are exposed to Po-210 all the time as it is part of the radioactive decay chain starting from U-238. Radioisotopes such as U-238, Th-232, or K-40 have half-lives of billion of years and can be found everywhere in our universe resulting in some amount of radioactivity even in the smallest quantities of matter. In addition to the gamma-ray and alpha-particle monitoring performed in the RadWatch UC Berkeley lab, we have performed so-called neutron-activation analysis experiments which allow us to measure trace amounts of non-radioactive matter in our environment and food. Of specific interest are trace metals, such as mercury or arsenic, which can be commonly found in food and fish. In keeping with the RadWatch mission, the goal is not to make the public fearful of food or our environment, but to make them aware of the world we are living in and to ultimately put the observation and risks of radiation in a proper context. All this information is available online on our Berkeley Radwatch website [15].

In order to further enhance awareness of radiation and its properties, and to combine efforts to raise awareness with education and outreach, we have established the Berkeley DoseNet program [21]. This program consists of a sensor network that is being developed across high- and middle schools in the Bay Area as well as well UC Berkeley and other high schools and research institutions in Japan, South Korea, and Sweden. The sensors are equipped with a Raspberry-Pi computer and initially set up with a simple radiation dosimeter. These dosimeters are loaned to the local school partners and collected data is made available to the public on the Dosenet website in 5-min intervals. The DoseNet team is currently collaborating with participating teachers to develop projects that students can work on with the data collected, either with their school's local data or with the data available across the network. These projects have two objectives: (1) Allow the students to "see" radiation in our environment and to learn important properties of radiation, e.g. the fact that it varies spatially and temporally or that it can be shielded or reduced by increasing the distance; (2) Enable a better understanding and appreciation of fundamental concepts in science and engineering such as uncertainties associated with observations and measurements, statistics and probability, and ultimately, risk. As the concept of risk is becoming ever more important in our modern and technological driven global society, it needs to be better understood by the public. The first objective addresses specifically the fear of radiation in the public as it can not be recognized with human senses. The second objective addresses the need to enhance more broadly the science and technology literacy of citizens. In parallel to expanding geographically, we are planning to upgrade and complement the radiation sensors with better radiation detectors and other sensors. We are in the process of integrating recently developed pocket-sized CsI(Na) scintillator-based detectors that enable us not only to register radiation, but also to measure the energy of radiation, e.g. to see the "color" of nuclear radiation. The "color" tells us about the origin and the type of materials, specifically the isotope, that emitted the radiation. Complementary sensors will include air-particulate and CO_2 sensors that will become part of the sensor package. We encourage schools to add digital weather stations to the sensor package, as it is quite interesting to study correlation between weather patterns and the quantities observed with the sensor package.

4.2 Science and Technology—Assess, Predict, and Minimize the Impact of Radiological Contamination

Our research within the context of radiological resilience at LBNL is currently engaged in four different scientific and technology domains, which address current needs in the evacuated areas in Fukushima Prefecture to ensure the safety of the population when they return and in the future. As of January 2016 about 85,000 people are still evacuated and many, particularly older people want to move back to their homes and communities [22]. The research areas we are currently pursuing in collaboration with scientists from JAEA aim at more effective means to map the contamination, at a better understanding and improved predictive power of environmental transport models, enhanced understanding and measurements of internal human radiation dose, and the removal of contamination from soil. While the focus initially is on the most abundant contaminant left (cesium) in the environment of Fukushima, the knowledge gained and technologies developed will provide significantly improved means in the aftermath of any radiological event in the future that is associated with the release of radioactive materials. The environment in Fukushima Prefecture represents a very different environment than, for example, the region of Chernobyl, as Fukushima Prefecture consists of large portions of forests and mountains with significant precipitation year-round, causing continuous changes in the contamination patterns. Figure 1 summarizes the four areas of research and their relationships. These activities are coordinated with the substantial efforts by the JAEA in Japan.

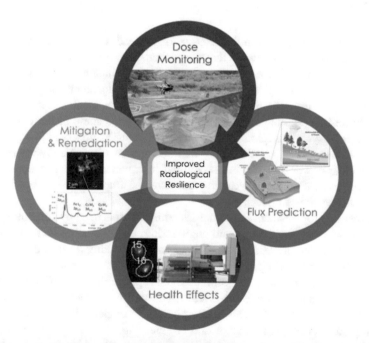

Fig. 1 The four main research areas being pursued initially as part of the new Institute for Resilient Communities. The goal is to enhance the effectiveness in monitoring and predicting radiological transport in the environment, to better understand and minimize the impact of radio-isotopes in the biospheres, particularly humans, and the remediation of these radiological materials, particularly cesium

4.3   Status and Path Forward

The aforementioned education, outreach, and research activities are the central pillars for the Institute for Resilient Communities. Based on these established activities we will expand our outreach and research activities locally in Berkeley and with our research and community partners and organizations in Japan and beyond. Reflecting the need to work with local communities we are actively collaborating with cities, such as Berkeley in the U.S. and Koriyama in the Fukushima Prefecture in Japan.

Complementary to the research, we will continue the RadWatch project's efforts, including near-realtime air monitoring and measurements of environmental and food samples, and Kelpwatch, as part of the outreach and educational efforts. As before and reflecting the importance of transparency in any of our activities to maintain public trust, we will publish all our measurements, procedures, and findings.

5   Summary

Recent events associated with the releases of radioactive materials and the recognition of the possibility of events that are associated with the release of radioactive materials to happen in the future represent major challenges for advanced and global societies. Radiological or nuclear events due to accidents have and will have an enormous socio-economical and political impact on local and global communities. While it is possible that such an event may have some health effects due to radiation, the psychological impact will be significant, as observed in Japan. While it is paramount to enhance the safety and security of the currently operating and future nuclear power plants, it is also critical to enhance means in responding and recovering from a possible event to minimize the impact of such events, i.e. to increase the resilience to such events.

The Institute for Resilient Communities addresses this need by combining natural and social sciences, technology and engineering, and education and outreach, and involves local communities, all in a global context. It addresses the need to improve the scientific understanding of the causes and impacts of such events. The education and outreach aspect aims to minimize the psychological effects through a better-informed public. While the initial focus will be on radiation, the goal is to establish programs to enhance science and technological literacy more broadly, including basic concepts in science and engineering. Data are collected and made available to recognize and appreciate the world we are living in, particularly the world we cannot see or physically feel. Local communities and schools are being involved to effectively introduce these concepts to the public and into schools. By providing such a framework, the Institute for Resilient Communities is a trusted resource to the public, media as well as decision makers, essential in the response to a radiological or nuclear event to minimize the effect and speed-up the recovery, i.e. enhancing resilience.

References

1. L. Labaka et al., Homeland Security & Emergency Management **10**(1), 289–317 (2013). doi:10.1515/jhsem-2012-0089
2. J.H. Kahan, Resilience redux: buzzword or basis for homeland security. Homel. Secur. Aff. **11**(2) (2015), https://www.hsaj.org/articles/1308
3. http://www.merriam-webster.com/dictionary/resilience
4. T.J. Vogus, K.M. Sutcliffe, Organizational resilience: towards a theory and research agenda, in *IEEE International Conference on Systems, Man and Cybernetics*, pp. 3418–3422 (2007). doi:10.1109/ICSMC.2007.4414160
5. L. Labaka et al., Enhancing resilience: implementing resilience building policies against major industrial accidents. Int. J. Critic. Infrastruct. **9**(1/2), 130–147 (2013)
6. M. Bruneau et al., A framework to quantitatively assess and enhance seismic resilience of communities. Earthq. Spectra **19**, 733–752 (2003)
7. J.H. Kahan et al., An operational framework for resilience. J. Homel. Secur. Emerg. Manage. **6**(1), 1–50 (2009)
8. Executive Office of the President, Presidential Policy Directive-8 (PPD8), February 2011, Washington, DC, 6
9. G. Pearson, A.T. Young, Technically speaking: why all Americans need to know more about technology. Committee on Technological Literacy, National Academy of Engineering; National Research Council (2002). ISBN: 0-309-51013-9
10. C. Snell et al., Assessment of the health impact of GM plant diets in long-term and multigenerational animal feeding trials: a literature review. Food Chem. Toxicol. **50**, 1134–1148 (2012)
11. Committee on "Genetically Engineered Crops: Past Experience and Future Prospects". Board on Agriculture and Natural Resources; Division on Earth and Life Studies; National Academies of Sciences, Engineering, and Medicine. doi:10.17226/23395 (2016). ISBN 978-0-309-43738-7
12. A.L. Van Eenennaam, A.E. Young, Prevalence and impacts of genetically engineered feedstuffs on livestock populations. J. Anim. Sci. **92**(10), 4255–4278 (2014). doi:10.2527/jas.2014-8124
13. D.H. Freeman, The truth about genetically modified food. Sci. Am. (2013)
14. M.S. Majumder et al., Substandard vaccination compliance and the 2015 measles outbreak. JAMA Pediatr. **169**(5), 494–495. doi:10.1001/jamapediatrics.2015.0384
15. https://radwatch.berkeley.edu/
16. https://kelpwatch.berkeley.edu/
17. V. Rossi et al., Multi-decadal projections of surface and interior pathways of the Fukushima Cesium-137 radioactive plume. Deep-Sea Res. I **80**, 37–46 (2013)
18. Woods Hole Oceanographic Institution, Fukushima Radioactivity Detected Off West Coast (2014), http://www.whoi.edu
19. Nuclear Regulatory Agency, Japan, Monitoring information of environmental radioactivity levels. Distribution map of seawater radioactivity around TEPCO Fukushima Dai-ichi NPP (2015), http://radioactivity.nsr.go.jp/en/
20. N.S. Fisher, "Evaluation of radiation doses and associated risk from the Fukushima nuclear accident to marine biota and human consumers of seafood. Proc. Natl. Acad. Sci. (2013). doi:10.1073/pnas.1221834110
21. https://radwatch.berkeley.edu/dosenet/map
22. http://www.japantimes.co.jp/news/2016/01/09/national/fukushima-nuclear-evacuees-fall-100 000/#.V9yRDJMrLUI

Part IV
Students Contributions

Ground Motion Prediction for Regional Seismic Risk Analysis Including Nuclear Power Station

Hiroyasu Abe

Abstract Ground motion simulation is one of techniques used to analyze seismic risk due to damage of structure and its effects on society. In this paper, the ground motion simulation using fault plane is used. Recently, ground motion simulation using fault model have been widely applied. Characterized fault model is conveniently used to model the heterogeneous slip distribution on fault plane, which divide the fault into two areas (asperity area and background area). More detailed model is needed to conduct probabilistic seismic risk assessment, which incorporate uncertainty in ground motion prediction. The model, however, is too simplified to model the complex characteristics of slip. In this paper, a stochastic model to simulate the slip distribution of fault plane is proposed for that purpose.

Keywords Seismic motion · Random field · Crustal earthquake · Fault model · Earthquake ground motion

1 Introduction

To discuss the safety of critical infrastructure such as nuclear power plants, seismic risk assessment is conducted. Usually engineered system consists of several facilities which are spatially distributed. Though the conventional risk assessment is mainly for a single facility, risk assessment for multi-facility is required. An example of spatially distributed system is a nuclear power station. In a nuclear site, several units are located. Additionally, sometimes several sites are located closely each other. For public, the information on the likelihood and possible amount of radioactive material release is needed and all the units may suffer from identical external events, multi-unit risk assessment is necessary. For that purpose, a technique to simulate spatially distributed ground motion probabilistically is needed.

H. Abe (✉)
Department of Nuclear Engineering and Management, The University of Tokyo, Tokyo, Japan
e-mail: h.abe.ju95th@gmail.com

© The Author(s) 2017
J. Ahn et al. (eds.), *Resilience: A New Paradigm of Nuclear Safety*,
DOI 10.1007/978-3-319-58768-4_17

221

Recently, ground motion simulation using fault model is widely used for ground motion simulation for a single site. In the fault model, a simplified characterized fault model was proposed and used for that purpose, and standardized method, e.g., 'recipe' [3], is proposed. The method, however, was proposed to calculate the average characteristics of ground motion. More detailed model is needed to conduct probabilistic seismic risk assessment, which incorporate uncertainty in ground motion prediction. Therefore, in this study, a probabilistic modeling of slip distribution focusing on crustal earthquake is proposed.

2 A Proposed Model to Simulate the Slip Distribution

Figure 1 is an example of actual slip distribution in the fault plane of West Off Fukuoka Earthquake in 2005. The spatial distribution of slip displacement exhibits stochastic characteristics. When two elements in the fault plane is closely located, slip displacement is similar, i.e., correlated. On the other hand, slip displacement is random when two elements are distant. These characteristics of slip displacement can be modeled using the spatial correlation model.

This correlation structure of the slip distribution is analyzed. The slip displacement at each element at Fig. 1 is denoted $y = \{y_1, ..., y_n\}$, where n is the number of element. Slip displacement is assumed to be distributed by the log normal distribution. Table 1 shows the average and standard deviation of logarithm of slip y in Fig. 1 obtained by the maximum likelihood estimate. Then, the correlation structure from slip distribution during actual earthquake is analyzed. Semivariogram r is used for that purpose. $r(Y_i, Y_j)$ is defined as follows:

Fig. 1 Slip distribution along fault surface of West Off Fukuoka Earthquake [1, 4]

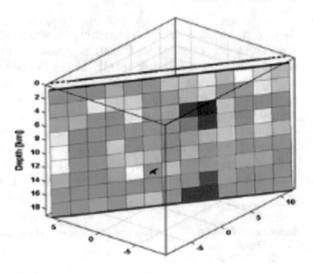

Table 1 Average and standard deviation of logarithmic slip of West Off Fukuoka Earthquake

Average	Standard deviation
3.90	0.87

$$r(Y_i, Y_j) = \frac{1}{2} E[(Y_i - Y_j)]^2 \tag{1}$$

where, Y_i and Y_j are the lengths of slip at the i-th and j-th element respectively. And, semivariogram '$r(Yi, Yj)$' is defined as follows:

$$r(Y_i, Y_j) = \sigma_Y^2 \times (1 - \rho_{Y_iY_j}) \tag{2}$$

where, σ_Y is standard deviation of fault plane and $\rho_{Y_iY_j}$ is the correlation coefficient between Yi and Yj.

In this study, this correlation is assumed to depend on h that is distance between two elements, and the correlation coefficient is defined as follows:

$$r(h) = \frac{1}{2N(h)} \sum_{i=1}^{N(h)} (Y(\mathbb{X}_{1i}) - Y(\mathbb{X}_{2i}))^2 \tag{3}$$

Provided that $N(h)$ is the number of pairs that fulfill (4) include $(\mathbb{X}_{1i}, \mathbb{X}_{2i})$.

$$h - \frac{\Delta h}{2} \leqq |\mathbb{X}_{1i} - \mathbb{X}_{2i}| \leqq h + \frac{\Delta h}{2} \tag{4}$$

where, Δh is assumed 1.34 km. Figure 2a is $r(h)$ obtained from Eq. (3). Figure 2b shows $N(h)$ for each bin. In Fig. 2a, fitted parabola that is determined from least-squares method is denoted.

Fig. 2 Semivariogram showing spatial correlation of slip obtained from slip distribution during West Off Fukuoka Earthquake

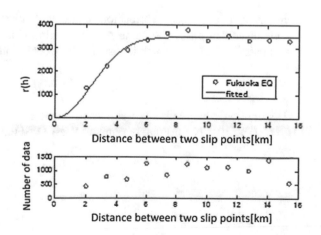

$$r(h) = b(1 - \exp(-ah^2)) \tag{5}$$

where, b is the constant equal to the dispersion σ_Y^2. a is estimated 0.0947, and b is 3.4819×10^3.

Simulation of slip distribution in fault plane is conducted by Monte Carlo Simulation. The slip of fault plane in i-th element is Y_i. $Y = [Y_1, Y_2, ..., Y_N]'$. Y follows logarithmic normal distribution. W is the normal random variable that is transformed from Y by Rosenblatt conversion as follows:

$$W = \Phi^{-1}(F(Y)) \tag{6}$$

where $\Phi^{-1}()$ is cumulative function of standard normal distribution, and $F(Y)$ is the cumulative distribution function of Y.

Z is decided by random number, and W is determined from that.

$$W = \Phi_w Z \tag{7}$$

W is determined from that. In this equation, Z is the stochastic variable vector that fulfill normal distribution, mean of 0 and standard deviation of 1. Φ_W is Eigenvector of covariance 'C_{WW}'. (8)

$$C_{WW} \Phi_W = \Phi_W \Lambda_W \tag{8}$$

Λ_W is the square matrix, diagonal element is characteristic number and the other is 0.

$$C_{WW} = \begin{bmatrix} \rho_{11} & \cdots & \rho_{N1} \\ \cdot & & \cdot \\ \cdot & & \cdot \\ \rho_{N1} & \cdots & \rho_{NN} \end{bmatrix} \tag{9}$$

In this equation, ρ_{WiWj} is correlation coefficient between Wi and Wj. In this study, it is premised that ρ_{WiWj} is equal to ρ_{YiYj}, and fulfill (5). So, it is premised that ρ_{YiYj} is a function of only h_{ij} which is distance between i and j.

$$\rho_{YiYj} = \exp(-ah_{ij}^2) \tag{10}$$

Distribution in fault is simulated under the condition of $Mw6.6$ ($M_0 = 9.0 \times 10^{18}$ N $\cdot$ m) that is same as West Off Fukuoka Earthquake.

Fig. 3 Geometrical relation
between fault plane and
receiver locations (A and B)

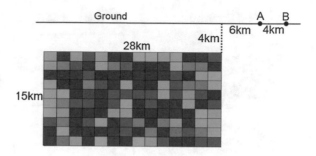

3 Ground Motion Simulation Using Proposed Model

Stochastic Green's function [2] method is used for ground motion simulation. The fault geometry and receiver location is showed in Fig. 3. S-wave velocity on surface is assumed to be 400 m/s in this study.

In Fig. 4a, samples of slip distribution is shown, while simulated ground motion (velocity time history) for respective case is shown in Fig. 4b. As shown in the figure, the temporal characteristics of velocity time history are different between samples. Maximum velocity increases if large slip appear between the hypocenter and the receiver as shown in the bottom figure in Fig. 4.

The simulated slip distribution was closer to real distribution than characterized fault model. However, the long slip area is scattered than actual case.

4 Conclusions

Multi-unit and multi-site probabilistic seismic risk assessment is needed to respond to the public concerns about offsite emergency response. In this paper, a stochastic fault rupture model is proposed for the purpose of multi-unit and multi-site probabilistic seismic risk assessment.

The simulated slip distribution was closer to real distribution than characterized fault model. However, the larger slip area scattered than actual case. It is needed to be improved in future study. For example, it would be needed that slip distribution is modeled by a different approach.

Acknowledgements This work is supervised by Professors Naoto Sekimura and Tatsuya Itoi.

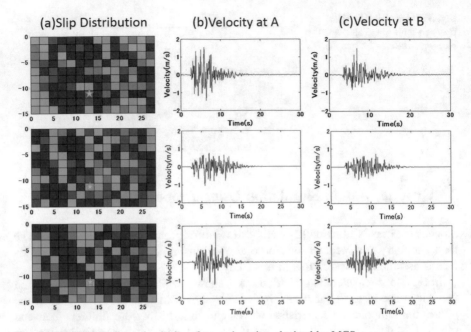

Fig. 4 Distribution slip and velocity of ground motion obtained by MCS

References

1. K. Asano, T. Iwata, Source process and near-source ground motions of the 2005. West Off Fukuoka Prefecture earthquake. Earth Planets Space **58**, 93–98 (2006)
2. D.M. Boore, Stochastic simulation of high-frequency ground motions based on seismological models of the radiated spectra. Bull. Seism. Soc. Am. **73**(6), 1865–1894 (1983)
3. Headquarter for Earthquake Research Promotion: Recipe for Predicting Strong Ground Motion by a fault plane of earthquake (2009). (In Japanese)
4. P.M. Mai, Finite-Source Rupture Model Database. http://equake-rc.info/SRCMOD/. Accessed 28 Feb 2015

Effects of Inelastic Neutron Scattering in Magnetic Confinement Fusion Devices

Ivana Abramovic

Abstract Components, surrounding the core of a magnetic confinement based fusion reactor, will be exposed to significant particle and heat fluxes that will cause severe (in many cases irreversible) damage of the components. In the D-T fusion reaction 80% of the energy is carried away by the 14 MeV neutrons and the rest by the emitted alpha particles. Motion of neutrons is not restricted by the present magnetic field, which is why the damage they cause by interacting with surrounding materials is to a large extent inevitable. The greater the neutron flux to the material the larger the damage, and shorter the lifespan of the reactor component taking the flux. In order to make fusion economically viable it is important to increase the lifespan of the components since they are costly to produce, replace and dispose of. It is the premise of this work that neutron inelastic scattering plays an important role in neutron transport in MCF systems. This reaction mechanism has been overlooked in neutron transport calculations. Planned work entails modeling of inelastic scattering using reaction codes and results compared with experiment where possible. Data obtained will be further used in neutron transport calculations and in damage analysis of various materials in order to establish how significant inelastic scattering is for the viability of fusion energy production.

Keywords Inelastic neutron scattering · Nuclear reaction modeling · Magnetic confinement devices · Neutron transport · Irradiation damage

1 Introduction

Nuclear fusion offers a prospect of an inexhaustible source of energy and promises a reduction of environmental impacts of worlds increasing energy demand. The fact that scientific and engineering challenges, which have to be surmounted, haven't fully been foreseen in the beginning of fusion research has greatly contributed to

I. Abramovic (✉)
Department of Nuclear Engineering, UC Berkeley, Berkeley, USA
e-mail: i.abramovic@berkeley.edu

© The Author(s) 2017
J. Ahn et al. (eds.), *Resilience: A New Paradigm of Nuclear Safety*,
DOI 10.1007/978-3-319-58768-4_18

establishing the reputation of fusion power as an elusive goal. Some of the key issues that still have to be addressed are: understanding of turbulent transport in magnetized plasma's, suppression of edge localized modes, realization of an effective fuel cycle based on tritium breeding, improvement of particle and energy exhaust systems, development of the materials able to handle extreme conditions in the reactor. Materials can rightfully be singled out as the most critical issue. Large heat loads and particle fluxes cause melting, sputtering, erosion, re-deposition, swelling and displacements of atoms from their lattice structure. This further results in fuel dilution, changes in size of the components, changing of properties of the material.

In general effects of incoming neutrons result from three types of interactions with the target material:

absorption (transmutation)
elastic scattering
inelastic scattering

The first reaction is responsible for creating transmutation products (such as He) that over time build up in the material. This consequently leads to deteriorating of material properties, making it swollen as the amount of helium increases, brittle as a result of accumulation of helium bubbles along grain boundaries, causing the change in thermal conductivity, activation of the material etc.

The second reaction is mostly responsible for damage expressed in terms of dpa-displacements per atom. Neutrons kick the lattice atoms out of their original position in the crystal lattice causing defects in the structure of the material by creating interstitials, substitutions and vacancies.[1] Accumulation of such defects can even lead to phase transitions.

The third reaction is in the focus of my research. Knowledge of it is limited [1] yet it is important for modeling of 14 MeV neutron interactions that originate from the D-T reaction. Change in the neutron energy spectrum along the propagation path will also be taken into account. Materials that are planned to be included in this analysis are the ones relevant for MCF: steel, tungsten, beryllium and carbon.

2   Description of the Actual Work

In order to address the problem of neutron transport and investigate the importance of inelastic neutron scattering both experimental and theoretical analysis will be conducted. Work can be divided into two conceptual parts. First part is related to nuclear reaction data evaluation while the second part deals with application of acquired knowledge to MCF systems.

---

[1]Types of defects in the lattice structure.

Experiments will be conducted with the Bay Area Neutron Group at the 88 inch Cyclotron at Lawrence Berkeley National Laboratory. First set of data will be collected in gamma–gamma coincidence measurements of inelastic neutron scattering on iron 56. Scattering cross sections for this element are inadequately known for transport calculations and although the level schemes of iron are well documented the accuracy of the cross section measurements needs to be improved by 10–15% in the relevant energy range of 0.5–20 MeV [1]. Reason for starting with iron 56, besides the fact that it is the major component of stainless and Euro-fer steels, is the fact that it has been singled out as a high priority isotope by the CIELO collaboration.

Measurements will take place in Cave 0 of the 88 inch Cyclotron facility. Four germanium detectors will be positioned roughly at an angle of 90° with respect to the incident neutron beam. Coincident gammas will be measured in order to allow the use of data for level scheme building. Once the data has been collected and analyzed it will be compared with calculations. Different nuclear models allow properties such as the reaction cross sections to be calculated. These calculations are extremely complicated and cumbersome which is why computer codes have been developed to perform them. State of the art code EMPIRE allows selection of options and adjustment of input parameters such that the calculations best fit the experimental conditions and allows loading of measured data for direct comparison between evaluated data, measured data and theoretical predictions. Once the data from upcoming experiments is analyzed the result will further be used in the process of data evaluation to reduce the currently existing uncertainties in the energy range of interest.

The same reaction code will later on be used for modeling the inelastic scattering through various materials in MCF devices. Thickness of the target is not an input parameter of EMPIRE however satisfactory results can be obtained by multiple runs at different incident energies simulating the change in neutron energy spectrum. Calculations will be weighted by the build up factor for materials of interest and further supplemented by calculations taking into account the incident neutron flux. If inelastic scattering turns out to be a significant mechanism in fusion reactor environment full MCNP calculations will be necessary for precise modeling, these calculations are out of the scope of this project.

3 Conclusions

Research outlined in previous sections has not yet been performed. Therefore no definite conclusions or results can be presented at this point. However research question that will be answered once the project is complete is how significant is the mechanism of inelastic neutron scattering for neutron transport in fusion reactor environment.

Acknowledgements Work planned has been made possible by the Bay Area Neutron Group in collaboration with UC Berkeley, Lawrence National Laboratory and University of Technology Eindhoven.

References

1. M.B. Chadwick et al., The CIELO collaboration: neutron reactions on H, O, Fe, U and Pu. Nucl. Data Sheets **118**, I-25 (2014).
2. Reaction code obtained from, http://www.nndc.bnl.gov/empire/
3. M. Victoria, Structural materials for fusion reactors. EPFL-CRPP Fusion Technology Materials, CH-5232 Villigen PSI, Switzerland
4. G.R. Satchler, *Introduction to Nuclear Reactions* (Oxford University Press, New York, 1990)

The Account of the Fukushima Daiichi Accident by the Plant Manager: A Source to Study Engineering Thinking in Extreme Situations

Aissame Afrouss

Abstract The concept of "engineering thinking in extreme situations" has been defined to make up for an epistemological lack in the field of Safety Studies. After the accident at the Fukushima Daiichi plant, official reports did not take an interest in analysing the conditions in which the recovery efforts had to be carried out. The description of the accident and its representation in the accident investigation reports convey these shortcomings. The Fukushima Daiichi plant manager Masao Yoshida testimony may allow us address them partly. Actually, the transcription of his hearings contains essential details and information to understand the sequence of events which took place after the 2011 Tōhoku earthquake and tsunami. This article intends to show the importance of studying this narrative, in order to highlight the relations between the Fukushima accident management and the concept of "engineering thinking in extreme situations".

Keywords Engineering thinking · Extreme situation · Yoshida testimony · Narrative analysis · Mental representation

1 Introduction

The Fukushima Daiichi accident is regarded as one of History's most important nuclear accidents, along with those at Chernobyl and Three Miles Island. Yet, the institutional accident investigation reports do not allow us to comprehend fully the complexity of the accident management by the operators. To address this problem, the concept of "engineering thinking in extreme situations" has been defined as *"engineering activities that are significantly impeded due to the lack of resources in the face of a societal emergency"* [1].

A. Afrouss (✉)
Centre for Research on Risks and Crises (CRC), MINES ParisTech/PSL-Research University, Sophia Antipolis, France
e-mail: aissame.afrouss@mines-paristech.fr

© The Author(s) 2017
J. Ahn et al. (eds.), *Resilience: A New Paradigm of Nuclear Safety*,
DOI 10.1007/978-3-319-58768-4_19

Masao Yoshida, the plant manager, has been heard by the Investigation Committee on the Accident at the Fukushima Nuclear Power Stations (ICANPS) on the plant's management after the earthquake and the actions undertook to deal with the accident. His account, made public in September 2014, is currently being edited in French [2].[1] The testimony enables us not only to address the silence of the reports, but to learn more about the factors affecting the decision-making and the action-taking during "extreme situations" as well. During such a situation, the actions taken by workers do not enable them to regain control of their production tool, and are regarded as responsible of an impending and irreversible damage [3].

This article therefore intends to demonstrate the importance of the Masao Yoshida first-hand account, and how its study contributes to the concept of "engineering thinking in extreme situations". In the first part of the article, some accident investigation reports and there content are presented. In the second part, correlations between the hearings of Yoshida and the narrative form are highlighted, and then, the significance of the manager's disclosures for "engineering thinking in extreme situations" is revealed.

2 The Accident Investigation Reports

The first part presents the four institutional reports used in this study and the bodies that produced them. Then, it sums their content up, with an eye to analyse the representation they make of the accident.

2.1 Description of the Institutional Report

Following the Fukushima accident, several bodies published reports about its causes. These reports also point out the lessons that are to be learned to enhance nuclear facilities safety. This paper discusses the reports issued by the Investigation Committee on the Accident at the Fukushima Nuclear Power Stations [4–7]. These documents have been written by two Japanese, an international and an American organisation. Obviously, they were drafted with different purposes and present different feedbacks and recommendations.

The Cabinet of Japan has decided to create The Investigation Committee on the Accident at the Fukushima Nuclear Power Stations (ICANPS) on 24 may 2011. The ten members (researchers, jurists…) are put under the direction of Yotaro Hatamura, professor emeritus of the University of Tokyo. The aim of the Committee is to

---

[1]The CRC—MINES ParisTech is publishing a French version of the entirety of the hearings of Yoshida. The first volume which contains the auditions of 22 and 29 July 2011 is available since March 2015. The auditions of 8 and 9 august are gathered in the second volume which was published in March 2016. The following hearings are to be found in the third volume.

suggest recommendations to limit the expansion of the damages of the accident and to prevent the recurrence of similar crises. The investigation should identify the causes of the accident and the causes of the damages it inflicted to Japan. The members of the Committee ambitioned to carry out a thorough investigation which outcome would satisfy every question about the accident, and whose results would remain valid for the next century. The final report is issued in July 2012.

The National Diet of Japan Fukushima Nuclear Accident Independent Investigation Commission (NAIIC) has been established via a dedicated act—named NAIIC act—on 30 October 2011. On 8 December 2011, the ten members of the Commission are designated by the Diet president. Kiyoshi Kurokawa, a doctor of medicine and former chairperson of the Science Council of Japan is appointed chairperson. The nine other members are scientists, legal experts and politicians. The investigation of the NAIIC focuses on the causes of the accident and of the damage of the accident. It also reminds how the stakeholders dealt with the accident, and points out the lack of efficiency of their responses in the face of the extreme situation. Finally, it suggests measures to be applied in order to prevent another nuclear accident from happening in Japan and to mitigate its possible consequences. The final report is published in September 2012.

One of The Nuclear Energy Agency's (NEA) tasks is to strengthen the legal, scientific and technological bases of the nuclear safety in the Organisation for Economic Co-operation and Development (OECD). In September 2013, it drew up a report on the lessons learnt from the Fukushima accident. This document has been written under the direction of the Director-General Luis Echávarri. It lists the efforts made by the OECD member countries to improve safety management following the accident, and gives general recommendations based on the main lessons learned.

The American National Academy of Science published a report sponsored by the United States Nuclear Regulatory Commission in July 2014. The study has been carried out by a committee consisting of various scientists and engineers with different competencies. The committee was led by Norman P. Neureiter. The aim of the report is to summarize the multiple causes of the nuclear accident and to suggest recommendations to enhance the American nuclear facilities' safety.

2.2 The Accident According to the Reports

The analysis of the recommendations given in the different reports highlights the flaws identified by the institutional bodies. This helps to understand their representation of the accident and the crisis management. In this article, the notion of "representation" refers to a mental map, that being the whole of causal, proximal and influence relations established by people in order to understand a problem or a problematic issue [8].

According to the four reports, the Fukushima Daiichi accident is due to a lack of preparation of TEPCO and the concerned institutions to deal with such an event. They reveal therefore that TEPCO staff lacked adequate training and appropriate skills to respond to emergency situations. They also point out the fact the communication between the workers and the authorities and the poor coordination of the emergency response centres has not permitted to react effectively.

Furthermore, the accident could have been avoided if the state of the art and the new safety concepts have been applied, especially the defence-in-depth concept. TEPCO and the Japanese institutions had not taken the appropriate measures to bring their facilities' safety up to current international standards. The reports underline also the necessity for all nuclear power plants to strengthen the defence-in-depth provisions, and to consider the occurrence of beyond design basis and multi-unit accidents.

Another recommendation involves the lack of independence of the Nuclear and Industrial Safety Agency (NISA). Although the nuclear regulatory body knew about some of TEPCO safety deficiencies, it did not face its responsibilities. The competencies, the involvement and the transparency of the NISA have been called into question. Consequently, Japan needed to deeply reform its nuclear facilities' regulation and monitoring system.

This short overview shows that the four reports do not bring renewed reflections on accident management. Instead, they only emphasise the need to strengthen concepts already acknowledged and to take larger margins to avoid potential accidents. New standards might be suggested and taking beyond design basis accidents is encouraged. To sum up, major accident management is regarded through the already existing organisation and resources.

However, during the accident, the operators found themselves in the face of a scenario that exceeds by far every known standard. The loss of electricity resources and the worsening of the site conditions point out the need to adapt to new and unexpected circumstances. The hearings of Masao Yoshida bring out a new consideration of the accident, giving specifications and details unfound elsewhere. His testimony enables a better understanding of the proceedings of the Fukushima crisis management.

3   The Importance of Yoshida's Testimony

This section analyses the hearings of Masao Yoshida and shows that the testimony he made can be considered as a narrative. A comparison is then made between the content of the institutional reports and the disclosures of the manager, binding these information and the concept of "engineering thinking in extreme situations".

3.1   The Yoshida Testimony: A Narrative of the Accident

The hearings of Masao Yoshida have been carried out by ICANPS. The Committee interrogated several political and technical actors, who were involved during the accident management. The manager has been summoned five times by the Commissions, between 22 July and 22 November 2011. These interviews lasted 28 h in total and primarily addressed the actions that were carried out in response to the accident. The transcript of the hearing was made public by the Japanese government on 11 September 2014 in the form of eleven documents.[2]

The content of these hearings can be regarded as a narrative of the nuclear accident. A narrative corresponds to an oral, written, drawn or ritualised representation of real or fiction events, arranged according to a chronological organisation and forming a consistent whole [9]. A life history is defined as *"a generic expression where one person tells their life or a part of their life to one or more interlocutors. This narrative may lead to a book, a recording or a film"* (Legrand, cited in Burrick [10]).

Yoshida's life history is produced during a "semi-structured interview". This kind of interview is an *"interaction close to conversation, thanks to the continuous adaptation of interrogations and interventions of the researcher to the ongoing exchange"* [11]. In such a narrative, the person interviewed—Yoshida—leads his discourse according to what he considers the investigator's expectations are [12]. The semi-structured interviews also encourage the narrator to digress and tell anecdotes [13].

These deviations from the main narrative plot are useful to add meaning to the story, by providing explanations and/or comparisons. In his account, via the addition of details and information, the narrator intends justifying the sequence of events: this guarantees the overall outline and the intelligibility of the narration. Yoshida selects the events he believes significant and establishes *"specific connexions to provide consistency"* [14].

Yoshida's account is obviously based on his own memories. The facts are then arranged according to the—necessarily subjective—point of view of a major stakeholder of the accident management. Even though the main plot of the account is led by the investigators, the overall meaning is instilled by the interviewee. The manager resorts to his own representational system to build a consistency between the facts reported.

The metaphors used and the reference to his states of mind reveals the complexity and the extent of the troubles that had to be dealt with during the decision-making. It is interesting to note the deviations to the main plot since they enable us to reveal the "absences" in the official investigation reports.

---

[2]These documents are available in Japanese in the following address: http://www.cas.go.jp/jp/genpatsujiko/hearing_koukai/hearing_list.html.

3.2   The Disclosures of the Manager

The analysis of the manager's testimony shows correlations between the accident management and the concept of "engineering thinking in extreme situations". Some of the information it contains is new, and can be divided into three aspects: factual, representational and operational.

The institutional reports present the succession of events in the Fukushima Daiichi plant with an a posteriori posture. They reflect a desire of comprehensiveness and explanation of all phenomena, notably from a technical point of view. This approach makes the reports relate some of the facts as they were deduced by simulation, especially when they concern chronological indications, as they were established by TEPCO. Yet, the Fukushima Daiichi onsite emergency response centre ignored about many of these phenomena until the simulations were performed. Indeed, certain phenomena were presumed to have happened only after the event, for example that a high concentration of hydrogen was the reason for the explosion of the first reactor building. Furthermore, Yoshida tells the investigators many times that he cannot recall some details or that he did not know about facts stated during the interview. This point is important, in particular to understand the uncertainty that reigned at the Fukushima site after the tsunami hit. It means that decisions must have been taken in the absence of knowledge about certain reactor parameters. Furthermore, the emergency response centre had to coordinate multiple tasks simultaneously, which cannot be effectively represented by a classic schema where events follow each other in chronological succession. The decisions and the actions that were taken, with a notable lack of resources, depended on one another. Many important factors came into play, including slow progress in completing tasks, the impact of events on the viability of the site, lessons learned from interventions at other plants, and the mobilization of resources for other activities.

From a representational point of view, some of the reports describe the Fukushima Daiichi reactors one by one. As for the manager's account, it reveals that the emergency response centre had to deal with the whole site at the same time, in order to avoid the deterioration of the different facilities. In addition, the evolution of the emotional state of Yoshida shows extreme complications in the handling of the situation. This can be proven by an excerpt of the hearings. After the earthquake, the concerned employees gathered in the anti-seismic building and established an informal emergency response centre. Since the tsunami warnings, this group considered the probability of an anomaly in the reactors' cooling. However it is the loss of the electric power resources following the tsunami that causes the distress of the group[3]: *"We're so dismayed that we're speechless. For the time being, we're quiet*

---

[3]The plant manager has to warn the authorities in case of an emergency situation, such as the loss of power resources. He then constitutes an emergency response center under his direction; in accordance with the article 10 of the Japanese act No. 156 on the Special Measures Concerning Nuclear Emergency Preparedness.

and we're tackling administrative tasks, as the declaration of loss of all AC power, the much talked about article 10. However, as I told you a bit earlier, as we carry out these administrative tasks, emotionally, we're devastated". The team finds itself, from this moment on, *"in the face of a catastrophe"* [2].

Devastation, nervousness, frustration and fear relate to an emotional anxiety which comes into play in the decision making and the action taking in an accident situation. The third aspect of the information found in the account of Yoshida is complementary to the others. It relates to the actions undertaken in the field. The manager underlines many times the complexity faced to see the tasks through to the end. This inability to have an effect on their work tool, paired up with the lack of understanding and the impatience of the off-site executives, contributes to the extreme situation experienced by the workers [2]. The attempts to vent the reactor 2, mentioned in few institutional reports, illustrate perfectly the gap between the field and the executives. TEPCO headquarters repeatedly issued orders to proceed with venting; although teams worked through the night, no workable solution could be found due to lack of suitable resources. Yoshida states his annoyance with this lack of understanding, which was perceived by the emergency response centre as a failure to acknowledge the work they were doing, and their ongoing efforts. This only served to increase the frustration of teams who could not make the unit respond to their actions. These strong emotions and the role they play in decision making must be taken into account, if we are to have a better understanding of how the accident was managed.

The analysis of the hearings permits to better take some factors inherent to the "extreme situation" concept into account. For instance, the uncertainty of the situation, the lack and inappropriateness of the available resources to deal with the accident, the social and hierarchy pressure, and the powerlessness facing the progressive decay of the facilities can be mentioned.

The factual lacks and the arrangement of facts in the reports do not convey accurately the complexity and the stakes of the crisis as faced by the workers and the onsite emergency response centre. Nevertheless, it should be made clear that the testimony is a reflection of how the manager recalls the story and depends particularly on his memory and the representation he makes of the whole situation. Thus, the account does not correspond to the objective succession of events, as they took place in the Fukushima Daiichi plant following the arrival of the tsunami.

4 Conclusion

The account of the accident by Yoshida is a valuable material to understand the Fukushima Daiichi accident management. The testimony makes up for some factual shortcomings and clarifies some information given by the institutional reports.

The information and the disclosures available in the narrative allow to establish a link between the Fukushima accident management and the concept of "engineering thinking in extreme situations". A more thorough analysis of this document should consequently showcase some pointers of this kind of engineering activities, performed in very hostile conditions with limited resources to mobilise. The analysis of some excerpts could for example show the influence of social pressure—embodied in the repeated demands of the Japanese Government—on the operations undertook to preserve the reactors' integrity.

The narrativisation of the nuclear catastrophe by Yoshida is also interesting to study. This research perspective should underline, for instance, how the manager relates the relation he maintained with the network of stakeholders in the accident's management, and the evolution of their interactions under the threat of physical, psychological and social destruction.

References

1. F. Guarnieri, S. Travadel, Engineering thinking in emergency situations: a new nuclear safety concept. Bulletin of Atomic Scientists **70**(6), 79–86 (2014)
2. F. Guarnieri (ed.), *L'accident de Fukushima Daiichi: le récit du directeur de la centrale*, vol. 1 (Presses des Mines, Paris, L'anéantissement, 2015)
3. S. Travadel, F. Guarnieri, «L'agir en situation extrême», *L'accident de Fukushima Daiichi: le récit du directeur de la centrale, volume 1, L'anéantissement*, ed. by F. Guarnieri (Presses des Mines, Paris, 2015)
4. Investigation Committee on the Accident at the Fukushima Nuclear Power Stations of Tokyo Electric Power Company. *Final Report* (2012)
5. The National Diet of Japan, *The Fukushima Nuclear Accident Independent Investigation Commission* (2012)
6. Nuclear Energy Agency, *The Fukushima-Daiichi Nuclear Power Plant Accident—OECD/NEA Nuclear Safety, Response and Lessons learnt*, NEA No. 7161 (2013)
7. The National Academy of Science, *Lessons Learned from the Fukushima Nuclear Accident for Improving Safety of U.S. Nuclear Plants* (The National Academies Press, Washington, DC, 2014)
8. S. Chaxel, C. Fiorelli, P. Moity-Maïzi, «Les récits de vie: outils pour la compréhension et catalyseurs pour l'action». *¿Interrogations? Revue pluridisciplinaire de sciences humaines et sociales* **17** (2014)
9. J.M. Adam, *Le récit*, P.U.F. «Que sais-je?» (Paris, 1996)
10. D. Burrick, «Une épistémologie du récit de vie» Recherches Qualitatives. Hors-Série **8**, 7–36 (2010)
11. S. Nossik, «Les récits de vie comme corpus sociolinguistique: une approche discursive et interactionnelle». *Corpus*, **10**, 119–135 (2011)
12. P. Brun, «Le récit de vie dans les sciences sociales». *Quart Monde* (2004)
13. M.C. Bernard, «La "présentation de soi": cadre pour aborder l'analyse de récits de vie.» *¿Interrogations? Revue pluridisciplinaire de sciences humaines et sociales* **17** (2014)
14. P. Bourdieu, L'illusion biographique. Actes de la recherche en sciences sociales **62–63**, 69–72 (1986)

On Safety Management Devices: Injunction and Order Use in Emergency Situation

Sophie Agulhon

Abstract This paper aims to introduce two main concepts regarding safety management which are injunction and order. In a first section, those two kinds of communication for action will be defined and distinguished through responsibility repartition criterion. Indeed, while injunction device involves addressee's commitment regarding action design, order device is a less complex one in which a specific authority is responsible of order content in a specific frame while the addressee is generally only responsible of the order content execution. To illustrate those concepts potential, injunction and order contribution to face an emergency situation will be demonstrated through local field management and Headquarter relationship analysis during a crisis exercise of major magnitude in a nuclear fuel cycle industry. As a general conclusion regarding safety management, one would note that injunction use ensures decision-making robustness by subjectivity mobilization, as challenging voices multiplication participates to solid evidence emergence thanks to cross-checking practices. Secondly, the specific result of this demonstration remembers one of the Fukushima-Daiichi management lessons, meaning that in a resilient system, Headquarter tends to communicate with Local Management Team through injunction.

Keywords Emergency situation management · Injunction use · Management devices and relationships · Order use · Nuclear safety

1 Introduction

According to CREAM methodology developed by Hollnagel [1], Human Reliability (HR) depends on three factors: human, technology and organization [2]. As this knowledge can possibly contribute to develop resilient systems, management devices

S. Agulhon (✉)
MINES ParisTech/CRC, Paris 8 Vincennes-Saint-Denis University/LED,
PSL Research University, Sophia Antipolis Cedex, France
e-mail: sophie.agulhon@hotmail.fr

© The Author(s) 2017
J. Ahn et al. (eds.), *Resilience: A New Paradigm of Nuclear Safety*,
DOI 10.1007/978-3-319-58768-4_20

use to overcome unexpected situations seems to be a relevant research topic to deepen. As Blau noticed, two kind of communication management can address to the people managed, depending on their "independence in the performance of [their] duty" [3] and impacting responsibility distribution. Those two kinds of communication are management devices that we would call order and injunction. Both of them are used by an authority demanding something from someone but with a different approach toward responsibility (which can imply notions such as liability or duty).

This contribution defines those two safety management devices called injunction and order and demonstrate their contribution to system recovery in an emergency situation. This "ongoing crisis in which conventional resources are lacking, but societal expectations are high" [4] was a particularly interesting case as explicitation processes and time acceleration effect emphasize how nuclear organizations deal with those issues.

2 Safety Management Devices Definition

2.1 What Is Injunction?

Injunction is a communication triggering action as the addressee should adapt his behavior regarding its message (conformity). This communication comes from an authority and is both binding and relying on its addressee subjectivity [5]; as the addressee is linked to the expected action or to its aim regarding responsibility criterion. Fundamentally, injunction implies a tension between what comes from oneself (autonomy) and what is implemented by external sources (heteronomy) [6]. This phenomenon affects one's identity as no one can predict how far a subject will integrate external things to his subjectivity [7] and experience.

By saying so, one would conclude right by stating that safety injunction is not always or completely defined in time, space and form as shown by its legal evolutions from British Equity system to its 19th to early 20th variations in the United States of America (Stewart 1895; Gregory 1898; Mc Murry 1961) cristalized in the Pullman strikes repression through the *Omnibus Indictment*, and to its actual uses demonstrating that the term 'injunction' has no fixed definition but is determined by its practice (Preston 2012, p. 5). That is why, prevention posters from Oak Ridge Laboratory dating from the Manhattan Project times are still quite relevant for any worker exposed to radiation sources, even though some military elements might have lost some sense since [8].

What is also interesting about injunction as a management device is that there is a wider array of potential issuers than in the order case. So far, three kinds of authorities have been identified as relevant to make an injunction.

The first authority observed is derived from the recognized power one has to direct someone else, such as in hierarchy case. This typical authority has been widely analyzed since management studies beginning, particularly with Henri Fayol description of administrative skills use [9].

The second authority defined comes from the legitimacy inherited from ones' function, in Weber sense [10]. So, experienced workers, specialists, inspectorate and auditors can also make injunctions. We chose here not to use the word "expert" as Blau showed that training purpose in organizations was mainly to make people experts in their respective domains; as we wanted to insist on the role idea which goes beyond knowledge.

The third authority observed results from a commitment. In this configuration, issuer and addressee are parts of the same community of interest and share an aim. That is why the issuer is legitimate to make an injunction and the addressee has to fulfill his duty as a group member. As one can guess, this is why safety culture development is encouraged in nuclear firms.

Finally, observing nuclear industry fieldwork shows that safety injunction use often implies an interesting labor division. Indeed, the issuer; or transmitter regarding its human or non-human status [11]; fixes goals that the addressee has to reach by defining himself means such as structures, equipment, workforce, and so on. So, safety injunction strength and weakness is its capability to rely on its addressee's experience by giving him some latitude to obtain a better individual contribution to safety. However, as nuclear industry also needs precision in several quality aspects, order as a management device can also be very helpful.

2.2   What Is Order?

Order is a time and space framed, oral or written binding communication, coming from an authority detaining a recognized power of direction over the addressee, to which the addressee must obey. In most of the cases, this authority is responsible of the given order result. Obedience and disobedience are not related to the autonomy-heteronomy tension derived from conformism but is a matter of dependence and independence balance. As a matter of fact, obedience in the kind of relationship previously described does not impact the addressee identity in the same way as injunction.

Indeed, as there are objective things showing the addressee's dependence and as the action expected is, apparently, not related to his own willingness, the subject is generally not easily questionable for his acts. As The Grapes of Wrath novel shows [12], when an expropriated farmer asked for who he should shoot to avenge his loss, the answer done by the mended man is that he is just following the owner orders who is just following the bank orders; and so on until the causal chain vanishes in the unknown, making the farmer's quest for a convener absurd.

Furthermore, orders are often combined with injunctions. Even in organizations when orders through short communications were openly favored such as in jail or in Christian schools during XVIIIth and XIXth centuries, Michel Foucault demonstrated the existence of another purpose than getting obeyed quickly.

What was at stake was to place bodies in a little world of signals to which an only and mandatory answer is attached. So, a daily-life order can also be combined with an injunction shaping prisoners and pupils' behavior, training them to react in the exact sense defined [13]. In this case, their individual contribution to performance tends to zero.

As subjects can be both commanded (when management makes them do something using order device) and governed (which means that management guides their actions and consequently modify their behavior by injunction device use) depending on management device choice, and because power relations are generally numerous and of various kinds [14], distinguishing how one is put under pressure and to what extend regarding his responsibility can be quite necessary to face all the expectations one is addressed in a particularly sensitive moment such as facing an emergency situation.

3 Safety Management Devices Contribution to System Recovery

3.1 Crisis Organization Context

In September 2014, a nuclear industry organized a major crisis exercise of 36 h that we will not try to analyze as such. Our demonstration will only focus on something out of all simulation aspects: the relationship between local and national level to manage an emergency situation.

Crisis mode is a simplified organization designed to save time. What should be remembered about this design is that:

- Local Emergency Management Team is responsible of field response to the crisis;
- While its national hierarchy (Headquarter) informs stakeholders and takes specific decisions like internal intervention force deployment;
- As this intervention force is composed of various specialists from other entities with no previous hierarchical link with Local Management Team but who will be placed under its command during field intervention.

As we explained earlier, an authority derived from the recognized power one has to direct someone else can possibly use injunction and order management devices. As time is lacking and precision necessary to get out of the crisis situation, one could have imagined that order would have been the main device used by Headquarter to lead the Local Management Team.

However, our observations note a different result which might clarify one of Fukushima-Daiichi management lesson regarding Yoshida and Prime Minister's coordination unit [15]: injunction can be used to handle uncertainty while order contributes to accelerate the recovery process.

3.2 Recovering with Injunction and Order Use

On the first day, a simulated tornado damaged the nuclear fuel cycle platform in the early afternoon. As no one knew exactly what were the consequences of such natural disaster on the plant, all actors tried to face the crisis in the best way they could think of. In this sense Becker's vision of enactment phenomenon, that is to say ways people find to cooperate for the moment to get to the next step in a specific occasion [16], began to appear.

During the mid-afternoon turn-over preparation conference call between Local Management Team and Headquarter, five issues were highlighted (in no preference order):

- Human assessment;
- Safety assessment;
- Production recovery conditions;
- International Nuclear Event Scale (INES) classification of the event;
- Plant workers evacuation.

As injunction use showed regarding the last point ("This needs to be addressed"); Local Management Team was clearly expected by Headquarter to solve those problems, though Headquarter also ordered "not to waste time" on INES classification.

On the field, as Local Team handles operational responsibility towards crisis management for legal and practical reasons, decision was made to prior human and safety assessment. So, rounds and competencies checking were organized to gather information on damages, assess risks and take back control on source terms. Workers evacuation was done during the night when Plant Management was sure no one would be carelessly exposed to danger.

On the second day, as reliable data were gathered, valuable technical solutions were found such as sprinkling devices and robot use to deal with the most risky situations. When it appeared that the intervention force would be sent in a relatively controlled environment, the Headquarter finally ordered to allocate the internal intervention force to spread uranium powder extraction, a relatively known action. As a consequence, the crisis exercise finished in the expected time and with no human loss due to National or Local effort for system recovery; which might not have been the case if previous decision had been confirmed to send the force right after the tornado instead of triggering its early warning mode for field checking support.

4 Results

First, injunction effectiveness to system recovery in national and local level management relationship has been demonstrated in several ways.

Injunction use contributed to data collect organization as the National level trusted Local Management Team ability to gather adequate means because of geographic position and responsibility repartition. But injunction use also contributed to recruit individual contribution to solution design such as sending a robot to a damaged building to prevent criticity peak consequences on intervention forces.

As a consequence, injunction reduced uncertainty and contributed to an effective internal force deployment through order. So this second management device could help, for its part, to solve the crisis in a clean-cut way.

Second, Headquarter injunction use in its relationship with local management allowed priority fixing, innovative choices but also, to a certain extent, contributed to limit errors due to omission or deny, as even the terrorism hypothesis has been considered. To put it in a nutshell, injunction contributed to an exhaustive situation assessment by cross-checking practices without penalizing field action.

5 Discussion

Choosing wisely between order and injunction management devices during the crisis participates in effective system recovery.

If resilience is a characteristic of a system with elastic behavior which can face disturbances [17]; that is to say a system able to partly absorb human experience through contextualization without rejecting all systemic aspects; knowing more about safety injunction reception could be an important step in for organizations dealing with high risks design.

But how could one understand that, in the one hand, injunction nature implies some result uncertainty because of the addressee regarding reliability criterion and, in the other hand that this device also effectively contributed to safety thanks to this same addressee?

Regarding management, qualifying devices, understanding their logics such as in order and injunction case might lead us beyond finding sole conditions of use.

Injunction use could be an empirical proof that systems are not only meant to be designed according to models [18], and in our case quite causal and narrow ones if we refer to Nancy Leveson and al. analysis: "this confusion of component reliability with system safety leads to a focus on redundancy as a way to enhance reliability, without considering other ways to enhance safety" [19].

But crossing an organizational approach with Professor Kyoko Sato's present book contribution regarding imaginaries referring to Castoriadis philosophy could also be adapted to complete our complex system understanding, as imaginaries are fundamentally out of modelling approaches but intertwined with their works: "reality and rationality" [20].

Acknowledgements The author gratefully acknowledges AREVA Safety Health Quality Environment Department regarding funding, data, analysis and diffusion aspects and UC Berkeley and MINES ParisTech Organizing Committee of the "International Workshop on Nuclear Safety: From accident mitigation to resilient society facing extreme situations".

References

1. E. Hollnagel, *Cognitive Reliability and Error Analysis Method* (Elsevier Science Ltd., 1998)
2. K. Furuta, K. Okano, T. Kanno, T. Morri, S. Shimizu, An incident reporting support system for airline cabin crew, in *24th European Safety and Reliability Conference*, Wroclaw, Poland, September 14–18, ed. by T. Nowakowski, M. Mlynczak, A. Jodejko-Pietruczuk, S. Werbinska-Wojciechowska (Taylor & Francis Group, 2015), pp. 1–8
3. P.M. Blau, The hierarchy of authority in organizations. Am. J. Sociol. **73**(4), 453–467 (1968) (The University of Chicago Press, 1968)
4. F. Guarnieri, S. Travadel, Engineering thinking in emergency situations: A new nuclear safety concept. Bull. Atom. Sci. **70**(6), 73–86 (2014)
5. S. Agulhon, F. Guarnieri, «L'injonction de sécurité comme dispositif organisationnel: le cas d'un atelier de retraitement de combustibles nucléaires», in *Faire l'économie de la dénonciation*, ed. by J.J. Perseil (L'Harmattan, 2015)
6. J. Spurk, *Une critique de la sociologie de l'entreprise: l'hétéronomie productive de l'entreprise, Logiques sociales* (L'Harmattan, Economie, 1998)
7. T.W. Adorno, How to look at television. Q. Film Radio Telev. **8**(3/Spring), 474–488 (1954)
8. S. Agulhon, D. Pecaud, F. Guarnieri, Rethinking nuclear safety management: Injunction as a meta-concept, in *24th European Safety and Reliability Conference*, Wroclaw, Poland, September 14–18, ed by T. Nowakowski, M. Mlynczak, A. Jodejko-Pietruczuk, S. Werbinska-Wojciechowska (Taylor & Francis Group, 2015), pp. 89–97
9. H. Fayol, *Administration industrielle et générale*, 1999 edn. (Broché, 1916)
10. M. Weber, *Economie et société: Les catégories de la sociologie, 1995 edn., Tome 1* (Agora, Pocket, 1921)
11. M. Akrich, «La construction d'un système socio-technique: esquisse pour une anthropologie des techniques», in *Sociologie de la traduction: textes fondateurs*, 2006 edn., ed. by M. Akrich, M. Callon, B. Latour (MINES ParisTech, Sciences sociales, Presse des Mines, 1989)
12. J. Steinbeck, *The grapes of wrath*, 2006th edn. (Penguin classics, Broché, 1939)
13. M. Foucault, *Surveiller et punir* (Editions Gallimard, 1975)
14. J.R.P. French, B. Raven, The bases of social power, in *Studies in Social Power*, ed. by M. Ann Arbor (Institute for Social Research, D. Cartwright, 1959)
15. S. Agulhon, F. Guarnieri, «L'injonction de sécurité comme dispositif de conquête de territoires organisationnels». Prospective et Stratégie **4–5**, 81–100 (Apors Editions, 2014)
16. S. Agulhon, C. Banaon, T. Lepers, M. Ndiaye, T. Nguyen, S. Sangkhavongs Pravong, The creative process: how sociological work really gets done—Rencontre avec Howard S. Becker. Faut LIRSA! (2014)
17. T. Kanno, Human-centered systems resilience, in *DEANS Forum Resilience Engineering 2013*, Paris, France, November 18–20 (2013)
18. A. Marchais-Roubelat, *La décision: Figures, symboles et mythes* (Bibliothèque Prospective, Broché, 2012)
19. N. Leveson, N. Dulac, K. Marais, J. Carroll, Moving beyond normal accidents and high reliability organizations: a system approach to safety in complex systems. Organ. Stud. **30**(2–3), 227–249 (2009)
20. C. Castoriadis, *L'institution imaginaire de la société* (Editions du Seuil, 1975)

The Water Neutron Detector

Alexandra (Sasha) Asghari

Abstract Information gathering and dissemination is a crucial aspect of a resilient society during and after a major disruptive event. Neutron detection is particularly important when spontaneously fissioning isotopes are present, such as following a severe nuclear accident. Historically, most neutron detectors have been based on helium-3. Yet because the supply of helium-3 has greatly diminished in the past decade, it is of international interest to develop non-helium-3 based neutron detectors. The Water Neutron Detector (WaND) provides an efficient, non-toxic, and non-flammable alternative detector method. The WaND system is currently under investigation for the nondestructive assay of spent nuclear fuel to quantify plutonium content.

Keywords Nondestructive assay · Neutron detection · Water Cherenkov · IAEA · Spent fuel monitoring

1 Introduction

1.1 Background

Without an adequate, assured supply of helium-3 (or an effective replacement), IAEA safeguards in particular (and worldwide safeguards in general) will be significantly impaired. The IAEA Workshop on Requirements and Potential Technologies for Replacement of helium-3 Detectors in IAEA Safeguards Applications.

The IAEA has been involved in Japan's efforts to plan and implement decommissioning of the Fukushima Daiichi Nuclear Power Station. For example, at the Government of Japan's request, the IAEA assessed the "Mid-and-Long-Term Roadmap towards the Decommissioning of TEPCO's Fukushima Daiichi Nuclear

A. (Sasha) Asghari (✉)
University of California, Berkeley, USA
e-mail: asghari@berkeley.edu

© The Author(s) 2017
J. Ahn et al. (eds.), *Resilience: A New Paradigm of Nuclear Safety*,
DOI 10.1007/978-3-319-58768-4_21

Power Stations Units 1–4" in an effort to enhance international cooperation and sharing of information and knowledge concerning the accident and the future decommissioning process. The review focused on the safety and technological aspects of decommissioning, radioactive waste management, control of underground water, and planning of the implementation of pre-decommissioning and decommissioning activities. A major aspect of decommissioning is removal of the spent fuel [1, 2].

Neutron detectors are an effective technology to search for signatures of fission because natural sources of neutrons are relatively rare compared to sources of other types of radiation, such as gamma. Neutron multiplicity counters can exploit the burst-like temporal signature of fission events to reveal information about the fissioning sample. For example, the multiplication, (α, n) rate, and effective plutonium mass may be determined. Helium-3-based neutron detectors comprise essentially all neutron detectors currently used by the IAEA, and the US has historically been the primary supplier [3]. The U.S. stockpile of helium-3 plummeted from $\sim$230,000 L in 2001 to $\sim$50,000 L in 2010 due primarily to an increase in radiation portal monitors to combat nuclear smuggling after 9/11 [4]. Because of the connection between the U. S. stockpile of helium-3 and the effectiveness of IAEA inspections, it is of national and international interest to develop non-helium-3 based neutron detectors.

Furthermore, the U.S. National Nuclear Security Administration, via the Next Generation Safeguards Initiative, has identified neutron multiplicity as a priority for nondestructive assay (NDA) of spent nuclear fuel. Methods for direct and accurate measurement of plutonium content in spent fuel requiring fewer unverified a priori assumptions about the fuel matrix are needed. Plutonium measurement in spent fuel using multiplicity counting is a technically challenging problem because the gamma flux from spent fuel quickly overwhelms most neutron detectors. Such measurements would aid in quantifying shipper/receiver differences, determining the input accountability value at reprocessing facilities, and provide quantitative input to burnup credit determination for repositories [5]. Knowledge of spent fuel plutonium content in the case of a nuclear accident would aid in better decision making based on scientific data for a more resilient society.

1.2   Novel Neutron Detector

The WaND (Water Neutron Detector) [6] is a non-helium-3 based neutron multiplicity counter under development at Lawrence Livermore National Laboratory. It is an efficient, stable, non-toxic, and non-flammable solution to some neutron multiplicity counting applications. Neutron multiplicity refers to the number of neutrons emitted per fission event and may be used to determine the effective plutonium mass in a plutonium-bearing sample or fingerprint special nuclear material, such as plutonium or uranium. The advantage of using a neutron multiplicity counter is that the analysis is non-destructive, has the possibility of being done on-site, and is relatively fast. The WaND system is composed of 1 m^3 of pure 18 MΩ deionized water doped

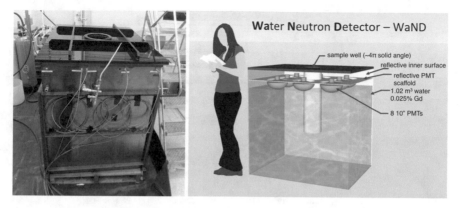

Fig. 1 A photograph (*left*) and 3D model created using Sketchup (*right*) of the WaND system. Note the person for scale

with 0.5% gadolinium-chloride ($GdCl_3$), contained within a stainless steel tank (121.9 cm × 91.4 cm × 119.4 cm). To protect the stainless steel tank from the corrosive water (due to the chloride content), the tank is coated with a baked-on layer of Teflon. Eight 10-in. diameter photomultiplier tubes are mounted on the top of the detector, looking into the detector volume. The inside of the tank is also lined with a 1.0 mm highly reflective (>99% in blue near UV) layer of GORE® DRP® material. Figure 1 shows a 3D model and a photograph of the detector.

The detection mechanism is multi-stepped. A neutron born from a fission event in the sample well must enter the water volume and thermalize. The thermalization distance is ∼35 mm for 1 meV neutrons. Once the neutron has thermalized, it will capture on a gadolinium nucleus. Gadolinium-157 (15% natural abundance) has the highest thermal neutron absorption cross-section of any stable isotope (259,000 barns), and natural gadolinium's thermal neutron absorption cross section is 50,000 barns. Upon capture, the target nucleus enters an 8 meV excited state, then de-excites by emitting a gamma cascade with a total energy of 8 meV. The gammas then Compton scatter electrons in the water, ejecting some at high energies. Electrons that are scattered with a kinetic energy above the Cherenkov threshold (250 keV [7]) will produce a ring of Cherenkov light, which is then detected by the photomultiplier tubes.

2 Results

Both gammas and neutrons produce a detector response in the WaND system. Figure 2 shows the spectral response of a 5.9 microCi cobalt-60 and a 0.82 microCi californium-252 source. The background spectrum is measured by performing a data run without sources. Then the background is statistically subtracted from a source run, leaving only the source contribution.

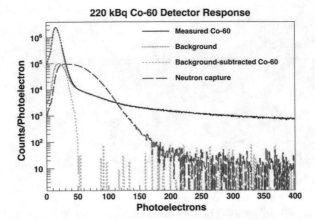

Fig. 2 The detector response spectrum from a 1 h run with a 5.9 microCurie cobalt-60 gamma and a 0.82 microCurie californium-252 neutron source [6]. The *solid black line* indicates the cobalt-60 spectrum prior to background subtraction. The *dotted blue line* is a no-source background spectrum. The *dashed green line* shows the statistical subtraction of the cobalt-60 source and background, which leaves the pure cobalt-60 detector response spectrum. The *dashed red line* shows a pure neutron californium-252 spectrum, also background subtracted

Cobalt-60 is used by national and international organizations as a standard candle for gamma rejection. In our detector, cobalt-60 is a proxy for unwanted low energy background gamma rays that may be associated with a source. Figure 2 shows that it is possible to remove nearly all of these gamma rays by applying a 50 photoelectron energy cut. The absolute neutron detection efficiency with this criteria is 28%, while only 1 part in 10^8 of the background remains. This translates to a sensitivity to ~ 20 to 30 mg of plutonium-240 by measuring the multiplicity distribution [8].

3   Conclusions and Future Work

The WaND system is currently under investigation for possible application to spent fuel monitoring. Spent fuel poses an especially difficult problem for neutron detection because the high intensity gamma field renders most neutron detectors useless. For example, scintillator-based neutron detector systems rely on pulse shape discrimination, placing severe limits on gamma background and neutron signal rate. Germanium or silicon-based detectors are small and susceptible to neutron damage. Boron-based systems such as BF_3 and ^{10}B tubes and planes present toxicity concerns and can be relatively inefficient. The WaND system is efficient, stable, non-toxic, and non-flammable. The effects of high gamma fields and how to mitigate them are currently under investigation. Future work includes experimentally measuring the multiplicity distribution of a plutonium source. The combination of plutonium sensitivity and gamma insensitivity creates a potential to directly measure the plutonium content in spent fuel. In the case of a severe nuclear accident, knowing the content nuclear contamination will aid decision makers in making science-driven decisions on how to better recover.

Acknowledgements The author would like to thank Karl van Bibber, Adam Bernstein, and Steven Dazeley for their mentorship and infinite patience. This material is based upon work supported by the National Science Foundation Graduate Research Fellowship under Grant No. DE-NA0000979.

References

1. International Atomic Energy Agency, Preliminary Summary Report: The IAEA International Peer Review Mission on Mid-and-Long-Term Roadmap Towards the Decommissioning of TEPCO's Fukushima Daiichi Nuclear Power Stations Units 1–4, Tokyo and Fukushima Prefecture, Japan (2015)
2. IAEA Team Completed Third Review of Japan's Plans to Decommission Fukushima Daiichi. International Atomic Energy Agency, 17 Feb 2015, Web
3. M.M. Pickrell, The IAEA workshop on requirements and potential technologies for replacement of helium-3 detectors in IAEA safeguards applications. J. Nucl. Mater. Manag. **41**(2), 14–29 (2013)
4. D.A. Shea, D. Morgan, *The helium-3 shortage: supply, demand, and options for congress, Congressional Research Service* (2010). www.crs.gov
5. S.J. Tobin, et al., Next generation safegaurds initiative research to determine the Pu mass in spent fuel assemblies: purpose, approach, constraints, implementation, and calibration. NIMA **652**, 73–75 (2011)
6. S. Dazeley, A. Asghari, A. Bernstein, N. Bowden, V. Mozin, A water-based neutron detector as a well multiplicity counter. NIMA **771**(0), 32–38 (2015). doi:http://dx.doi.org/ 10.1016/j. nima.2014.10.028
7. G.S. Mitchell, R.K. Gill, D.L. Boucher, C. Li, S.R. Cherry. doi:10.1098/rsta.2011.0271 (Published 28 November 2011)
8. S. Dazeley, et al., Performance characterization of a water-based multiplicity counter, in *55th INMM Annual Meeting in Atlanta, GA. 2014* (for a Journal or Transactions Summary. Trans. Am. Nucl. Soc. **98**, 1200 (2008))

Human Error and Defense in Depth: From the "Clambake" to the "Swiss Cheese"

Justin Larouzée

Abstract After the Fukushima accident, a new concept of nuclear safety arouse: engineering thinking facing extreme situations. One of the specificity of emergency situations being a rise of social demand on engineering process, safety scientist have to make an anti-dualist move in order to improve collaboration between social scientists and engineers. In this aim, this article studies a case of efficient collaboration: the Swiss Cheese Model (SCM) of accidents. Since the early 1990s, SCM of the psychologist James Reason has established itself as a reference in the etiology, investigation or prevention of accidents. This model happened to be the product of the collaboration between the psychologist and a nuclear engineer (John Wreathall). This article comes back on the journey of the SCM and its fathers. It is based on an exhaustive literature review of Reason's work and interviews of Reason and Wreathall carried out in 2014. The study suggests that the success of the model is not so much due to appropriation of the work of the psychologist by the industrial community but to a complex process of co-production of knowledge and theories. To conclude, we try to highlight ways that should encourage, in the future, such collaborative ways of working.

Keywords Swiss cheese model · James Reason · John Wreathall · Coproduction · Collaboration

1 Introduction

The Fukushima nuclear accident that occurred in Japan, March 2011, and its aftermath reinforced the need of theoretic and pragmatic studies over industrial and social resilience. Since there is no end in sight to the accident, it also raised the issue of engineering thinking facing extreme situation [1]. Defined as *"engineering activities*

J. Larouzée (✉)
Centre for Research on Risks and Crises (CRC), MINES Paristech/PSL Research University, Sophia Antipolis, France
e-mail: Justin.larouzee@mines-paristech.fr

© The Author(s) 2017
J. Ahn et al. (eds.), *Resilience: A New Paradigm of Nuclear Safety*,
DOI 10.1007/978-3-319-58768-4_22

that are significantly impeded due to a lack of resources in the face of a societal emergency", this new concept of nuclear safety insist on the link between engineering processes and social contingencies. In this paper we therefore try to shade a light on collaborative process where social scientist and engineers come to work together (more than side to side) in order to seek determinants of successful collaborations. The paper focuses on an historical case: the (so called) *Swiss Cheese Model* of accidents. Since the early 1990, the Swiss Cheese Model (SCM) of the psychologist James Reason has established itself as a reference model in the etiology, investigation or prevention of industrial accidents. Its success in many fields (transport, energy, medical) has made it the vector of a new paradigm of *Safety Science*: the organizational accident. A comprehensive literature review of Reason's work leads us to consider the SCM as the result of a complex (and poorly documented) collaboration process between the fields of research and industry; human sciences and engineering sciences. In a dualistic premise where research and industry would be two entities interacting but still separable, this collaboration would be understood as the appropriation of research work by the industrial world. However, the complexity of the genesis of the SCM forces an overcoming of this dualism to bring out a process of "co-production" of knowledge. As part of this research, the two main "fathers" of the SCM: James Reason (psychologist and theorist of human error) and John Wreathall (nuclear engineer) where interviewed by the author. These meetings shed a new light on a prolific era for the Safety Sciences field. We therefore hope to keep from a retrospective bias that tends to smooth and simplify facts. This chapter deals with the induced effects of the collaboration between a psychologist and an engineer in terms of models production. In the first section, we briefly present the two "fathers" of the SCM and the social and historical context in which their collaboration took place. In the second section, we focus on the effects of this collaboration over their intellectual and scientific productions. Note that prior knowledge of the SCM, its theoretical foundations and its main uses is requested (see, for example Larouzée et al. [2]).

2 The Fathers of the Model

This section presents the two fathers of the SCM. Reason a psychologist of human error and Wreathall a nuclear engineer. After presenting their backgrounds (Sects. 1 and 2), we present the social and industrial context in which they were brought to meet and work to create the first version of the SCM (Sect. 3).

2.1 James Reason, the Psychologist

Reason gets a degree in psychology at Manchester University in 1962. He then works on aircraft cockpit ergonomics for the (UK) Royal Air Force and the US

Navy before defending a thesis on motion sickness at Leicester University in 1967. Until 1976, he works on sensory disorientation and motion sickness. In 1977 he becomes professor of psychology at Manchester University. In 1977, Reason makes a little action slip that will impact his scientific career. While preparing tea, he began to feed his cat (screaming with hunger). The psychologist confused the bowl and teapot. This was of great interest to him and he started a daily errors diary. That's how he started a ten years research on human error which resulted in a taxonomy (1987). After he became a referent on the issue, he was a keynote speaker in various international conferences on human error. During one of these conferences, he met John Wreathall, nuclear engineer, with who Reason built working relationship and "*strong intellectual communion*" (in his words). On their collaboration will be drawn the first version of the SCM. Since then, Reason kept working on human and organizational factors in many industrial fields.

2.2   John Wreathall, the Engineer

John Wreathall studies nuclear engineering at London University, undergraduate in 1969; he gets a masters' degree in systems engineering in 1971. Later he studies an Open University course "Systems Thinking, Systems Practice" based on Checkland's models of systems. This option brings the young engineer to human factors and systems thinking. From 1972 to 1974 he works on the British nuclear submarine design which allows him to access confidential reports on HRA by Swain. From 1976 to 1981, Wreathall works for the CEGB (English energy company), first as design reviewer for control systems then as an engineer on human factors in nuclear safety. As an acknowledged expert he was brought to participate in conferences organized by NATO and the World Bank called *Human Error* (book "*Human Error*" by Senders and Moray is the only published product from the 1981 conference of the same name). After meeting Reason there, they both started professional collaborations on accident prevention models (including SCM). His interest in the human factor brought him to several leading functions where he worked on human factor. Most of his works also were funded by the nuclear industries in the USA, Japan, Sweden, the UK and Taiwan, and by the US Nuclear Regulatory Commission.

2.3   Meeting and Collaboration, a Particular Context

Industrial and research community's interest for human factors is nothing new in the mid-1980s. By the 1960s, development of the nuclear industry and modernization of air transport stimulates many research programs (e.g. Swain 1963; Newell and Simon 1972; Rasmussen 1983; quoted by Reason [3]). Researches then were mostly conducted under the '*human error*' paradigm. The 1980s were marked by a series of industrial accidents (Three Mile Island, 1979; Bhopal, 1984, Chernobyl and Challenger, 1986; Herald of Free Enterprise and King's Cross Station 1987; Piper Alpha, 1988). Investigations following these accidents brought the Safety community to question the understanding of accidents solely based on operator's error. In this scientific, industrial and social context, NATO and the World Bank funded many multidisciplinary workshops on accidents. The first one was held in Bellagio, Italy, 1981. It received the name of "*first human error clambake*".

At Bellagio's Clambake, Reason and Wreathall met. This fortuitous meeting led them to become (in Wreathall words) "*social friends*". Indeed, according Wreathall, "*intellectual communion was quick with Reason but also with other researchers in vogue on the issues of human error at the time. Swain, Moray, Norman*". Reason and Wreathall started corresponding and met at different conferences during the 1980s. Both took commercial projects for industrial groups such as British Airways and US NRC in which they employed each other as professional colleagues. At that time Reason was ending his taxonomy of unsafe acts. He started writing a book on human error aimed to his cognitive psychologist peers. The *Safety Culture Decade* context and choice of reducing first chapter's size brought him into writing a chapter on industrial accidents. Therefore, he intended his book to both the research and the industrial world (he progressively became familiar with thanks to his Wreathall & Co's joint missions as well as others). To communicate his new vision of *organizational* accidents, Reason called on his friend Wreathall to help to design a simple but effective model that would be included in the 7th chapter of *Human Error*. This model was to become, ten years later, the famous SCM.

3 Birth and Growth of the SCM

Section 2 has presented the two SCM's fathers, their backgrounds and the context in which they were brought to meet. This section focuses on their collaboration from 1987 (when the writing of *Human Error* begun) to 2000 (publication of the latest SCM version). We first look back at the discovery and exploitation by Reason of the nuclear field (Sect. 3.1). We then explicit the shift that the psychologist made from fundamental to applied research (Sect. 3.2). Section 3.3 is devoted to the percolation of defense in depth into the SCM. Finally, we look at the developments which led the Wreathall and Reason's early accident model, to become, in 2000 the famous and widely used SCM (Sect. 3.4).

3.1 Reason, Human Error and NPPs

In the late 1970s Reason is still far from the nuclear power plant (NPP) control rooms. Yet this industrial field will be one of the most influential for its work. In

1979, the TMI incident operates an awareness of the influence of local workplace conditions on the operator's performance. While Charles Perrow sees in TMI the advent of a *normal accident*, Reason finds the first level of his taxonomy: distinction between *active* and *latent* errors. In 1985, Reason and Embrey publishes *Principles Human factors relating to the modeling of human errors in abnormal condition of nuclear power plants and major hazardous installations*. One year later, the Chernobyl disaster provides an unfortunate case study. Reason introduces a new distinction between errors and violations in his taxonomy. In 1987, he publishes an article in *British Psychological Society* bulletin devoted to Chernobyl errors' study from a theoretical perspective. In 1988, he publishes *modeling the basic tendencies of human operator error*, thus introducing an error model which allows modeling the human behaviour of problem solving (the Generic Error Modeling System, GEMS). Reason's cognitive models were then based on observations in NPPs control rooms as case study of human behavior.

The development of distinction between accidents theories based on active or latent errors and violations, is strongly linked to the development of nuclear energy and its safety culture. From 1979 to 1988, Reason uses accident investigations and gets used to the field and its culture. For all that, his productions remains designed to his peers. A turning point is met when the observation process becomes a collaborative one and that Reason's psychologist work mingles with the engineering one of Wreathall.

3.2 From Fundamental to Applied Research

1987 represents a break in Reason's work [2]. After studying everyday errors for ten years, Reason holds a major contribution to his discipline with the taxonomy of unsafe acts [3, p. 207]. He publishes the *Generic Error Modelling System* ([4] Fig. 1a), a combination of his classification with the *Skill Rule Knowledge* model of Danish psychologist Rasmussen [5]. It presents the types of human failures linked with the specificities of a given activity. This theoretical cognitive model still belongs to the field of psychological research (model quoted 192 times).

The same year, Reason works on a chapter of *Human Error* dedicated to industrial accidents and designed for security practitioners. He has the backing of his friend Wreathall. Reason says he looked for a manner of *"showing people what our work was about"*. Wreathall talks in these terms of the genesis of the first model *"during an exchange in a pub (the Ram's Head) in Reason's home town (Disley Cheshire, England), we have drawn the very first SCM on paper napkin. Initially, James saw the organizational accident as a series of "sash" windows opening or closing thus creating accident opportunity"*. Wreathall allowed the psychologist to combine his accident theory (resident pathogens metaphor; [6]) and his error taxonomy with a pragmatic model of any productive system.

The shift over, the cognitive and theoretical model changed into a descriptive and empirical one (Fig. 1b). The book *Human Error* received a warm welcome by

(a)   **(b)**

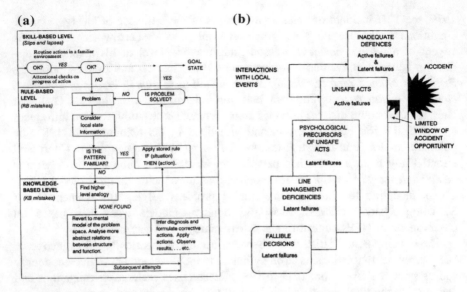

Fig. 1 Reason's taxonomy backed **a** at the cognitive SRK model by Rasmussen produces a theoretical model; **b** at the Wreathall's productive system's model produces an effective descriptive model

both research and industrial communities (quoted 8604 times). Reason became a Wreathall & Co' director and continued his work related to industries "*he supported psychological dimensions of the reports produced by the firm. As early as 1991* according to Wreathall, *James was familiar with the engineering community and became conductor of the various works made by Wreathall & Co', especially for the American nuclear domain*". Reason will remain a part-time collaborator of Wreathall & Co' and then WreathWood Group until he retired in 2012.

3.3   The Defense in Depth Contributions

The engineer's contribution goes beyond the pragmatic modeling of a productive system. Wreathall's training and experiences with the British submarines nuclear reactor and CEGB NPPs' safety gave him specific defense in depth[1] thinking. When

---

[1]Early 1960, the military 'defense in depth' concept is introduced into the US nuclear safety policies. It concerns the hardware and construction design (fuel and reactor independent physical barriers containment). The TMI incident extends it to human and organizational dimensions. In 1988, an International Atomic Energy Agency working group publishes an issue entitled Defense in depth in nuclear safety [7] which establishes defense in depth as a doctrine of nuclear safety. Doctrine based on three concepts: barriers (implementation of physical protection systems), defensive lines (structural resources and organizational security), and levels of protection

he designed the first SCM, Wreathall chose a representation of superimposed plates. These plates evokes defense in depth's *levels of protection*. Reason then explains each plate's failure using his taxonomy and understanding of organizational accidents. The *Swiss cheese* nickname and representation is late. Still it's rooted in the first graphical choice. Wreathall's contribution overtakes engineering understanding of a system: it carries the defense in depth thinking.

Defense in depth is clearly mentioned in an early SCM version ([3, p. 208]; Fig. 2a).[2] It incorporates an accidental *trajectory of accident opportunity* which provides information on respective contributions of the psychologist and the engineer. On the left hand, the white plates represent the organizational (managerial level) and human failures (unsafe acts): contribution of the psychologist. On the right hand, gray plates represent defense in depth as a block (set of defenses ensuring the system's integrity): it's the engineer contribution. Human variability may confuse the engineer (which partly explains the historical *human error* understanding of accidents). On the other hand, technical and organizational sides of safety often confuse academic researchers. In the SCM, disciplines collaboration is used to display the complex interactions between humans and technology and therefore, emergent properties of system's security (Fig. 2b). Finally, the differences in graphical complexity between the theoretical and empirical models are to be noted. In the next section, we will argue that the success of the SCM also lies in the choice to simplify the drawing in a heuristics release.

3.4 SCM Evolutions

Reason and Wreathall kept working together and using the SCM within Wreathall & Co's reports. A little after 1993 Wreathall suggests replacing "*latent error*" (referring to organizational failures) by "*latent conditions*". This change acknowledges the fact that efficient decision at a given time may have negative outcomes at another time or place in the system but these decisions may not be wrong at the time —they are just made under uncertainty. In addition to these semantic changes, SCM graphically evolves (over 4 times in the 1990s). Its use reached many sectors such as energy or transportation [11]. During 1990s, Rob Lee, director of the Australian Bureau of Air Safety Investigation, suggested representing gaped barriers as Swiss

---

(Footnote 1 continued)

(arrangement of barriers and defensive lines according to structured objective regarding the potential event's gravity).

[2]If the original version labels defense in depth (Fig. 2a), the 1993 French translation (by an academic) changes the label for «défenses en série» (serial defenses). Loss of sense due to field sensitivities' manifestation.

(a)

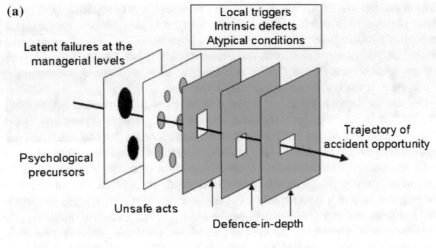

(b)

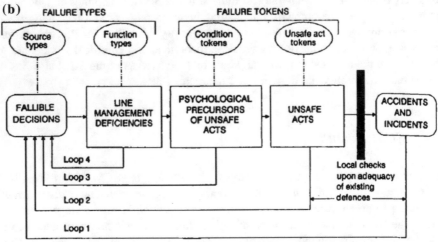

Fig. 2 a The accident causation model published in 1990 explicitly introduced the defense in depth concept. **b** A more complex representation showing the interactions between human and technical dimensions of the system

cheese slices [9]. The idea attracted Reason, then working on a new SCM version for the *British Medical Journal* ([8], Fig. 3). This was a landmark article (quoted 3442 times) and in 2003 Reason was appointed *Commander* of the British Empire for his work on patient safety. The SCM was born. Its simplicity and empirical pragmatism made it the vector of a new paradigm of Safety: *the organizational accident.*

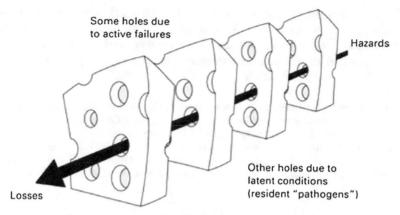

Some holes due
to active failures

Hazards

Other holes due to
latent conditions
(resident "pathogens")

Losses

Successive layers of defences, barriers and safeguards

Fig. 3 SCM version where the cheese slices represent a system's defenses [8]

4 Discussion

A detailed study of the SCM is both simple and complex. Simplicity comes from the abundance of sources. This model has been widely quoted and Reason is a prolific author (149 publications; [10]). Complexity arises from the nature of the model's origin: a collaborative and poorly documented work between distinct but interactive worlds, research and industry. Meeting the two fathers of the SCM was a great help, it surely helps preventing from retroactive bias.

This study was guided by intuition that the success of SCM lays (mostly) in its simple graphical representation. If it is undeniable that *Swiss cheese* representation has played a role in the socialization process of Reason's work, it actually seems it has mostly caused theoretical and methodological pitfalls [11]. A second hypothesis was that success of the model was the result of the appropriation of research findings by industry. It emerges that it is more the appropriation of industrial experience by the academics and long term collaboration that gave the SCM its empirical pragmatism, likely to encourage its use and spread. If Reason and Wreathall's meeting was helped by a favorable social and industrial context (Safety Culture decade and human error clambakes), their collaboration stood thanks to a mutual will of convergence. We note the importance of backgrounds and early life experiences that led Reason working in aviation community and Wreathall meeting systemic thoughts and human factors early in his studies. This shared background guaranteed sensitivity and brought a common language to the two: a collaboration prerequisite. Finally, more than simply causing their meeting, the social demand at that time (industry funding many research programs) also allowed the evolutions of the model. Through various research programs and industrial demands, the SCM was used and shaped.

The SCM took time to evolve and meet industrial (and in a way, social) demand. As we tried to demonstrate here, the essence of its efficiency is cross-disciplinary background and collaboration. We must now use these assets as a mean to address extreme situation so one can operate quick, innovative and pragmatic solutions when unfortunately faced with it.

Acknowledgements My thanks to Professor Reason for his hospitality, on 7 January, 2014, and to Mr. Wreathall for coming and meeting us at MINES ParisTech Centre of Research on Risk and Crises, on 10 October, 2014. Their answers provided valuable insight to our work. Great thanks to Professor Frank Guarnieri for his unconditional support and precious advices. Thank you to Professor Joonhong Ahn for hosting the International Workshop on Nuclear Safety: *From Accident Mitigation to Resilient Society Facing Extreme Situations*, at the University of Berkeley, California, March 2015.

References

1. F. Guarnieri, S. Travadel, Engineering thinking in emergency situations: A new nuclear safety concept. Bulletin of Atomic Scientists **70**(6), 79–86 (2014)
2. J. Larouzee, F. Guarnieri, D. Besnard, *Le modèle de l'erreur humaine de James Reason.* Papiers de Recherche du CRC, MINES ParisTech, 44 pp. (2014)
3. J. Reason, *Human Error* (Cambridge University Press, 1990)
4. J. Reason, *Generic error-modelling system (GEMS): a cognitive framework for locating common human error forms.* New Technol. Hum. Error **63** (1987)
5. J. Rasmussen, *Skills, rules, and knowledge; signals, signs, and symbols, and other distinctions in human performance models.* IEEE Trans. Syst. Man Cybern. **SMC-13**(3) (1983)
6. J. Reason, *Resident pathogens and risk management.* W.B. Workshop Safety Control and Risk Management (1988)
7. INSAG, *Defense in depth in Nuclear Safety.* International Nuclear Safety Advisory Group Report (INSAG) IAEA (1996)
8. J. Reason, Human error: models and management. BMJ **320**(7237), 768–770 (2000)
9. J. Reason, E. Hollnagel, J. Paries, *Revisiting the "Swiss Cheese" Model of accidents.* ECC No. 13/06 (2006)
10. J. Larouzee, F. Guarnieri, *Fond bibliographique Sir James Reason – une vie dans l'erreur.* Papiers de Recherche du CRC, MINES ParisTech, 12 pp. (2015)
11. J. Larouzee, F. Guarnieri, *Huit idées reçues sur le(s) modèle(s) de l'erreur humaine de Reason.* Revue d'Electricité et d'Electronique (2014)

Criticality Safety Study for the Disposal of Damaged Fuels from Fukushima Daiichi Reactors

Xudong Liu

Abstract This paper summarizes our previous works on neutronics analysis for the disposal of damaged fuels from Fukushima Daiichi reactors. Three major stages have been identified for the criticality safety assessment after disposal. In order to evaluate the criticality safety for certain repository conditions and engineered barriers designs, neutronics models have been defined for different stages, and numerical results have been calculated by a Monte-Carlo code MCNP. For stages when fissile nuclides in the damaged fuels remains in the vicinity of the engineered barriers, the neutron multiplicity (k_{eff}) for a canister containing fuel debris surrounded by buffer was calculated over the leaching time. For the stage when fissile nuclides originated from multiple packages deposit in far-field host rocks, the critical masses for uranium depositions were studied for various rock types and geometries. The methodology presented in the present paper could be further improved and utilized to assist the repository system design and criticality safety assessment in the future.

Keywords Criticality safety · Geologic disposal · Damaged fuels · Fukushima accident · Radioactive waste management

1 Introduction

The accident at the Fukushima Daiichi Nuclear Power Station in March 2011 generated damaged fuel in three crippled reactors, containing nearly 250 metric tons of uranium and plutonium along with fission products, minor actinides, and other materials such as fuel cladding, assemblies, and in-core structural material [1]. The damaged fuels will have to be disposed of in a deep geological repository. For a prospective repository, a criticality safety assessment (CSA) should be performed to ensure that the repository system including the engineered barriers and far-field

X. Liu (✉)
Department of Nuclear Engineering, University of California, Berkeley, CA, USA
e-mail: xdliu@berkeley.edu

© The Author(s) 2017
J. Ahn et al. (eds.), *Resilience: A New Paradigm of Nuclear Safety*,
DOI 10.1007/978-3-319-58768-4_23

geological formations remains sub-critical for tens of thousands to millions of years. For various repository concepts, CSA is considered to include three major stages in a chronological order: (1) the stage before package failure, (2) the stage after package failure, while fissile nuclides remain within the engineered barriers, and (3) the stage in which fissile nuclides originated from multiple packages deposit in far-field host rocks.

This paper summarizes our previous works [2, 3] on neutronics analysis for the disposal of damaged fuels from Fukushima Daiichi reactors, during the three stages in CSA. Current understanding about the conditions of the damaged fuel is very limited, and the location and design of the repository have not been determined. Therefore, the primary objective of our study is to establish a consistent methodology to evaluate the criticality safety for certain repository conditions and engineered barriers designs. The methodology could be further improved and utilized to assist the repository system design and criticality safety assessment in the future. For stages (1) and (2), neutronics analysis for the engineered barrier region consisting of a single waste package containing damaged fuel debris, failed overpack and the buffer materials [2] will be reported in Sect. 2. For stage (3), our study on the criticality conditions for uranium depositions in geological formations resulting from geological disposal of damaged fuel [3] will be reported in Sect. 3.

2   Neutronics Analysis on Engineered Barrier System Containing Damaged Fuel Debris

2.1   Model and Assumptions

The repository is assumed to be in a water-saturated reducing environment. The neutronics model consists of a canister containing fuel debris from Fukushima Daiichi Unit 1 reactor and the buffer surrounding the canister. Because there is no current design for the disposal system for the damaged fuels, the composition and dimension of the canister and buffer are assumed based on the design for spent fuel disposal [4]. The damaged fuel is assumed to be disposed of after 50 years of cooling. The fuel composition after the accident was calculated by burnup code ORIGEN, which was reported in [1]. Gaseous, soluble, and volatile neutron absorbing nuclides in the fission products (such as Xe and Cs) might have been separated from the fuel and released during and after the accident [5]. Therefore, in this study, only physically and chemically stable, and strongly neutron absorbing nuclides in fission products are considered, which include Gd, Nd, Sm, Rh, and Eu isotopes.

The present work considers six nominal time steps for neutronics analysis: the emplacement time (t = 0), the canister failure time (t = T_f), and four steps during the dissolution of debris particles (t = T_f + 0.2 T_l, t = T_f + 0.4 T_l, t = T_f + 0.6 T_l, and t = T_f + 0.8 T_l). At t = 0, the canister only contains fuel debris. The failure time

Fig. 1 Schematic layout of the neutronics model of the engineered barrier system containing damaged fuel debris

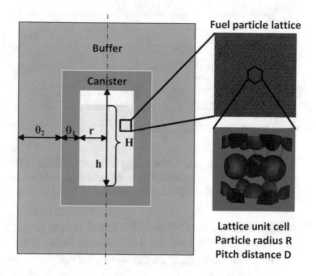

(T_f) of the carbon steel canister is assumed to be 1000 years. After canister failure, water fills the canister, and the canister is modeled as a porous medium with porosity of 0.3. The geometry of neutronics model for the damaged fuel debris at different time steps have been built based on our literature review on defueling process for the Three Mile Island (TMI) accident [6]. A hexagonal lattice of spherical fuel particles is assumed. The pitch distance between particles is assumed to be either (1) make particles contact each other or (2) make the particles lattice fully fill the canister. In the leaching steps, the released materials from the damaged fuel particles is assumed to be either (a) removed from the canister-buffer system, or (b) be homogeneously mixed with the corroded canister. Combinations of the above variations makes four cases: case 1a, case 1b, case 2a, and case 2b. The schematic layout of the MCNP model is shown in Fig. 1. The engineered barriers consist of a carbon-steel canister surrounded by buffer (a mixture of bentonite and silica sand). The canister is filled with spherical fuel particles in a hexagonal lattice. The unit cell of the lattice is shown in the right bottom of Fig. 1. More detailed descriptions about the model parameters can be found in [3].

2.2 Summary of Numerical Results

The numerical results were calculated by a Monte-Carlo code MCNP [7], and are shown in Fig. 2, where the neutron multiplication factor k_{eff} is plotted against the nominal time steps for various combinations of cases and initial loadings. Note that the time axis only represents the order of the time steps and does not represent the actual time. The failure time (1000 years) should be several orders of magnitude

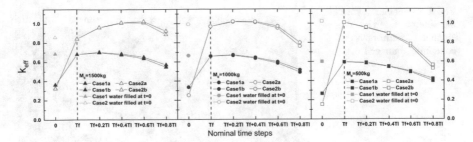

Fig. 2 Calculated k_{eff} for various cases versus time. Solid points represent case (1), hollow points represent case (2). Red points represent case (a), and blue points represent case (b). Squares, circles, and triangles represent initial loading of 500 kg, 1000 kg, and 1500 kg, respectively

smaller than the leach time. The green points in three figures in Fig. 2 represent cases assuming the canister is filled with water at time zero.

The major findings from the numerical results include, (a) the calculated neutron multiplication factor (k_{eff}) is sensitively dependent on assumptions related to moderation, (b) the carbon steel canister plays an important role in reducing the potential for criticality, (c) the maximum k_{eff} of the canister-buffer system could be achieved after a fraction of fissile nuclides been released from the canister, and (d) under several assumptions, the maximum k_{eff} of the canister-buffer system could be principally determined by the dimension and composition of the canister, not by the initial fuel loading.

3   Conditions for Criticality by Uranium Deposition in Water-Saturated Geological Formations

3.1   Model and Assumptions

From previous discussions, in stage (3), a plume of uranium-bearing groundwater originated from multiple canisters containing damaged fuels could form a uranium deposition in geological formations in the far-field. The deposition could locate in either porous or fractured rock. Because the size of the uranium deposition in porous rock is of the order of the grain size of those rock is much smaller than typical neutron mean-free-path, we can consider that uranium deposition in porous rock is homogeneously mixed with rock and water in a neutronics model. For deposition in the fractured rock, two different configurations have been considered for the mixing between uranium deposition and water in the fracture (i.e. fully mixed or fully separated).

Figure 3 shows the schematic of the MCNP model, in which the spherical core is filled with one of the three different geometries (shown right), surrounded by the one-meter-thick rock as reflector. The combination of rock, water, and heavy metal is expressed by two independent variables: void volume fraction (VVF) and

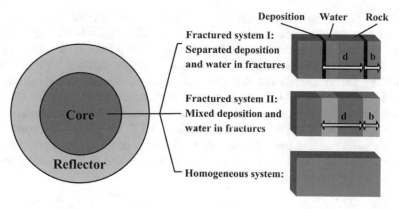

Fig. 3 Three geometries for the MCNP simulations: (1) fractured system I, (2) fractured system II, and (3) homogeneous system

heavy-metal volume fraction (HMVF). For the heterogeneous systems, the VVF is given by b/d, representing the averaged fracture volume fraction, or the fracture porosity in rock. For the homogeneous system, VVF represents the void space fraction that is filled with water and heavy metal precipitations, equivalent to the porosity of a porous rock. The HMVF is defined in a similar way, representing the volume fraction of heavy metal precipitations in the entire core. The volume fraction of the solid-phase of the rock then equals to (1-VVF), and the water volume fraction is given by (VVF-HMVF). By definition, the HMVF must be smaller than VVF, because the volume of precipitation cannot exceed the available void space in the rock.

Two types of host rocks are considered in the present study: average sandstone and magnetite-hematite-bearing pelitic gneiss containing 15% iron. For the heterogeneous systems, the fracture aperture takes values of 0.1, 0.2, 0.5, 1.0, 2.0, 3.0, 4.0, 5.0, and 10.0 cm. For given compositions and geometry for rock and heavy metal, calculations have been first performed for various VVF and HMVF parameters, assuming that the mass of heavy metal in the core is 250 MT, which is the total mass of the damaged fuels form three reactor cores. The discrete k_{eff} results have been used to generate a k_{eff} contour plot by interpolation. By defining a nominal sub-criticality criterion $k_{eff} < 0.98$, the super-critical region can be determined in the parametric space. Within the super-critical parameter range, MCNP calculations have been conducted to obtain the critical mass of heavy metal deposition. More detailed descriptions about the model parameters can be found in [2].

3.2 Summary of Numerical Results

The numerical results for the effective neutron multiplication factor k_{eff} for the deposition containing 250 metric tons of uranium are shown in Fig. 4 (a) and (b) for two types of host rocks. The contour line in red color, referred to as the critical contour line, indicates the nominal criticality criterion, $k_{\text{eff}} = 0.98$. The triangular region results from the fact that the HMVF cannot be greater than VVF. In either case of rock, the k_{eff} value tends to be greater for a greater value of VVF (i.e., to the right along the horizontal axis). A maximum k_{eff} is observed as HMVF increases for a fixed VVF. If the VVF is 0.094 or smaller for sandstone Fig. 4 (a) and 0.265 for iron-rich rock Fig. 4 (b), then the uranium deposition is always subcritical. We call this threshold VVF as the minimum critical VVF hereafter. The comparison between sandstone and iron-rich rock shows importance of rock compositions. For the iron-rich rock, the likelihood of criticality event would be significantly smaller because iron strongly absorbs neutron.

Fig. 4 **a** k_{eff} contour plot for fractured sandstone with fractured geometry I. **b** k_{eff} contour plot for iron-rich rock with fractured geometry I

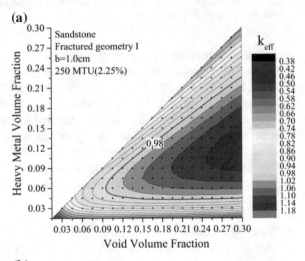

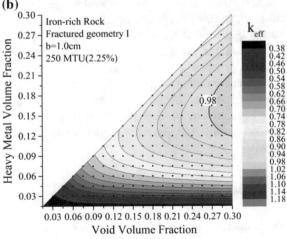

Fig. 5 a Critical mass contour plot for fractured sandstone. **b** Critical mass contour plot for homogeneous sandstone. The values in the figure and in the side-bar scale are logarithm of MT of uranium included in the system

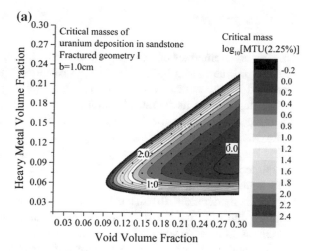

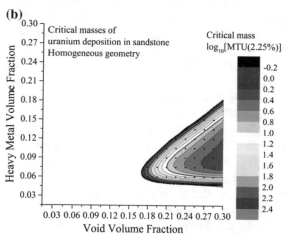

The minimum critical VVF can be found similarly for every combination of rock type, geometry, and a certain mass of uranium deposition. For both rock types with fractured I geometry, the minimum critical VVF becomes the smallest at aperture $b = 1.0$ cm. Figure 5(a) and (b) show the contour plots for the critical mass for sandstone comparing two geometries as indicated within the figures. The boundary of the plot is extracted from the red contour line from the k_{eff} results Fig. 4 (a) and (b). The values for critical masses are shown in a logarithm scale in unit of metric tons of uranium, and the contour lines for 1, 10, and 100 metric ton are shown in black, red, and blue, respectively.

From the numerical results, we can conclude that, the k_{eff} for the deposition become greater with (1) smaller concentrations of neutron-absorbing materials in the host rock, (2) larger porosity of the host rock, (3) heterogeneous geometry of the deposition, and (4) greater mass of uranium in the deposition.

4 Conclusions

This paper summarizes our previous works on neutronics analysis for the disposal of damaged fuels from Fukushima Daiichi reactors. Three major stages have been identified for the criticality safety assessment after disposal.

For stages when fissile nuclides in the damaged fuels remains in the vicinity of the engineered barriers, the k_{eff} for a canister containing fuel debris surrounded by buffer was considered over the leaching time. Based on literature review, the fuel debris has been modeled as a hexagonal lattice of spherical fuel particles. Based on the numerical results, the following key observations can be made: (a) the calculated neutron multiplication factor (k_{eff}) is sensitively dependent on assumptions related to moderation, (b) the carbon steel canister plays an important role in reducing the potential for criticality, (c) the maximum k_{eff} of the canister-buffer system could be achieved after a fraction of fissile nuclides been released from the canister, and (d) under several assumptions, the maximum k_{eff} of the canister-buffer system could be principally determined by the dimension and composition of the canister, not by the initial fuel loading. Future works in this area are planned to apply the present approach for damaged fuels from Unit 2 and Unit 3, to consider more modes for release from the canister, such as leaching of the damaged fuels by reducing the radius of each fuel particle, to consider buffer swelling or collapsing due to degradations, and to develop detailed models to connect the models for single canister with models for the deposition from multiple canisters. The dependence on model parameters, such as fuel particle radius, need to be further examined. We will also investigate the option of using backfilling materials to control criticality in the engineered barrier design.

For the stage when fissile nuclides originated from multiple packages deposit in far-field host rocks, the critical masses for uranium depositions were studied for various rock types and geometries. The analysis has been made for two kinds of rocks by considering a finite system with three different geometries, containing various masses of uranium. The three different geometries include heterogeneous (fractured I and II) and homogeneous systems. The exploration was performed to find optimized combinations of geometry, fracture aperture and the model parameter HMVF, to give the minimum rock porosity (VVF) for criticality. The numerical results show that: the k_{eff} for the deposition become greater with (1) smaller concentrations of neutron-absorbing materials in the host rock, (2) larger porosity of the host rock, (3) heterogeneous geometry of the deposition, and (4) greater mass of uranium in the deposition. After the present analysis, we conclude that various far-field critical configurations are conceivable for given conditions of materials and geological formations. Whether any of such critical configurations would occur in actual geological conditions remains unanswered. To answer this question, we need to extend the present study into the following directions. First, from the neutronics point of view, a more "realistic" fractured system with both the fracture orientation and size randomly distributed is suggested. Second, we need to perform the mass transport analysis to explore whether such a configuration obtained by neutronics analysis is likely to be occurred in geological formations.

Acknowledgements This study was carried out under a contract with METI (Ministry of Economy, Trade and Industry) of Japanese Government in fiscal year of 2012 as part of its R&D supporting program for developing geological disposal technology.

References

1. K. Nishihara, H. Iwamoto, K. Suyama, *Estimation of Fuel Compositions in Fukushima-Daiichi Nuclear Power Plant*. JAEA-Data/Code 2012–018, Japan Atomic Energy Agency (2012)
2. X. Liu, J. Ahn, F. Hirano, Conditions for criticality by uranium deposition in water-saturated geological formations. J. Nucl. Sci. Technol. **52**(3), 416–425 (2014)
3. X. Liu, J. Ahn, F. Hirano, *A Criticality Safety Study for the Disposal of Damaged Fuel Debris, Proc. IHLRWM 2015*, Charleston, South Carolina, April 12–16 (2015) (Accepted)
4. Japan Atomic Energy Agency (JAEA), *Report on Technology Development for Direct Disposal of Spent Nuclear Fuel, Project for Technology Investigation for Geological Disposal*, (in Japanese) (2014), http://www.enecho.meti.go.jp/category/electricity_and_gas/nuclear/rw/library/2013/25-11-1.pdf
5. *Reflections on the Fukushima Daiichi Nuclear Accident*, pp. 51–84, JOONHONG AHN, CATHRYN CARSON, MIKAEL JENSEN, KOHTA JURAKU, SHINYA NAGASAKI, and SATORU TANAKA Ed., Springer (2014)
6. *Three Mile Island Nuclear Station Unit II Defueling Completion Report*, GPU (1990)
7. X-5 Monte Carlo Team, MCNP - Version 5, Vol. I: Overview and Theory, LA-UR-03-1987, (2003)

Rational and Non-rational Influence in a Time-Constrained Group Decision Making

Dipta Mahardhika, Adrián Agulló Valls, Taro Kanno
and Kazuo Furuta

Abstract When humans make decisions, they tend to rely on the heuristic approaches, instead of considering all available facts. When humans need to make decisions as a group, this tendency also seems true. However, there are some additional mechanisms that can only be observed in the group level, which are influence and conformity. Understanding these mechanisms and their process patterns is necessary to interfere and manipulate a group decision making in order to make a good group decision. This is particularly critical in emergency situations where decision making needs to be done under time and risk pressure. This paper proposes a model of group decision making process using I-P-O model, emphasizing the influence process in the group. Besides, this paper also explains an analysis towards a group decision making experiment in laboratory setting. The discussion process was observed to find the influence pattern among the members.

Keywords Group decision making · Group conformity · Influence · Personality

1 Introduction

One important aspect in resilience engineering is how an organization responds to changes that happen from either within or outside the organization. When a change happens, the organization needs to shift from its usual routine into a new action. Changing from the routine into a new condition requires some decision makings.

D. Mahardhika (✉) · T. Kanno
Department of Systems Innovation, University of Tokyo, Tokyo, Japan
e-mail: dipta@cse.t.u-tokyo.ac.jp

A.A. Valls
Industrial Management Engineering, Universitat Politècnica de Valencia, Valencia, Spain

K. Furuta
Resilience Engineering Research Center, The University of Tokyo, Tokyo, Japan

© The Author(s) 2017
J. Ahn et al. (eds.), *Resilience: A New Paradigm of Nuclear Safety*,
DOI 10.1007/978-3-319-58768-4_24

279

In the modern world, most of that decisions should be made by a group instead of a single individual. As pointed out by Stasser and Dietz-Uhler [1] even though group is far from perfect, their choice, judgment, and solutions are generally better than individuals.

Individual decision making has been studied intensively in many fields, including psychology, economics, politics, and so on [2, 3]. Group decision making, however, is relatively a newer and less matured subject compared to the studies of individual decision making. Group decision making is more complex, since it is both involving the information processing by its member individually and the social process between the members. This social process is not just including information exchange between the members but also how each member influences on other members' agreement towards the group decision.

Reaching a consensus in a group decision making is not just the norm of the group, but also triggered by human's natural response against cognitive dissonance. Cognitive dissonance happens when group members discover that they do not agree with other members [4]. This is such an unpleasant state that people are motivated to take steps to reduce it [5]. Some of the steps done by the members are changing their own position (conform) or trying to change other members' position (influence). These processes sometimes happen unconsciously because the nature of social influence is very complex [6, p. 221].

In this paper, the dynamics of conforming and influencing process in group decision making is observed. Even though conformity and influence are affected by many situational characteristics [6, p. 211], in this paper it is argued that in general, it can be divided into rational and non-rational influence. This research aims to find the pattern of rational and non-rational influence in several short decision making discussions.

2 Theoretical Perspective

To make a good decision, group members should thoroughly and carefully consider as much information as possible relevant to the problem. When a person makes a decision as an individual, however, a heuristic approach is usually used due to human's limited cognitive ability. One often relies on simple, fast, and easy-to-access heuristics [7]. This situation also happens when a group makes a decision. Even though there are more resources to process information, group members still need to rely on heuristics such as other people's opinion instead of information from actual sources.

To illustrate the group decision making process and observe the social influence, input-process-output model is used. The model is explained in the following subsection.

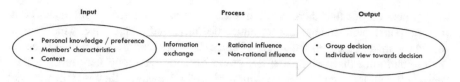

Fig. 1 The IPO model of influence in group decision making

2.1 Conceptual Model

As shown in Fig. 1, the inputs consist of three entities. Personal knowledge or preference is the individual's knowledge relevant to the problem being discussed, and also their personal preference regarding the decision. The difference in this part will trigger the discussion process until a consensus is reached. Members' characteristic is the personal traits that will determine the dynamics of the discussion process, mainly the non-rational one. Some examples of this characteristic factors are introversion/extroversion and self-esteem [6, p. 214]. The context defines the situational characteristics of the decision making. For example, in an emergency situation, usually the group is pressured to make a unanimous decision in a relatively short time.

In the process, the members exchange information, influence, and conform to each other. In this model, 'information exchange' is different from 'rational influence' due to its neutral property. In 'information exchange', the information is exchanged merely for letting the others know about it. Meanwhile, 'rational influence' is described as a process when a person use information or logical analysis to convince other people towards a certain decision. Rational influence is used because naturally people are motivated to interpret and perceived the reality as accurate as possible [8]. Other factors besides rational influence that can affect conformity are defined as non-rational influence. One example of this is a persistent statement of a person without providing any argument, or when someone emphasizing the time constraint to force other members to conform, or the pressure that exists due to the opinion of the majority.

The output of the group decision making process are the group decision and the individuals' view towards it. Even though the group has reached a consensus, it does not mean that all members also privately agree with the decision. This will be explained in the next subsection.

2.2 Social Response in Group Decision Making

Like all other social processes, a group decision making is involving social influences—interpersonal activities that change other people's thoughts, feelings, or behaviors [6, p. 203]. In a group decision making, Nail and MacDonald [9] (as cited in [6]) pointed out that there are five social responses related to influence: compliance, conversion, independence, anticonformity, and congruence. That categorization is made based on the status of agreement between the individual decision (before and after discussion) and the group decision. A group may reach a consensus after all the members agree with it *publicly*. However, they may or may not *privately* agree with it. If they agree both publicly and privately, it is called conversion. Otherwise it is called compliance. However, if they have agreed even before the discussion, it is called congruence. Independence is when a person disagrees with the group decision at all time. Anticonformity is when a person initially agree (or neutral) but afterward disagree with the group.

3 Experiment, Data, and Analysis

3.1 Experiment Design

The purpose of the experiment is to find the pattern of rational and non-rational influence in regard to three different responses of group decision making: compliance, conversion, and congruence. The data were taken from a previous research about mutual belief model [10]. In that experiment, 21 groups of size 3, 4, and 5 persons were asked to give ranks to 15 items (such as oxygen, water, rope, etc.) based on their priority related to a fictional moon-survival story [11]. They were required to provide their own rank before discussion (IIR—Individual Initial Rank), publicly agreed rank (GR—Group Rank), and individual rank after discussion (IRR—Individual Revised Rank). In this experiment, only congruence, compliance, and conversion can be observed. Independence cannot be observed because by definition, this type of response is disagreeing with the group decision at all time, while in the experiment, the final group decision must be unanimous. For the similar reason, anticonformity cannot be observed as well. The discussion was conducted for 15 minutes, and they have to reach a unanimous decision within that time. The whole discussion process was recorded by both video and audio.

It is assumed that if the IIR between members are diverse, they will have more discussion. In the same manner, it is also assumed that if a person has a higher congruence, this person is regarded to be influential to the group because his/her initial answer remains close to the final group answer. By using these two assumptions, all participants were then mapped based on their diversity and congruence.

From the mapping, several participants' data were chosen, and their videos were observed to find the conformity pattern in the discussion. The detail of the mapping and video observation is explained in the following subsections.

Fig. 2 Distribution of members based on congruence and diversity score

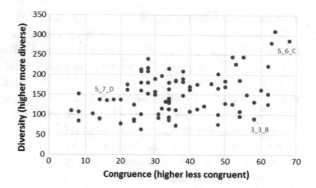

3.2 Congruence and Diversity Distribution

As mentioned previously, if a member agrees to the group's decision even from before the discussion, this type of response is called congruence. By comparing IIR and GR, a score can be assigned to the degree of congruence. Next, by comparing IIR of members in a same group, a score can be assigned to the degree of diversity. The calculation is shown by the following formulas (m = ID of participant, i = item number, g = ID of group where participant m belongs to, p = partner in the same group). The result of the mapping can be seen in Fig. 2.

$$\text{Congruence}_m = \sum_{i=1}^{15} \left| IIR_{i,m} - GR_{i,g} \right| \tag{1}$$

$$\text{Diversity}_m = \sum_{i=1}^{15} \left(\sum_p \left| IIR_{i,m} - IIR_{i,p} \right| \right) \tag{2}$$

In the graph, different conditions regarding congruence and diversity experienced by each member can be seen. For example, participant 5_6_C had a close answer to the group decision, but she faced a very diverse opinion from her partners. Meanwhile participant 3_3_B did not face such a diverse opinion. From this graph, three data were chosen arbitrarily from three different areas. The videos from these three participants were then analyzed further.

3.3 Analysis of Influence

In the current progress of the research, the video of group 5_6, 5_7, and 3_3 have been observed. The videos was transcribed and the discussions were analyzed. The protocol analysis aims to separate rational influence and non-rational influence, and to see if there is a pattern of emergence of the influence, such as time-wise emergence. For each utterance in the discussion, two tags were assigned. The first is the item related to the utterance, and the second is whether the utterance is an influence or not. If it is considered by the researcher as an influence, then it would be decided whether it is a rational or non-rational influence.

One example of rational influence is as follows: for the item oxygen, someone says, "The amount 2.5 kg is not sufficient". The logical reason may be strong or weak but at least the participant shows a fact-based argument. On the other hand, an example of non-rational influence is as follows: for the item pistol, someone says, "Pistol is really unimportant", without providing any argument. The statement that says whether an item is important or not is the expected decision, therefore when someone mention their preference about a decision without any argument, it is considered as a non-rational influence.

The result of the analysis towards group 5_6 is shown in Fig. 3. In the figure, one dot represents one utterance. This group mostly discuss about compass and oxygen, 30 and 32 utterances respectively. For compass, 17 utterances were rational while for oxygen it was 24. The group was focusing on those two items for most of its rational influence. When looking at the distribution of IIR for each item in that group, oxygen on the other hand is the item with the least diverse IIR. Meanwhile, the item with the most diverse IIR, the heating unit, was discussed only in 5 utterances and 2 of them are rational influence.

As can be seen from Fig. 3, in the first half of the discussion they intensively used rational influence, while in the latter half non-rational influence was more dominant. One might think that the second half is probably the decision stage after they discussed the matters in the beginning. However, most of the other items appeared in the second half. It means that for these items they conformed to each other with few rational influences or even not at all. This happened regardless of the difference of IIR of these items. One possible factor is the time pressure. Reaching the latter half they realized that they do not have much time to make a decision. At that moment, they started to use the heuristic approach by conforming or influencing other member even without a logical reason. Most of the time in the latter half, some of the members mentioned their rank for some items and then other people conformed to it, or provided alternative rank without any argument.

However, a different pattern was found in group 3_3. In the beginning, rational influence was not so intensive. When the content of the discussion was checked, it is found out that they began intensively using rational influence only after they found a difference of opinion against a certain rank or item.

Another fact that was observed from the discussion is the domination of the discussion. In group 5_6, the discussion was dominated by two of the members. One other member almost never spoke or gave any opinion. When their IIRs were

Fig. 3 Distribution of rational and non-rational influence of group 5_6

compared to the expert rank [11], this least contributive member is on the other hand, had the closest rank to the expert rank. However, she did not try to influence other members to follow her rank. In some of the items, she converted or complied to the group rank. In the end the group score fell down and got further from the expert rank. Her IRR also became bad, and quite far from her IIR. However, in group 5_7, the opposite situation was found. The member with best IIR was also the most talkative and the group score also become good, close to his initial rank.

4 Preliminary Findings

Even though at the current stage of research the supporting data are not enough and more of the videos in the data need to be analyzed, there were several interesting findings that can lead to further investigation or elaboration. From the analysis explained in the previous section, three preliminary results were found.

The first finding is that group members sometimes do not realize what is important to discuss and what is not. Moreover, in a time-constrained decision making situation where members are pressured to make a decision as fast as possible, they may skip the orientation process of understanding the differences of the topics. As found in the analysis, group 5_6 discussed more about oxygen regardless of its low difference of IIR, while discussed less about a heating unit regardless of its high difference of IIR. This was triggered simply because oxygen was mentioned very early by one of the members, while nobody mentioned about a heating unit until the middle of the discussion. Similar situation was also found in group 5_7, even though not as extreme as in group 5_6.

The second finding is about the pattern of rational influence and non-rational influence. It is found that when the time pressure is higher (in this case, reaching the end of discussion time) members tend to use non-rational influence than rational influence. Since they do not have enough resources (time) to consider all facts, they may choose to rely on their partners' preference. When it comes to this situation, whose preference will be used will depend on various things such as the members' personality. However, sometimes when time pressure does not exist *and* they feel that the difference of opinion is not so significant, they still use non-rational influence. Such a case was found in the beginning of the discussion of group 3_3. They did that probably to save resources, since argumentation requires resource (time, energy, and so on).

The third finding is the effect of the members' personality on the decision. As found in the analysis of group 5_6, the member with the best answer is unfortunately the least contributive member. On the other hand, the answers of the dominant members were not so good. The opposite situation happened in group 5_7. Since the groups relied on the heuristic approach for most of the items, then the dominant members were more influential regardless the level of accuracy of their answer.

Human decision making behavior under stress has been studied quite extensively [12]. However, there seems to be more to study about decision making under stress for group situations. The interaction and influencing process between the members, related to time or members' personality still need to be explored further. Such studies can later improve group decision making, particularly in an emergency situation.

5 Conclusions

In this paper, a model of group decision making has been proposed. The model emphasizes the aspect of individual influence towards each other in making a group decision. In the current stage of research, some issues were found. This paper wants to highlight that in a group setting, humans also tend to rely on a heuristic approach in decision making, just like in an individual setting. However, there is a mechanism that was not found in individual decision making, which are influence and conformity. Further research need to be done to ensure that group decision making in emergency situation will produce a good decision.

References

1. G. Stasser, B. Dietz-Uhler, Collective choice, judgment, and problem solving, in *Blackwell Handbook of Social Psychology: Group Processes Hogg/Blackwell*, ed. by M.A. Hogg, R.S. Tindale (Blackwell, Malden, MA, 2001), pp. 31–55
2. R.A. LeBoeuf, E.B. Shafir, Decision making, in *The cambridge handbook of thinking and reasoning*, ed. by K.J. Holyoak, R.G. Morrison (Cambridge University Press, New York, 2012), pp. 301–321
3. K.L. Mosier, U.M. Fischer (eds.), *Informed by knowledge: Expert performance in complex situation* (Psychology Press, New York, 2011)
4. D.C. Matz, W. Wood, Cognitive dissonance in groups: The consequences of disagreement. J. Pers. Soc. Psychol. **88**, 22–37 (2005)
5. L. Festinger, *A theory of cognitive dissonance* (Stanford University Press, Stanford, CA, 1957)
6. D.R. Forsyth, *Group Dynamics*, 6th edn. (Wadsworth, Cengage Learning, Belmont, CA, 2010)
7. D. Kahneman, *Thinking, fast and slow* (Farrar, Straus and Giroux, New York, 2011)
8. R.B. Cialdini, N.J. Goldstein, Social influence: compliance and conformity. Annu. Rev. Psychol. **55**(1974), 591–621 (2004)
9. P.R. Nail, G. MacDonald, On the development of the social response context model, in *The science of social influence: Advance and future progress*, ed. by A.R. Pratkanis (Psychology Press, New York, 2007), pp. 193–221

10. D. Mahardhika, T. Kanno, K. Furuta, *"Study of Mental Subgrouping Pattern in Extended Mutual Belief Model", in Symposium on Systems and Information* (Society of Instrument and Control Engineer, Okayama, 2014)
11. NASA—Langley Research Center, "Survival! Exploration: Then and Now," *NASA & Jamestown Education Module*, (2006). [Online]. Available: http://www.nasa.gov/audience/foreducators/topnav/materials/listbytype/Survival_Lesson.html
12. K. Starcke, M. Brand, Decision making under stress: a selective review. Neurosci. Biobehav. Rev. **36**(4), 48–1228 (2012)

Evaluation of Optimal Power Generation Mix Considering Nuclear Power Plants' Shut-Down Risk

Hiromu Matsuzawa, Ryoichi Komiyama and Yasumasa Fujii

Abstract After Fukushima nuclear power plant accident, resilience engineering has emerged as a new paradigm of risk management, and the design of resilient energy system is getting more and more important. Energy model analysis based on mathematical programming contributes to discussing how to implement resilience into energy system by identifying quantitative suggestions. In this paper, as an example of such analysis, the authors try to derive possible appropriate measures to enhance electricity supply system resilience to successive nuclear power plants' shut-down risk. The model developed in this paper is a dynamic power generation planning model, which considers nuclear power plants' shut-down risk stochastically and identifies resilient capacity expansion in Japan from 2012 to 2030 under the uncertainty of the risk from a quantitative perspective. This resilient capacity expansion includes the necessity of alternative power resources and demand response compensating for supply capacity loss due to nuclear power plants' shut-down, considering economic constraints. Simulation results successfully show the need for these measures in the capacity expansion. Importantly, the suggestion is not like a future prediction but a normative image of the system through the comprehensive incorporation of forecasted future parameters and scenarios. The more detailed the parameters and the scenarios are, the better image can be obtained. Learning from past accidents and updating our scientific knowledge base will detail the parameters and the scenarios and make energy model analysis very effective. It will tell us how to make resilient energy system.

Keywords Energy model analysis · Resilient energy system · Power generation planning · Nuclear power plants' shut-down · Stochastic dynamic programming

H. Matsuzawa (✉) · Y. Fujii
Department of Nuclear Engineering and Management, The University of Tokyo, Tokyo, Japan
e-mail: matsuzawa@esl.t.u-tokyo.ac.jp

R. Komiyama
Resilience Engineering Research Center, The University of Tokyo, Tokyo, Japan

© The Author(s) 2017
J. Ahn et al. (eds.), *Resilience: A New Paradigm of Nuclear Safety*,
DOI 10.1007/978-3-319-58768-4_25

289

1   Introduction

Fukushima nuclear power plant accident has highlighted insufficient preparedness for unexpected events in many systems. Electricity supply system is one of them. By the shut-down of nuclear power plants and thermal plants in Kanto region and Tohoku region, rolling blackout was taken place to compensate for supply capacity shortage, which was the first time in Japan after the World War II. In addition, the Japanese government issued the restriction of electricity use against the large-lot electricity users in summer 2012 for the same reason. These policies and voluntary demand side management prevent massive blackout, but Japanese society had suffered heavy social and economic damages. Therefore, in the future power generation planning, preparation for power plants' successive shut-down should be considered so that social and economic damages caused by the shut-down will be the smallest. That is to say, implementation of seismic resilience into electricity supply system is necessary.

Resilience in this context refers to the adaptive capacity of a system to absorb changes and to maintain its functionality. From a quantitative perspective, resilience can be enhanced by the following three measures: "Reduces failure probability," "Reduced consequences from failures," and "Reduces time to recovery." Furthermore, enhancement measures of seismic resilience can be classified according to the following four properties [1]:

- Robustness: strength, or the ability of elements, systems, and other units of analysis to withstand a given level of stress or demand without suffering degradation or loss of function
- Redundancy: the extent to which elements, systems or other units of analysis exit that are substitutable, i.e., capable of satisfying functional requirements in the event of disruption, degradation, or loss of functionality
- Resourcefulness: the capacity to identify problems, establish priorities, and mobilize resources when conditions exit that threaten to disrupt some element, system, or other unit of analysis; resourcefulness can be further conceptualized as consisting of the ability to apply material (i.e., monetary, physical, technological, and informational) and human resources to meet established priorities and achieve goals
- Rapidity: the capacity to meet priorities and achieve goals in a timely manner in order to contain losses and avoid future disruption

Hence, seismic resilience enhancement measures in electricity supply system can be presented in accordance with these four properties. For example, robustness in electricity supply system can be enhanced by improving power plants' earthquake resistant capacity. Alternative power resources enhance redundancy. Demand side management enhances resourcefulness. Then, restoration plans enhance rapidity. Of course, it must be noted that these example are illustrative only. Many other researches which contribute the enhancement of seismic resilience also are classified into the four properties. The classification gives systematic understanding of the way various research activities contribute the enhancement of seismic resilience. And then, by measuring the four properties quantitatively, comprehensive coordination of those researches can be achieved.

Energy model analysis using mathematical programming is an effective tool measuring 4Rs (robustness, redundancy, resourcefulness, and rapidity) quantitatively. Then, in this paper, the authors try to present quantitative analytical framework to discuss possible appropriate measures to implement seismic resilience into electricity supply system, using simple energy model analysis. The energy model used in this paper is a dynamic power generation planning model, which incorporates the modelling of some measures in accordance with 4Rs and assesses optimal capacity expansion strategy under successive nuclear power plants' shut-down risk. It must be understood that the output of the energy model analysis should not be like the future prediction. Its major concern is not to forecast a likely future image of the energy system, but rather to derive a normative future image of the system through the comprehensive incorporation of forecasted future parameters and scenarios. The goal of this paper is to stimulate discussions about seismic resilience enhancement by the normative image obtained through energy model analysis and present its usefulness in that it can provide quantitative suggestions.

2 Dynamic Power Generation Planning Model

2.1 Mathematical Formulation

The dynamic power generation planning model mathematically expresses nuclear power plants' shut-down risk as yearly stochastic transitions of nuclear power plants availability, and identifies the optimal capacity expansion strategy under the uncertainty of the risk. This strategy is denoted by stochastic dynamic programming [Eqs. (1) and (2)] and minimizes the expected total system cost necessary from y to the expiration of an analytical period. This formulation highlights that it can consider the uncertainty of both nuclear power plants' shut-down and recovery from their disruptions, so it can express the preparation for risk and adaptive methods to disruptions, which corresponds to redundancy and resourcefulness of the system.

As an analytical period, this paper assumes form 2012 to 2030, and power plants considered are thermal power (coal, LNG steam turbine (ST), LNG combined cycle (CC) and oil), nuclear power, hydro power including pumped type, and stationary sodium-sulfur battery. Exogenous variables about them are based on [2], and fuel price is set based on [3]. Concerning the installed capacity of coal-fired power plants, LNG ST power plants, oil power plants, nuclear power plants and hydro power plants, the maximum upper limit is assigned due to political, geographical or some other reasons. Regional scope is the whole region of Japan and the electricity market is assumed as a monopoly market. Annual power demand is expressed by four representative load curves of each season in 2012, and it does not change until 2030. Problem formulation is described as follows.

Endogenous variables:

$V_y(K_y, i_y)$	expected total system cost ($)
$C_y(K_y, i, dK_y)$	yearly system cost incurred in year y ($/year)
K_y	capacity mix of power plants and power storage facilities (GW)
$Kp_{y,p}$	capacity of p-th type of power plant in year y (GW)
$dKp_{y,p}$	newly constructed capacity of p-th type of power plant in year y (GW)
$X_{p,t,d,y}$	output of p-th type of power plant in day d at time t and year y (GW)
$Ks1_{y,s}$	kW capacity of s-th type of power storage facility in year y (GW)
$Ks2_{y,s}$	kWh capacity of s-th type of power storage facility in year y (GWh)
$dKs1_{y,s}$	newly constructed kW capacity of s-th type of power storage facility in year y (GW)
$dKs2_{y,s}$	newly constructed kWh capacity of s-th type of power storage facility in year y (GWh)
$Cha_{s,d,t,y}$	input of s-th type of power storage facility in day d at time t and year y (GW)
$Dis_{s,d,t,y}$	output of s-th type of power storage facility in day d at time t and year y (GW)
$Ss_{s,d,t,y}$	stored energy of s-th type of power storage facility in day d at time t and year y (GWh)
$Save_{d,t,y}$	electricity demand saving in day d at time t and year y (GWh).

where $p \in \{1: \text{Nuclear}, 2: \text{Coal}, 3: \text{LNG CC}, 4: \text{LNG ST}, 5: \text{Oil}, 6: \text{Hydro}\}$, $s \in \{1: \text{Pumped hydro}, 2: \text{NAS battery}\}$, $d \in \{1, 2, ..., 4\}$, $t \in \{1, 2, ..., 24\}$, $y \in \{0, 1, ..., 18\}$, i_y: state of nuclear power plants availability in year y ($i_y = \{0: \text{nuclear power unavailable}, 1: \text{nuclear available}\}$).

2.1.1 Objective Function

Objective function is the discounted total cost considering all the possible yearly state-transitions about i_y from 2012 to 2030, which corresponds to $V_0(K_0, i_0)$ in Eq. (1). Discount rate in this paper is assumed as 3%. As the initial state in dynamic programming, the existing capacity in 2012 is given and nuclear power plants are assume to be available [Eq. (4)].

$$V_y(K_y, i_y) = \min_{dK_y} \left\{ C_y(K_y, i_y, dK_y) + \exp(-\gamma) \sum_{i_{y+1}=0}^{1} P(i_y \to i_{y+1}) V_{y+1}(K_{y+1}, i_{y+1}) \right\}$$

(1)

$$V_{19}(K_{19}, i_{19}) = 0$$

(2)

$$K_{y+1} = K_y + dK_y + dec_y \tag{3}$$

$$(K_0, i_0) = (kini, 1) \tag{4}$$

$$K_y = \left(Kp_{y,1}, Kp_{y,2}, \ldots, Kp_{y,6}, Ks1_{y,1}, Ks1_{y,2}, Ks2_{y,1}, Ks2_{y,2}\right) \tag{5}$$

where, dec_y: decommission capacity, $kini$: existing capacity in 2012, γ: discount rate, $P(\cdot)$: state transition probability

$$C_y\left(K_y, i_y, dK_y\right) = \sum_{p=1}^{6} \left(g_p \times pf_p \times dKp_{y,p} + \frac{365}{4} \sum_{d=1}^{4} \sum_{t=1}^{24} pv_{p,y} \times X_{p,t,d,y} \right) \\ + \sum_{s=1}^{2} CS_{s,y} + \frac{365}{4} \sum_{d=1}^{4} \sum_{t=1}^{24} Csave_{d,t,y} \tag{6}$$

$$CS_{s,y} = gs1_s \times pfs1_s \times dKs1_s + gs2_s \times pfs2_s \times dKs2_s + pfs3_{s,y} \times \frac{TCha_{s,y}}{cycle_s} \tag{7}$$

$$TCha_{s,y} = \frac{365}{4} \sum_{d=1}^{4} \sum_{t=1}^{24} Cha_{s,d,t,y} \tag{8}$$

where g_p: annual fixed charge rate of p-th type of power plant (capital recovery factor), pf_p: unit fixed cost of p-th type of power plant ($/kW), $pv_{p,y}$: unit variable cost of p-th type of power plant ($/kWh), $CS_{s,y}$: annual cost of s-th type of power storage facility, $gs1_s$: annual fixed charge rate for power component of s-th type of power storage facility, $pfs1_s$: unit fixed cost for power component of s-th type of power storage facility (cost for kW capacity, $/kW), $gs2_s$: annual fixed charge rate for energy component of s-th type of power storage facility, $pfs2_s$: unit fixed cost for energy component of s-th type of power storage facility (cost for kWh capacity, $/kWh), $pfs3_{s,y}$: unit fixed cost for consumable material of s-th type of power storage facility($/kWh), $cycle_{s,y}$: maximum recharge times of s-th type of power storage facility, $TCha_{s,y}$: annual total charged electricity of s-th type of power storage facility(kWh/year).

The fourth item in right hand side of Eq. (6) is demand saving cost, which is mathematically modelled as the penalty cost incurred by economic loss based on [4]. This paper assumes typical demand curves (Fig. 1) where reference price is P_0 and reference demand power load in day d at time t and year y is $load_{d,t,y}$. According to this curve, the promotion of energy saving from reference point cause the escalation of electricity price, and eventually, the integral of the demand curve from reference demand $load_{d,t,y}$ to curtailed demand $load_{d,t,y}$ minus $Save_{d,t,y}$ corresponds to the penalty cost, which is formulated in Eq. (9) or (10). Variable $Save_{d,t,y}$ is endogenously determined through cost minimization considering the cost competitiveness towards the capacity expansion cost. This modelling methods of supply shortage depends on reference price P_0 and price elasticity. In this paper,

Fig. 1 Electricity demand
curve and energy saving

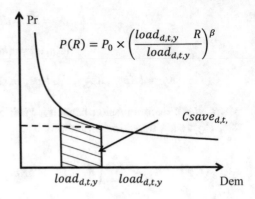

P_0 is assumed as the shadow price of electricity demand in dynamic power generation planning model that does not consider the nuclear power plants' shut-down risk, and price elasticity is assumed as 0.1%. From systemic view, the modelling of demand saving corresponds to that of resourcefulness.

$$Csave_{d,t,y} = \int_0^{Save_{d,t,y}} P(R)dR = \int_0^{Save_{d,t,y}} P_0 \times \left(\frac{load_{d,t,y} - R}{load_{d,t,y}}\right)^{-\frac{1}{\beta}} dR$$

$$= \frac{\beta}{1-\beta} \times load_{d,t,y} \times P_0 \times \left\{\left(\frac{load_{d,t,y} - Save_{d,t,y}}{load_{d,t,y}}\right)^{\frac{\beta-1}{\beta}} - 1\right\} \quad (\beta \neq 1) \tag{9}$$

$$Csave_{d,t,y} = \int_0^{Save_{d,t,y}} P(R)dR = \int_0^{Save_{d,t,y}} P_0 \times \left(\frac{load_{d,t,y} - R}{load_{d,t,y}}\right)^{-\frac{1}{\beta}} dR$$

$$= load_{d,t,y} \times P_0 \times \log\left(\frac{load_{d,t,y} - Save_{d,t,y}}{load_{d,t,y}}\right) \quad (\beta = 1) \tag{10}$$

where β: price elasticity, P_0: reference price, $load_{d,t,y}$: power demand in day d at time t and year y (GWh).

2.1.2 Constraints

The dynamic power generation planning model minimizes total expected cost under various technical constraints employing linear programming technique. The constraints considered in this paper are based on [2], and formulated as follows:

(a) *Available capacity constraint of the plants*

$$X_{p,t,d,y} \leq avp_{d,p}(i_y) \times Kp_{y,p} \tag{11}$$

where $avp_{d,p}(i_y)$: availability factor of i-th type of power plant.

(b) *Power demand and supply balances*

$$\sum_{p=1}^{6} X_{p,t,d,y} + \sum_{s=1}^{2} (Dis_{s,d,t,y} - Cha_{s,d,t,y}) = load_{d,t,y} - Save_{d,t,y} \tag{12}$$

(c) *Constraint on installable capacity*

$$Kp_{y,p} \leq KPmax_{y,p} \tag{13}$$

$$Ks1_{y,s} \leq KS1max_{y,s} \tag{14}$$

$$Ks2_{y,s} \leq KS2max_{y,s} \tag{15}$$

where $KPmax_p$: maximum installable capacity of p-th type of power plant in year y (GW), $KS1max_{y,s}$: maximum installable kW capacity of s-th type of power storage facility in year y (GW), $KS2max_{y,s}$: maximum installable kWh capacity of s-th type of power storage facility in year y (GWh).

(d) *Constraint on load following capability of the plants*

$$X_{p,t+1,d,y} \leq X_{p,t,d,y} + increase_p \times avp_{d,p}(i_y) \times Kp_{y,p} \tag{16}$$

$$X_{p,t+1,d,y} \geq X_{p,t,d,y} - decrease_p \times avp_{d,p}(i_y) \times Kp_{y,p} \tag{17}$$

where $increase_p$: maximum output increase rate per unit time of p-th type of power plant, $decrease_p$: maximum output decrease rate per unit time of p-th type of power plant.

(e) *Charge and discharge balances of energy storage technology*

$$Ss_{s,d,t+1,y} = (1 - sd_s) \times Ss_{s,d,t,y} + \left(\sqrt{eff_s} \times Cha_{s,d,t,y} - \frac{Dis_{s,d,t,y}}{\sqrt{eff_s}} \right) \tag{18}$$

where sd_s: self-discharge rate of s-th type of power storage facility, eff_s: cycle efficiency of electricity storage of s-th type of power storage facility.

(f) *Available capacity constraint of the battery technology*

$$Cha_{s,d,t,y} + Diss_{s,d,t,y} \leq us1_{s,d} \times Ks1_{y,s} \tag{19}$$

$$Ss_{s,d,t,y} \leq us2_s \times Ks2_{y,s} \tag{20}$$

$$Ss_{s,d,t,y} \leq m_s \times us1_s \times Ks1_{y,s} \tag{21}$$

where $us1_s$: kW availability factor of s-th type of power storage facility, $us2_s$: kWh availability factor of s-th type of power storage facility, m_s: energy storage capacity per generation capacity of s-th type of power storage facility (kWh/kW).

2.2 Nuclear Power Plants' Shut-Down Model

In this paper, nuclear power plants' shut-down risk is mathematically expressed as yearly stochastic transitions of nuclear power plants availability based on [5]. The methodologies are as follows.

Suppose *normal* state means nuclear power plants are available and *accident* state means unavailable, disruption rate λ, the rate of disruption occurrence per time step, is formulated by the reliability of nuclear power plants availability R [Eq. (22)]. If disruption rate λ is independent from time steps, disruption density function $f(y)$, which means the rise rate of unreliability $1 - R$, is denoted by Eqs. (23) and (24). Mean time between disruptions (MTBD), the expected mean time nuclear power plants can continue their operation, is obtained [Eq. (25)], and mean time to recovery (MTTR), the expected mean time nuclear power plants restart their operation after disruption, is formulated in a similar way through the definition of recovery rate μ. State transition probability between *normal* state and *accident* state, which corresponds to unreliability $1 - R$, is determined by MTBD and MTTR. This paper assumes MTBD and MTTR as 30 years and 2 years respectively. In this model, MTBD and MTTR represents the robustness and the rapidity of the system.

$$\lambda = -\frac{1}{R}\frac{dR}{dy} \tag{22}$$

$$R = \exp(-\lambda y) \tag{23}$$

$$f(y) = \frac{d(1 - R)}{dy} = \lambda \exp(-\lambda y) \tag{24}$$

$$MTBD = \int_0^\infty yf(y)dy = \frac{1}{\lambda} \tag{25}$$

$$MTTR = \frac{1}{\mu} \qquad (26)$$

$$P(0 \rightarrow 1) = 1 - \exp(-1/MTTR) \qquad (27)$$

$$P(1 \rightarrow 0) = 1 - \exp(-1/MTBD) \qquad (28)$$

where λ: disruption rate, R: the reliability of nuclear power plants' availability, $f(y)$: disruption density function, $MTBD$: mean time between disruptions (year), $MTTR$: mean time to recovery (year).

2.3 Calculation Algorithm

To solve strictly this model formulated using stochastic dynamic programming needs a lot of computations because of the high dimensionality of K_y, and calculation is difficult due to computational constraints. This problem is called "the curse of dimensionality". Therefore, as an approximate solution method to stochastic dynamic programming, cutting planes method [6] is adopted in this paper. This method uses the convexity of $V_y(K_y, i_y)$, which characteristic is due to the linear programming technique employed in Eqs. (1)–(28), and $V_y(K_y, i_y)$ is approximated as a set of hyper planes defined at sample points K_y^* on the function described in Fig. 2 and Eq. (29). The approximation allow us to solve this model at each time step while it is necessary to solve at one time considering all the time steps during analytical period if you want to solve this model strictly. It achieves a lot of computational saving, and can successfully address the curse of dimensionality. The detailed algorithm of cutting planes method is shown in Fig. 3.

$$V_y\left(K_y, i_y\right) \geq V_y\left(K_y^*, i_y\right) + \frac{\partial V_y\left(K_y^*, i_y\right)}{\partial K_y}\left(K_y - K_y^*\right) \qquad (29)$$

where K_y^*: a sample point defined on $V_y(K_y, i_y)$.

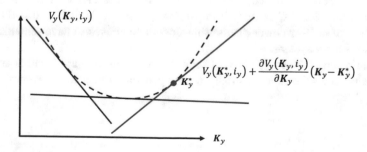

Fig. 2 Approximation by cutting planes method

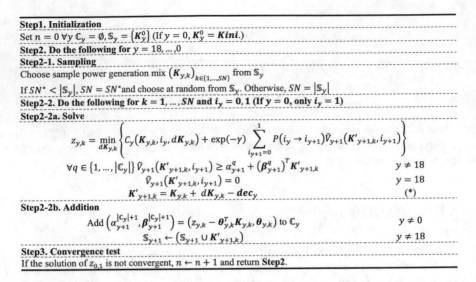

Step1. Initialization

Set $n = 0$ $\forall y$ $\mathbb{C}_y = \emptyset$, $\mathbb{S}_y = \left\{K_y^0\right\}$ (If $y = 0$, $K_y^0 = \textbf{Kini}$.)

Step2. Do the following for $y = 18, \ldots, 0$

Step2-1. Sampling

Choose sample power generation mix $\left(K_{y,k}\right)_{k \in \{1,\ldots,SN\}}$ from $\mathbb{S}_y$

If $SN^* < |\mathbb{S}_y|$, $SN = SN^*$ and choose at random from $\mathbb{S}_y$. Otherwise, $SN = |\mathbb{S}_y|$

Step2-2. Do the following for $k = 1, \ldots, SN$ **and** $i_y = 0, 1$ **(If** $y = 0$, **only** $i_y = 1$**)**

Step2-2a. Solve

$$z_{y,k} = \min_{dK_{y,k}} \left\{ C_y\left(K_{y,k}, i_y, dK_{y,k}\right) + \exp(-\gamma) \sum_{i_{y+1}=0}^{1} P\left(i_y \to i_{y+1}\right)\tilde{V}_{y+1}\left(K'_{y+1,k}, i_{y+1}\right) \right\}$$

$$\forall q \in \{1, \ldots, |\mathbb{C}_y|\} \; \tilde{V}_{y+1}\left(K'_{y+1,k}, i_{y+1}\right) \geq \alpha_{y+1}^q + \left(\beta_{y+1}^q\right)^T K'_{y+1,k} \qquad\qquad y \neq 18$$

$$\tilde{V}_{y+1}\left(K'_{y+1,k}, i_{y+1}\right) = 0 \qquad\qquad y = 18$$

$$K'_{y+1,k} = K_{y,k} + dK_{y,k} - \textbf{dec}_y \qquad\qquad (*)$$

Step2-2b. Addition

$$\text{Add}\left(\alpha_{y+1}^{|\mathbb{C}_y|+1}, \beta_{y+1}^{|\mathbb{C}_y|+1}\right) = \left(z_{y,k} - \theta_{y,k}^T K_{y,k}, \theta_{y,k}\right) \text{ to } \mathbb{C}_y \qquad\qquad y \neq 0$$

$$\mathbb{S}_{y+1} \leftarrow \left(\mathbb{S}_{y+1} \cup K'_{y+1,k}\right) \qquad\qquad y \neq 18$$

Step3. Convergence test

If the solution of $z_{0,1}$ is not convergent, $n \leftarrow n + 1$ and return **Step2**.

Fig. 3 Cutting planes method algorithm

where n: iteration number, K_y^0: initial sample point (arbitrarily set except in case $y = 0$), $\mathbb{C}_y$: constraint space in year y, $\mathbb{S}_y$: sample point space in year y, $K_{y,k}$: chosen sample points, SN: sampling number, SN^*: maximum sampling number, $\tilde{V}_{y+1}$: approximated function of V_{y+1}, $\theta_{y,k}$: shadow price of equation (*) in step 2-2a.

3 Results and Discussion

The algorithm shown in Fig. 3 allow us to simulate any scenarios by solving the problem forwardly from $y = 0$ to $y = 18$. This paper presents two representative scenarios. The first scenario (Scenario 1) assumes the shut-down does not happen during the time period. In the second scenario (Scenario 2), the shut-down happens in 2026 and it recovers 2028. In addition to the two scenarios, reference case, where the shut-down risk is zero, is calculated.

Figure 4 shows estimated capacity mix in reference case and Scenario 1. In scenario 1, the LNG CC capacity is expanded at larger scale. It can be said that the uncertainty of nuclear power plants' shut-down encourages to have redundancy in electricity supply system.

Figure 5 shows the comparison of supply capacity in Scenario 1 and Scenario 2, and Fig. 6 shows the estimated daily power generation dispatch in summer 2026,

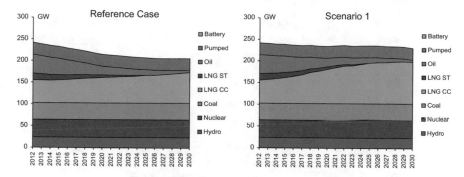

Fig. 4 Estimated capacity mix from 2012 to 2030

Fig. 5 Power supply capacity

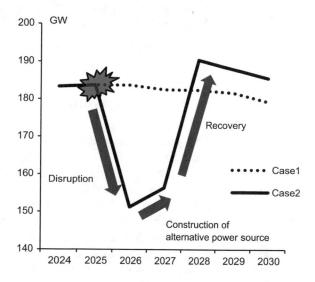

when nuclear power plants' shut-down happens. Immediately after disruption, demand saving happens at peak time (11:00–18:00), and then, the construction of alternative power source can been seen in 2027 to compensate for the supply capacity loss. These policies highlight the resourcefulness of the system to addresses the nuclear power plants' shut-down. It should be noted that the construction of alternative power source minimizes the resilient triangle expressed in Fig. 5 under economic constraints.

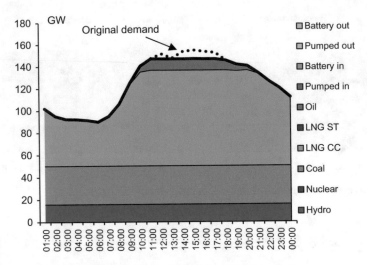

Fig. 6 Daily power generation dispatch in 2026 (Scenario 2)

4 Conclusions and Future Works

This paper presents a quantitative analytical framework to assess seismic resilient enhancement measures in electricity supply system, referring to the four essential properties; robustness, redundancy, resourcefulness and rapidity. The dynamic power generation planning model considering nuclear power plants' shut-down risk successfully derives possible appropriate measures to enhance resilience of the system from a quantitative perspective, and gives systemic understanding of the system's seismic resilience. Simulation results show that nuclear power plants' shut-down risk creates a need for redundancy in power generation planning and that the decrease of electricity supply capacity caused by nuclear power plants' shut-down can be compensated for by demand saving and construction of alternative power source. In other words, it can be compensated for by the system's resourcefulness.

As mentioned in introduction, the simulation results highlight a normative image of the system through the comprehensive incorporation of forecasted future parameters and scenarios. However, the model developed in this paper is a work in progress and not simple enough to do so. To get better understanding of seismic resilience, our future works consists in refining the model and consider such parameters and scenarios. With respect to the refinement of parameters and scenarios, it depends on technical and social researches. For example, MTTR can be shortened through enhancement of quake-resistance and tsunami protection of nuclear power plants. It can also shortened through good design of social systems where effective decision making is done even under extreme situations, and effective decontamination actions. The value of MTBD can be assessed more precisely by geologically-based study. Price elasticity, which determines the resourcefulness of the system, will be lowered aggressive introduction of emergency power source although cost-effectiveness of their installation should be considered, of course. Assessment on these policies through energy model analysis will give the potential contributions and benefits of them. Concerning scenarios planning and model refinement, the design of systems has a key role. Social activities in today's world are supported by highly complex and interdependent system, and so, risks surrounding us are very systemic. Fukushima nuclear power plant accident is such a kind of risk. Therefore, targeted system should be comprehensive enough to consider their inter-relations although the system considered in this paper is limited within electricity supply system. In addition, more kinds of risks should be considered because our challenges to be dealt with now in this world are not only nuclear usage.

Finally, to address above requirements, the most important seems to learn effectively from accidents and update our social scientific knowledge base. The implement of what we have learned into energy model makes it more sophisticated and the model will tell us how to design resilient systems.

Acknowledgements This work was supported by JST Strategic Basic Research Programs RISTEX, Resilience Analysis for Social Safety Policy.

References

1. M. Bruneau et al., A framework to quantitatively assess and enhance the seismic resilience of communities. Earthquake Spectra **19**, 733–752 (2003)
2. R. Komiyama, Y. Fujii, Long-term scenario analysis of nuclear energy and variable renewables in Japan's power generation mix considering flexible power resources. Energy Policy **83**, 169–184 (2015)
3. International Energy Agency (IEA), *World Energy Outlook 2013* (OECD, Paris, France, 2013)
4. H. Matsuzawa, R. Komiyama, Y. Fujii, Analysis of energy system resilience to disaster in Kanto Region using stochastic programming. Proc. Conf. Energy Econ. Environ. **34**, 223–226 (2015)
5. Y. Uchiyama, Y. Hatano, K. Okajima, *Social Risk in Energy System* (CORONA Publishing Co., Ltd., 2012), pp. 70–74
6. Z.L. Chen, W.B. Powell, Convergent cutting plane and partial-sampling algorithm for multistage stochastic linear programs with recourse. J. Optim. Theory Appl. **102**, 497–524 (1999)

A Hybrid Finite Element and Mesh-Free Particle Method for Disaster-Resilient Design of Structures

Naoto Mitsume, Shinobu Yoshimura, Kohei Murotani
and Tomonori Yamada

Abstract The MPS-FE method, which is a hybrid method for Fluid-Structure Interaction (FSI) problems adopting the Finite Element method (FEM) for structure computation and Moving Particle Semi-implicit/Simulation (MPS) methods for free surface flow computation, was developed to utilize it in disaster-resilient design of important facilities and structures. In general free-surface flow simulation using the MPS method, wall boundaries are represented as fixed particles (wall particles) set as uniform grids, so the interface of fluid computation does not correspond to the interface structure computation in the conventional MPS-FE method. In this study, we develop an accurate and robust polygon wall boundary model, named Explicitly Represented Polygon (ERP) wall boundary model, in which the wall boundaries in the MPS method can be represented as planes that have same geometries as finite element surfaces.

Keywords Fluid-structure interaction · Finite element method · Moving particle Semi-implicit/simulation method · Mesh-free particle method · MPS-FE method

1 Introduction

Large-scale facilities such as petrochemical and nuclear power plants, and tsunami evacuation facilities built along coastal regions are vulnerable to water-related disasters. The resulting damage to equipment and instruments has the potential to cause catastrophic harm to people and local society. However, it is economically difficult to completely prevent the effects of disasters of extreme severity. We need quantitative measures to minimize the damages and loss. Detailed numerical computations treating the disasters and damages as Fluid-Structure Interaction (FSI) problems involving free surface flow give us a measure for disaster-resilient design of important structures.

N. Mitsume (✉) · S. Yoshimura · K. Murotani · T. Yamada
Department of Systems Innovation, University of Tokyo, Bunkyō, Japan
e-mail: Mitsume@save.sys.t.u-tokyo.ac.jp

© The Author(s) 2017
J. Ahn et al. (eds.), *Resilience: A New Paradigm of Nuclear Safety*,
DOI 10.1007/978-3-319-58768-4_26

We have developed a hybrid method for FSI problem with free surface, named the MPS-FE method [1, 2]. This method adopts the Moving Particle Semi-implicit/Simulation (MPS) [3] method, a mesh-free particle method, for free surface flow computation because of its robustness in long-term analyses with moving boundaries, and Finite Element Method (FEM) for structure computation because of its high accuracy and reliability. The method combines the advantages of both methods and achieves efficiency and robustness. These two methods are coupled with a partitioned coupling approach, i.e. the conventional serial staggered (CSS) scheme [4], which can set different time step sizes for the fluid and structure computations.

The conventional MPS-FE method [1], in which MPS wall boundary particles and finite elements are overlapped in order to exchange information on fluid-structure interfaces, has difficulty in dealing with complex shaped fluid-structure interfaces, because the wall particles have to be set in an orthogonal and uniform grid manner. In addition, forces on fluid-structure interfaces are not balanced when the pressure on the walls is calculated in the conventional MPS-FE method. As the next step, we adopted existing polygon wall boundary models [5, 6], which can treat a wall boundary as an arbitrary plane, and improved the MPS-FE methods so that the fluid and structure interfaces are consistent. However, the existing polygon wall boundary models cannot satisfy the pressure Neumann or the slip/no-slip boundary conditions, so these cause instability near the boundaries and deteriorate the accuracy.

In this study, we developed a new polygon wall boundary model for fully explicit algorithms (Explicit-MPS [7]: E-MPS), called the Explicitly Represented Polygon (ERP) wall boundary model [8] to compose more accurate MPS-FE method. The ERP model is formulated such that it satisfies the pressure Neumann boundary condition and the slip/no-slip boundary condition, without requiring the generation of virtual particles or treating angled edges as exceptional cases. Moreover, the ERP model eliminates the problem of force imbalance on the boundaries, which occurs in conventional models.

2 ERP Wall Boundary Model for Explicit-MPS Method

Regarding the wall boundary treatments, research has made greater progress for the SPH method [9], which is one of mesh-free particle methods. The repulsive-force model [10] has been developed in order to avoid penetration of fluid particles across wall boundaries. Although this model is relatively easy to implement, it causes the instability of fluid particles near wall boundaries because the boundary conditions are not satisfied. On the other hand, the ghost (mirror) particle approach [11] is widely used to satisfy the boundary conditions on walls. In this approach, virtual particle is generated at reflectional position across the wall of each fluid particle. These mirror particles are given pressure and velocity values so that the pressure Neumann boundary condition and slip/no-slip condition are satisfied. However, this approach has several problems, including a high computational cost caused by the need to generate virtual particles, and leakage of particles at the angled edges of surfaces.

Regarding polygon wall boundary models in the MPS method, Harada et al. [5] derived the force exerted on a fluid particle from a wall from impulse-momentum relationship at the particles near the wall. This force modeling can be classified as repulsive-force model, so Harada's model has the same problem of the instability and strange behavior of fluid particles near the wall. Yamada et al. [6] focused on the E-MPS method and they proposed another formulation. Although this is a natural expansion of the MPS differential operator models, the model causes excessive pressure oscillations.

The ERP model is based on the mirror particle approach and can satisfy the boundary conditions on walls without virtual particles, and it is versatile enough to treat arbitrarily shaped boundaries and arbitrary movements. The ERP model adds a repulsive force adaptively only when the boundary conditions are not satisfied near the angled edges. The ERP model has the following characteristics:

- Wall boundaries are represented explicitly.
- Generation of virtual particles and the need to make special adaptations for angled edges are not required.
- The pressure Neumann boundary condition and the slip/no-slip condition on the walls are satisfied.

3 Verification of the ERP Model

To verify the accuracy of the ERP model applied to the E-MPS method quantitatively, we analyzed a hydrostatic pressure problem in a rectangular vessel. The numerical results were compared with the theoretical solution

$$p = \rho |g| h$$

where ρ is the fluid density, g is the gravitational acceleration vector, and h is the depth of the static water surface. The initial configuration, in which the depth of the rectangular vessel is 0.1 m and the width is 0.04 m, is illustrated in Fig. 1. In the E-MPS computation, weak compressibility causes vertical vibrations of the fluid surface. To reach the static state as quickly as possible, we chose a relatively high value for the kinematic viscosity. The conditions used in the analysis are listed in Table 1, in which l^0 is the initial particle spacing.

Figure 3 shows the pressures of fluid particles computed by (a) the ERP model, (b) the ERP model using only the repulsive force, (c) Harada's model, and (d) Yamada's model. These are the results at the 200,000th step, at which the pressure field can be regarded to be in a steady state. In Fig. 2, snapshots obtained

Fig. 1 Hydrostatic pressure:
initial configuration

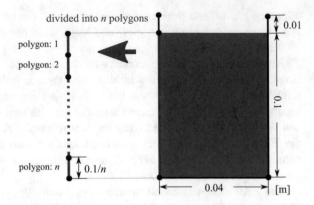

Table 1 Hydrostatic
pressure: analysis conditions

Time step width	5.0×10^{-5} (s)
Number of particles	4000
Particle spacing	1.0×10^{-3} (m)
Effective radius	2.91^{0} (m)
Fluid density	1.0×10^{3} (kg/m^3)
Kinetic viscosity	1.0×10^{-4} (m^2/s)
Gravitational acceleration	9.8 (m^2/s)
Sound speed coefficient	9.44 (m/s)
Repulsive coefficient	1.0×10^{7} (N/m^3)

Fig. 2 Hydrostatic pressure:
visualization of results

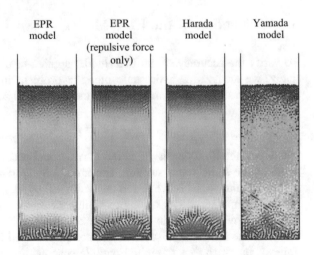

by each models at the 200,000th step are shown; the pressure on the fluid particles
is indicated by color (unit [N/m^2], min: 0, max: 1,000).

As indicated in Fig. 3b, c, the results from using only the repulsive force show
the same tendency as those from Harada's pressure gradient model, which is one of

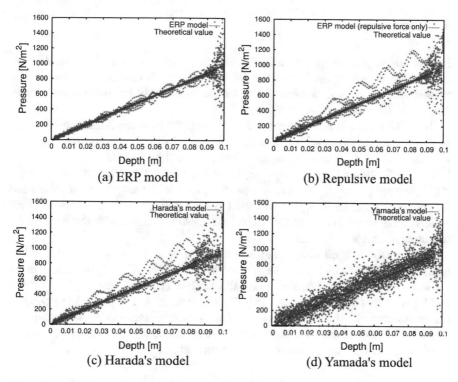

Fig. 3 Hydrostatic pressure: pressure on particles

the repulsive force models as we mentioned. Both of the results exhibit two strange lines that indicate pressures that are higher than the theoretical values. These results indicate that the pressures on the particles in contact with polygons have not been evaluated correctly, as shown in Fig. 2. Therefore, the model using only the repulsive force encounters the same problems as are found with Harada's model. On the other hand, Fig. 3d shows that Yamada's pressure gradient model results in a disturbed pressure field that has a wide dispersion compared to those of the other methods.

Unlike the existing polygon wall boundary models, the ERP model does not have the problems that occur with Harada's and Yamada's models, and it obtains a better pressure distribution that is in agreement with the theoretical solution. The results of the ERP model are also in better agreement with the theoretical solution. This is because the pressure gradient in the ERP model satisfies the pressure Neumann boundary condition, whereas the pressure gradient obtained using the existing polygon wall boundary models do not satisfy it rigorously. The results of the polygon wall boundary model involving the ERP model, however, have highly dispersed pressures near the bottom of the vessel, because the accuracy deteriorates at the angled edges of polygons. The influence of the edges can be reduced by enhancing the spatial resolution by using smaller particles.

4   Conclusions

In this study, we developed and verified the Explicitly Represented Polygon (ERP) wall boundary model for the E-MPS method. It can deal with arbitrarily shaped boundaries and movements, and it can accurately impose boundary conditions for free-surface flow analysis.

The ERP model is formulated so as to satisfies the pressure Neumann boundary condition and the slip/no-slip boundary condition, without requiring the generation of virtual particles or treating angled edges as exceptional cases.

For verification of the proposed model, we conducted simulations for a hydrostatic pressure problem. The results were compared with the theoretical values and the results of other models. We confirmed that the boundary conditions of the ERP method were appropriately modeled, and the E-MPS method with the ERP model can achieve adequate accuracy.

Now, we have been developing a more accurate and robust MPS-FE method applying the ERP model, and a large-scale parallel MPS-FE analysis system using the large-scale parallel code for the E-MPS method (HDDM_EMPS [12, 13]) and FEM (ADVENTURE_Solid [14, 15]). This system makes it possible to conduct robust simulations of three-dimensional fluid-structure interaction. An example is shown in Fig. 4, which is a three-dimensional simulation of dam break and an elastic column with constraints on the bottom surface.

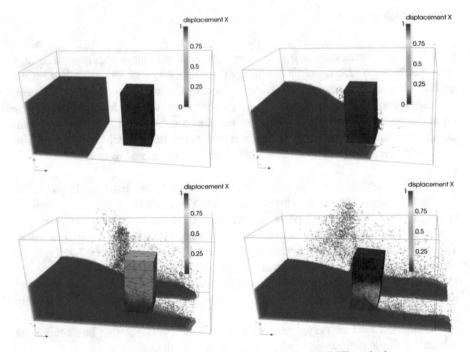

Fig. 4 Example of three-dimensional FSI simulation using the MPS-FE method

Acknowledgements The present research was partly supported by the JST CREST project "Development of a Numerical Library Based on Hierarchical Domain Decomposition for Post Petascale Simulation".

References

1. N. Mitsume, S. Yoshimura, K. Murotani, T. Yamada, MPS-FEM partitioned coupling approach for fluid-structure interaction with free surface flow. Int. J. Comput. Methods, **11**(4), 1350101 (16 pages) (2014)
2. N. Mitsume, S. Yoshimura, K. Murotani, T. Yamada, Improved MPS-FE fluid-structure interaction coupled method with MPS polygon wall boundary model. Comput. Model. Eng. Sci. **101**(4), 229–247 (2014)
3. S. Koshizuka, Y. Oka, Moving-particle semi-implicit method for fragmentation of incompressible fluid. Nucl. Sci. Eng. **123**(3), 421–434 (1996)
4. C. Farhat, M. Lesoinne, Two efficient staggered algorithms for the serial and parallel solution of three-dimensional nonlinear transient aeroelastic problems. Comput. Methods Appl. Mech. Eng. **182**(3), 499–515 (2000)
5. T. Harada, S. Koshizuka, K. Shimazaki, Improvement of wall boundary calculation model for MPS method. Trans. Jpn. Soc. Comput. Eng. Sci. (20080006) (2008) (in Japanese)
6. Y. Yamada, M. Sakai, S. Mizutani, S. Koshizuka, M. Oochi, K. Muruzono, Numerical simulation of three-dimensional free-surface flows with explicit moving particle simulation method. Trans. Atomic Energy Soc. Jpn. **10**(3), 185–193 (2011). (in Japanese)
7. A. Shakibaeinia, Y.C. Jin, A weakly compressible MPS method for modeling of open-boundary free-surface flow. Int. J. Numer. Meth. Fluids **63**(10), 1208–1232 (2010)
8. N. Mitsume, S. Yoshimura, K. Murotani, T. Yamada, Explicitly represented polygon wall boundary model for the explicit MPS method. Comput. Particle Mech (Submitting)
9. L.B. Lucy, A numerical approach to the testing of the fission hypothesis. Astron. J. **82**, 1013–1024 (1977)
10. J.J. Monaghan, Simulating free surface flows with sph. J. Comput. Phys. **110**(2), 399–406 (1994)
11. J.P. Morris, P.J. Fox, Y. Zhu, Modeling low Reynolds number incompressible flows using SPH. J. Comput. Phys. **136**(1), 214–226 (1997)
12. http://adventure.sys.t.u-tokyo.ac.jp/lexadv/
13. K. Murotani, M. Oochi, T. Fujisawa, S. Koshizuka, S. Yoshimura, Distributed memory parallel algorithm for explicit MPS using ParMETIS. Trans. Jpn. Soc. Comput. Eng. Sci. Paper No. 20120012, (2012) (in Japanese)
14. http://adventure.sys.t.u-tokyo.ac.jp
15. S. Yoshimura, R. Shioya, H. Noguchi, T. Miyamura, Advanced general-purpose computational mechanics system for large-scale analysis and design. J. Comput. Appl. Math. **49**, 279–296 (2002)

Lack of Cesium Bioaccumulation in Gelatinous Marine Life in the Pacific Northwest Pelagic Food-Web

Delvan R. Neville and Kathryn A. Higley

Abstract Bioaccumulation of cesium with increasing trophic position is well known across nearly every ecosystem for most organisms. In the marine environmental, typical (concentration ratios (Bq/kg in tissue: Bq/kg in seawater) range from 50–100 in lower trophic levels to 300–10,000+ for apex predators. Recent surveys of 7 gelatinous organisms off the coast of Oregon ranging in trophic position from 1.0 to 3.0 revealed a concentration ratio maximum of 12.5 and typical concentration ratios no higher than 4.4. The implications on human diets and ecosystem shifts for large radiocesium releases are discussed.

Keywords Cesium bioaccumulation · Marine life · Pacific Ocean · Pelagic food-web

1 Introduction

Accumulation of radiocesium with increasing trophic position has been well known since the 1960s. Most radioecological assessments at that time were driven by nuclear weapons testing, and concerned with human food safety in Western diets rather than ecosystem effects, which come from trophic levels of ~ 2.7 (e.g. pandalid shrimp, forage fish) up to and exceeding 4+ (e.g. older large tunas) [6]. In pelagic marine environments like those sampled in this study, the preferential retention of Cs with respect to K has been shown to increase by a factor of 2.4 per trophic step [3]. Recommended concentration ratios (CRs) for Cs vary by species to some degree, from 50 in crustaceans to 100 in fish [1], although observed CRs in higher predators can range much higher (~ 200–300 in albacore, 400–500 in sharks analyzed alongside the gelatinous organisms reported here). Cephalopod molluscs, however, break this trend with CRs typically closer to 10–20.

In the years following the release of substantial quantities of Cs-137 and Cs-134 (among other radionuclides) from reactors at the Fukushima Daiichii Nuclear Power

D.R. Neville (✉) · K.A. Higley
Oregon State University, Corvallis, USA
e-mail: dnevill@gmail.com

© The Author(s) 2017
J. Ahn et al. (eds.), *Resilience: A New Paradigm of Nuclear Safety*,
DOI 10.1007/978-3-319-58768-4_27

311

Station, this overall focus on Western diets has lead to surprises relative to expectations. For example, tea plants at the time of the release did not yet have new leaves (those harvested), eliminating foliar uptakes as a pathway. Nonetheless, some tea leaves harvested the following season still showed substantial radiocesium uptake. As gelatinous organisms are essentially wholly absent from Western diets, there has been far less study on cesium retention. Thus, there are surprises in store with regards to human dietary concentrations of radiocesium in jellyfish. It is hoped that this survey of various gelatinous organisms will help serve as a first approximation of expected outcomes in dietary jellyfish such as *Rhizostomatidae* and *Stomolophidae*.

2 Methods

Organisms were collected off the coasts of Oregon and Washington either directly from surface trawls or via dip nets. Collections are from 2012–2014, conducted as part of research cruises conducted by the National Marine Fisheries Service. After identification, samples were kept frozen until processed. Before radio-analysis, samples were first baked to dryness at 100 °C, before being carefully dry ashed at a maximum temperature of 450 °C and a maximum temperature increase of 100 °C per hour. Recovery of cesium through ashing was verified to be unity, within the range of counting uncertainty ($\sigma = 4.074\%$), by processing IAEA-414 freeze dried fish tissues [5]. Samples were analyzed for Cs-137 concentrations on a high-purity germanium γ spectrometer. The 72.5 mm diameter, 68 mm long closed-end coaxial detector has 70% relatively efficiency and 2.0 keV resolution [full-width at half-maximum] at 1.33 meV and 1.0 keV at 122 keV. To account for differences in detection geometry arising from differing volumes of ash, samples of known activity were counted at various fill volumes and a weighted least-squares fit for the absolute efficiency based on the fill volume was produced. Uncertainties in count rates, mass, geometry-altered efficiency, and chemical yield (on the basis of the yield using the IAEA-414 standards) were propagated.

3 Results and Discussion

Initial analyses of 1000 g by wet weight (standard sub-sample weight for other phyla being analyzed) yielded no detectable concentrations. Increasing this to the entire mass of each collection yielded detectable levels in only 1 of all collections despite wet weights up to and exceeding 10,000 g. Table 1 reports the results of radio-analysis for the 7 collections reported here. However, the minimum detectable activity (MDA) as based on the Currie limit[1] and reported in terms of Bq/kg

---

[1]The Currie limit is based on a decision level that produces a 5% false positive rate, and defines the minimum activity necessary at such a decision level to expect a 5% false negative rate.

Table 1 Activities/MDAs of analyzed organisms

Sample ID	Species	Composition	Cs-137 Act Bq/kg fw	Cs-137 rel err (1σ)	Wet mass (g)	Collected
JP1	*Pleuro brachia sp.*	Bulk	0.0125	38.86%	3246	July 2013
JS1	*Salpa fusiformis*	Bulk	0.0044	*<MDA*	2706	July 2013
JE1	*Phacellophora camtschatica*	Individual	0.0036	*<MDA*	5726	July 2013
CAR1	*Carenaria sp.*	Bulk	0.0044	*<MDA*	871	July 2013
JT1	*Thetys vagina*	Bulk	0.0060	*<MDA*	3041	July 2013
JBig	*Phacellophora camtschatica*	Individual	0.0108	*<MDA*	10,458	July 2013
C42	*Chrysaora fuscescens*	Bulk	0.0038	*<MDA*	9880	October 2013

Most samples consisted of bulk samples of multiple individuals. Phacellophora were large enough to analyze as individuals. MDAs calculated as the Currie limit e.g. activity for 5% false negative rate given detection limit

constrain the maximum possible bio-accumulation for these organisms to well beyond the bounds of those seen practically all other marine biota. Assuming a surface water concentration of 1 mBq/kg, 6 of the 7 samples have practically no bio-accumulation whatsoever: MDAs for bulk samples of the salps *Salpa fusiformis* and *Thetys vagina*, the medusae *Chrysaora fuscescens* and the two individual medusa *Phacellophora camtschatica* yield CRs of just 3.6–4.4. The mollusc *Carenaria* featured a CR no higher than 4.4, remarkably low compared to that recommended for most other non-cephalopod molluscs of 60 [1], which may be due to its dietary inclusion of gelatinous prey such as thaliacians and relative jelly-like body composition: it features only a vestigial shell a few mm in size and extremely soft translucent tissue, and by appearance outside of the water can be easily misidentified as a torn fraction of *Salpa fusiformis*.

The only sample with both sufficient biomass and bio-accumulation to be directly quantified still had resulting concentrations yielding a remarkably low CR. The small ctenophore *Pleurobrachia* had a CR of just 12.5 (σ = 38.85%), producing a total number of counts only barely above the decision level.

With regards to trophic level [6], these results are extremely surprising. All of these organisms, by trophic level, are at least comparable with forage fish such as sardine and smelt, with the latter having 10 times as much bio-accumulation at the maximum possible here. Further, as all but one of these are maximum levels constrained by MDAs, the actual CRs could be much lower...as ionoconformists it is not out of the question to consider a CR of 1 for the lower trophic position samples. In the case of *Phacellophora*, it occupies a trophic position above 3.0, specializing in predating on the gelatinous organisms like those sampled here. This puts it on par with young albacore and bluefin tuna in terms of trophic level, which have CRs of 200+ [2, 4].

One implication of this result is that, at least with regards to food safety, dietary jellyfish are likely to remain safe for consumption even in cases of substantial contamination. Drinking water standards for Cs-137 are several orders of magnitude below those for most foods owing to the much larger quantities of water consumed by a human. For example, the EPA in the US has a drinking water standard of 7.4 Bq/L (approximately 7.4 Bq/kg) for Cs-137, whereas the FDAs derived intervention level for Cs-137 in food products is 1,200 Bq/kg. Given the CRs presented here even for the highest trophic positions of gelatinous organisms, jellyfish should remain a relatively safe food product even when harvested in water well of 100 times past the safe drinking water standards.

Another implication of these results is one that has been mirrored in other areas of anthropogenic effects on marine ecosystems: jellyfish blooms. In the event of a radiocesium release of a large enough magnitude to produce population level effects in that ecosystem, the doses to gelatinous organisms can be expected to be tens to hundreds of times lower than that to crustaceans, fish and even macroalgae. Although no in-depth studies have been conducted on the radiosensitivity of gelatinous organisms, there is nothing to suggest they would be especially radiosensitive to offset this greatly decreased dose. Because of their extremely simple immune system that lacks the an equivalent "weak-link" of bone-marrow, and an intestine that is replaced directly by mitosis rather than the "weak-link" of crypt cells, the general indication is that they would show at least moderate radioresistance. One might expect, then, that ecosystem-affecting anthropogenic radiocesium releases might lead to jellyfish blooms and increased competition with forage fish much, much as been observed from anthropogenic eutrophication and apoxia (albeit for a different set of sensitivity/resistance mismatches).

4   Conclusion

Gelatinous organisms seem to show remarkably low retention of radiocesium when compared to their competitors in the food web. Even with 24 h counting periods, several kilograms of sample tissue to concentrate and a relatively high-efficiency HPGe in a low-background environment, only one of all gelatinous samples collected exceeded the detection threshold for Cs-137. CRs were no higher than 4.4 and may very well have been 1.0 for some of the organisms collected. Jellyfish may thus serve as a preferential food source during periods of high levels of radiocesium contamination, and may experience more population growth due to lack of competition in the event contamination levels are sufficient to produce population level effects in the ecosystem. More work on radiosensitivity of jellyfish and the means by which they maintain such low levels of retention are needed in the future.

References

1. IAEA, *Sediment Distribution Coefficients and Concentration Factors for Biota in the Marine Environment* (International Atomic Energy Agency, Vienna, 2004), p. 95
2. D.J. Madigan, Z. Baumann, N.S. Fisher, Pacific bluefin tuna transport Fukushima-derived radionuclides from Japan to California. Proc. Nat. Acad. Sci. U.S.A. **109**(24), 9483–9486 (2012). doi:10.1073/pnas.1204859109
3. A. Mearns, D.R. Young, R. Olson, H. Schafer, Trophic structure and the cesium-potassium ratio in pelagic ecosystems. CalCOFI Rep. **XXII**(1980) (1981)
4. D.R. Neville, J. Phillips, R.D. Brodeur, K.A. Higley, Trace levels of Fukushima disaster radionuclides in East Pacific Albacore. Environ. Sci. Technol. **48**(9), 4739–4743 (2014)
5. M.K. Pham, J. La Rosa, S.-H. Lee, P.P. Povinec, International Atomic Energy Agency (IAEA). Report on the Worldwide Intercomparison IAEA-414, Radionuclides in Mixed Fish from Irish Sea and the North Sea; IAEA: Vienna, Austria, 2004; IAEA/AL/145, IAEA/MEL/73, http://nucleus.iaea.org/rpst/Documents/al_145.pdf (2004)
6. J.J. Ruzicka, R.D. Brodeur, T.C. Wainwright, Seasonal food web models for the Oregon inner-shelf ecosystem: investigating the role of large jellyfish. Cal Coop Ocean Fish Invest Rep **48**(1), 106–128 (2007)

RadWatch Near-Realtime Air Monitoring (Natural Radioactive Backgrounds and Outreach)

Ryan Pavlovsky

Abstract Radioactive backgrounds establish the limit of sensitivity in detection systems for the general search scenario, and set the reasonably unavoidable dose limits for members of the public. Measurement of NORM isotopes in the air provides a unique opportunity to serve the dual goals of capturing temporal/ meteorological NORM variations as well as coordinate public outreach/education of NORM exposure. The RadWatch Near-Realtime Air Monitor (RAM) stores meteorological and high resolution spectroscopy data as a function of time from six stories above UC Berkeley Campus. This data is served hourly to the public via radwatch.berkeley.edu/airsampling to demonstrate, not only the existence of NORM, but also the large variations observed in radioisotope air concentration. Clarity and transparency in this education effort are paramount, and complement the urgency of a 'realtime' system. In the future RadWatch will expand to interactive, networked devices to broaden the scope and engage the public.

Keywords NORM · Background radiation · Radon · Low dose · Outreach

1 Introduction

Currently there exists a major deficit in public knowledge about nuclear technologies and nuclear science. A resilient society will include the education of the public and will be crucial to the success of nuclear technologies. The RadWatch program seeks to fill those gaps by providing data in context for background radiation measurements around the Bay Area. Here transparency and clarity

R. Pavlovsky (✉)
Etcheverry Hall, University of California, Berkeley, USA
e-mail: rp@berkeley.edu

© The Author(s) 2017
J. Ahn et al. (eds.), *Resilience: A New Paradigm of Nuclear Safety*,
DOI 10.1007/978-3-319-58768-4_28

317

measurements/data are paramount. A branch of the RadWatch activities, the Near-Realtime Air Monitor (RAM) is crucial to public outreach efforts and serves the dual purpose of providing high resolution spectroscopy data to analyze temporal and meteorological inputs on variations in the natural radioactive background.

2 Near-Realtime Air Monitor Detection Scheme

The air monitor that we have constructed forces 21SCFM of air through a 4″ diameter HEPA filter which is continuously assayed by a high resolution gamma spectrometer (HPGe). The FPAE-102 filter collects 99.99% DOP 0.3 μm particles. The mechanically cooled, n-type HPGe detector has superior energy resolution and efficiency, even at ∼ 10 keV energies. The detector is sensitive in the 3–3000 keV energy range. Spectra are reported every hour to the public to promote transparency (Figs. 1 and 2).

Fig. 1 RAM system, acquisition and weather station electronics

Fig. 2 Weather station with 6 parameters of local meteorological data, about 25 m above street level

3 Radioisotopes of Interest

Isotopes reported by RAM are primarily selected by total contribution to the spectra. These lines are NORM materials, composed of ^{40}K, ^{238}U, ^{232}Th or their daughters. Specifically thoron and radon daughters are of interest because they provide almost 2/3 of the natural dose that healthy individuals receive. Other isotopes are of public interest after events such as Fukushima Dai'ichi which include residual fallout or TENORM. The RAM is an improvement over the previous revision of the sample based air monitors that we employed just after Fukushima. The measurements for the time period just after Fukushima were mainly focused on the detection of ^{134}Cs and ^{137}Cs, the results are provided in Fig. 5 (Figs. 3, 4 and 6).

Isotope	Origin
Bi214	Naturally Occurring(U238)
Pb212	Naturally Occurring(Th232)
Tl208	Naturally Occurring(Th232)
K40	Naturally Occurring
Cs134	Reactor
Cs137	Technically Enchanced/Naturally Occurring

Fig. 3 The isotopes reported by the RAM system to the web. These are significant in that they are NORM or isotopes related to Fukushima releases

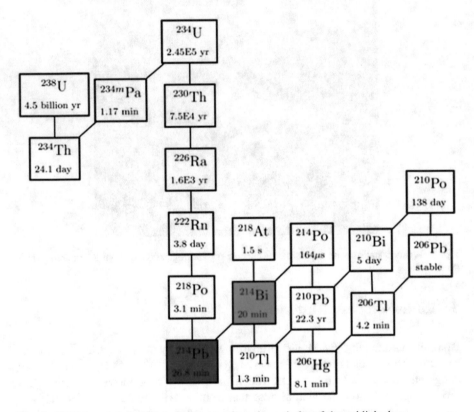

Fig. 4 238U decay chain. These decays constitute the majority of the public's dose

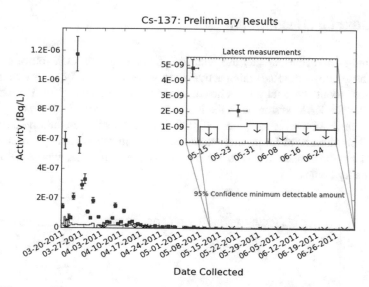

Fig. 5 ^{137}Cs and [134]Cs discrete filter air sampling just after Fukushima with first revision air sampler. The RAM system would have the ability to continuously sample events of this kind, capturing the transient response

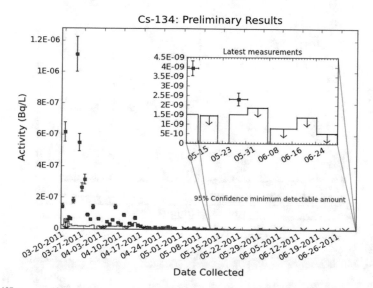

Fig. 6 ^{137}Cs and [134]Cs discrete filter air sampling just after Fukushima with first revision air sampler. The RAM system would have the ability to continuously sample events of this kind, capturing the transient response

4 Outreach (Heading A)

Accessibility and transparency are vital to establishing trust in an outreach effort. The website allows the RadWatch activities to scale accessibility, by providing data in context about the naturally radioactive world we are living in. The strongest context for the RAM system is in the long time history over which we are tracking isotopes. In the future, a computed NORM dose-rate-variance could be used as an unavoidable dose band for putting excess dose risk in perspective. Transparency is imposed by the immediacy with which results are published, while maintaining good quality control.

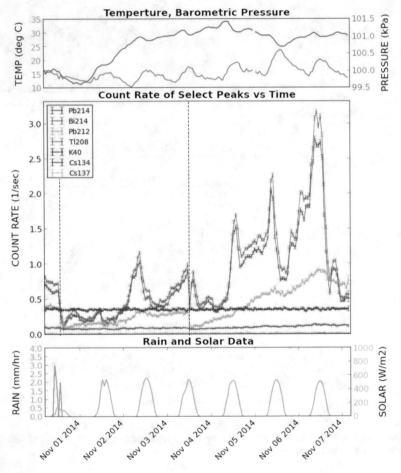

Fig. 7 [*Middle*] Net peak counts rate for 1 week. The natural background lines vary by as much as factors of 25. Observation of this variance is considerable. *Dashed black lines* indicate the filter exchange times. [*Top* and *bottom*] Meteorological data collected from the weather station just above the monitor. Currently the air monitor is down for upgrades to allow for deduction of weather trends

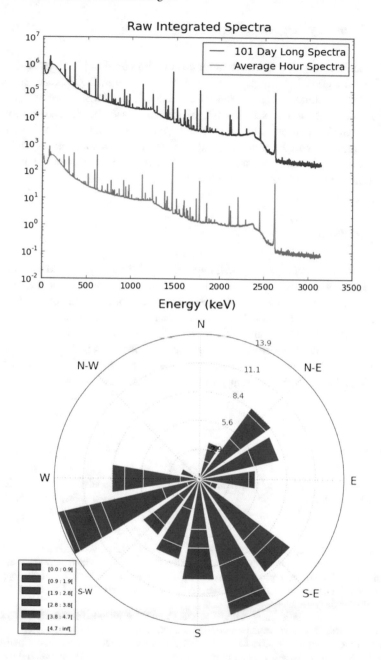

Fig. 8 [*Top*] A wind rose measured from the weather station. The most probable direction for wind was W-SW at about 14% of the time with gusts up to 4.7 m/s+. [*Bottom*] Integrated and average spectra from 100d period

5   Preliminary Results/Status

The air monitor provided data with a total downtime of about 72 h over the Feb to Dec 2014 time period. A sample of this data is provided in Fig. 7. One can observe factors of 25 variation in certain gamma ray lines. This defines the natural band of variation for these products in the air. More work is to be done to remove systematics from the data, with the caveat that isotope selection is a bandwidth filter. Planned upgrades will help achieve the optimum time resolution for radon concentration in the air. The optimum is defined by the rate kinetics with some simple assumptions about the isotope collection (Fig. 8).

6   Conclusions

A radiologically resilient society will depend on public data provided in context and transparently. We have demonstrated the operation of a high resolution spectroscopy system with accessible data to provide the public with the facts that the world is radioactive and that they concentration of natural radioactivity in the air varies considerably. The goal of this research is to eventually calculate an estimate of current dose rate. Concentration of background isotopes in the environment are incredibly useful for putting nuclear technologies and events in context, allowing the public and scientific community to weigh the risks and benefits associated with this field.

Reference

1. NCRP Report No. 160: Ionizing radiation exposure of the population of the United States 2009. J. Radiol. Prot. **29**, 465

Incorporating Value Discussions into High Level Radioactive Waste Disposal Policy: Results of Developing Fieldwork

Rin Watanabe

Abstract The disposal of high level radioactive waste has fared no better in Japan since its legislation in 2000 than in most countries grappling with the same problem. This research aims to contribute to realizing some form of disposal in Japan, by suggesting ideas for an improved institutional scheme of policy making. This scheme concerns value judgments in decisions of technology use. Historically, implementing agencies have allowed limited debate on issues of value. What would develop if values previously neglected were given a chance to become a technical option of their own for the disposal of high level radioactive waste? This question has been taken out to the field, in the form of group interviews of young citizens. Here, the details and preliminary discussion of the fieldwork are described, as a temporary result of this study. A final section is dedicated to discuss the possible contributions of this study to the consideration of engineering resilience.

Keywords High level radioactive waste disposal · Policy making · Value discussions · Citizen interviews

1 Introduction

1.1 High Level Radioactive Waste Disposal in Japan

The disposal of high level radioactive waste has fared no better in Japan than in most countries grappling with the same problem. Since 2002, two years after legislation announced that Japan's waste would be vitrified, stored for 30–50 years, then disposed in a geologic repository within its territory, Japan has embarked on a process of 'local solicitation', asking municipalities to apply appropriate sites for further investigation. Compensatory money and services sparked near calls for application nationwide, but nothing has been officially registered.

R. Watanabe (✉)
University of Tokyo, Tokyo, Japan
e-mail: watanabe@esl.t.u-tokyo.ac.jp

J. Ahn et al. (eds.), *Resilience: A New Paradigm of Nuclear Safety*,
DOI 10.1007/978-3-319-58768-4_29

Moves to fix problems within the set framework, such as finding visual means of explaining the official position, and elaborating on the compensatory measures, have been studied and adopted in abundance. There have also been moves to shift—less from the localities, more to the national government—the political weight of applying to the siting process. All have ended to no avail as of today.

Meanwhile, opposition against the present method of disposal of high level radioactive waste (and of its by-production) has been repeatedly put forth by the concerned public in the form of public comments on policy drafts and opinions gathered at public hearings. These opinions have been presented both prior to [1] and after [2] legislation, but close inspection reveals that it is hard to say they have been given a substantial response [3]. Much of the official publications and discussion events have dismissed other methods of disposal on grounds of their political and technical challenges to realization, arguing instead on the safety of geologic disposal, or what can be done for localities accepting site exploration.

This apparent evasion of debate occurs not necessarily because geologic disposal cannot withstand counterarguments, but rather, because the choice of geologic disposal relied heavily on politically top-level selection based on prior international debate and engineering perceptions. This means that, in the Japanese context, it is difficult to give a logical explanation why the choice of disposal method was made; what criteria were considered, whether the choice of criteria or their relative significance were democratically or scientifically warranted. Albeit mention in the present policy of securing a margin for the 'future generations' to reform the present policy, doubts arise whether anyone can recognize a "betterment" coming along when there has been no attempt to substantiate what "better" actually means.

Such a history suggests, both in practicality and in democratic theory, that more radical reconsideration of policies is due.

1.2 Overarching Aims

This research aims to contribute to realizing some form of disposal in Japan, by suggesting an improved institutional scheme for radical (re)consideration of high level radioactive waste disposal policies. The improvisation will center on problems of how decision making agendas have been limited/assigned to certain actors over time. This direction will lead further to questions such as, "How should the public participate?" or "What is the role of certain specialists in policy making?" which are both questions echoed from related studies [4, 5].

To reconsider institutional schemes in this view, the history of decision making on high level radioactive waste disposal will be reviewed with a mind to discern what agendas have been heretofore assigned to which actors, while consciously inquiring how agendas and actors could or should be defined. These queries are of importance, since determining ways in which agendas/actors could be categorized, and being able to position existing issues within the spectrum of these categories (e.g. from the technically general to the specific, from national to local, etc.) is fundamental to understanding what agendas have been "mis-assigned" in past policies; what has been assigned perhaps against a more desirable order of agendas/actors at some point in time.

1.3 Research in Development: Incorporating Values into Decisions of Technology Use

This study is in process (most of the understanding arrived at this point has been formed as the author's graduate thesis [3], and is expressed in the first sub-section of the introduction), and a working hypothesis has been derived: One problem of prior institutional schemes is that citizens have been deprived of debate concerning overall value judgments of technology use (upstream engagement), and instead have been expected to participate in judging technical, specific issues (downstream issues). Since it can be assumed that the characteristic of citizen actors lies is in their citizenship and not their engineering expertise (or the lack of it), it follows that devising a means to ask citizens to discuss values concerning such issues would be a better combination of agendas/actors. This is a point shared with researchers working for the Science Council of Japan's on issues of high level radioactive waste disposal policy [6].

It is worthy of noting here that the categories of actors employed here does not split society into engineers and citizens, but rather assumes an abstract notion of citizens being those thinking as a politically independent member of society (including vocational engineers, researchers etc.), with the other end of the scale being so-called specialists, engineers, or non-"value-thinkers"; those who have the know-how of solutions to predefined problems based on their expertise. It is not necessarily that "the non-engineers' view was not considered" but rather that "those of the public who spoke as citizens were dismissed" or "agendas which concerned value judgments were put aside" in the debate concerning high level radioactive waste disposal policy of Japan.

This working hypothesis has been put to test. Using chances granted by funds from Tokai village, a municipality with a long relationship with the nuclear industry, a method for consulting people in matters of value regarding disposal related technology has been devised and conducted. The details are introduced in the subsequent sections.

2 Methods of Incorporating Values into Discussion of Technology Use

2.1 Objectives and Methods

The objective of the fieldwork at Tokai village is to attempt to consult people on issues of values which influence decisions on technology use (as concerns this presentation). To have achieved this, the value-laden issue needs to be defined, then the context of the technological problem conveyed, then the problem consulted. The results of the consultation can then be analyzed by the researchers and translated back into its implications for discussions of technology use. This process needs to sufficiently convey the context of the discussion, and be appropriately defined (if the issue is too abstract in terms of value, for example if people were asked to discuss political values each endorses, this would likely not work any better than asking them to talk about the appropriate thickness of waste containers).

The method of consultation devised here takes the form of a group interview. During August 2014 to February 2015, 18 people were interviewed in various groups, each session lasting around 90 min. Participants were asked to relax and to discuss freely "what you think is a desirable means of 'disposal' of high level radioactive waste" without worrying about present technical limitations. A less than 5 min explanation of what high level radioactive waste is, what it looks like, how much radioactivity it holds and how long that lasts, how much of it Japan has, and its present designation was given at certain timings using the figures shown (Fig. 1).

One important characteristic of these interviews was that the targeted age group of the participants was limited to "young" people, set as being between 16 (a high school freshman) to 34–35. This target was meant as a current attempt of listening to the "future generation".

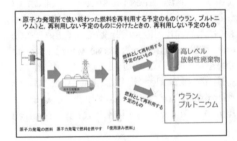
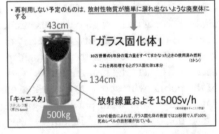
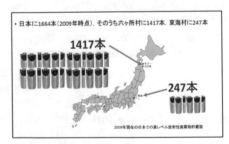
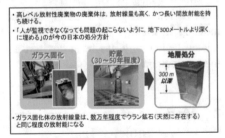

Fig. 1 The 4 slides handed to participants during the group interviews

2.2 Results and Discussion

Some of the frequently observed opinions obtained through the group interviews are listed above (Table 1). Many of the participants from varying groups raised questions about reducing and re-using the waste, and some mentioned that such efforts to fundamentally ameliorate the situation should be continued after the waste has been stored in a repository. There were mixed feelings about putting hopes in future technology development, but not many were in favor of giving it up altogether. Questions about what sort of place would seem fit for the storage of waste drew answers that places isolated from the human environment and places which humans have little relation to even into the future were considered better, revealing that the general idea of a geologic repository (for storage/final disposal) may not be far off. The seabed and outer space, though controversial within the groups, was also raised as a possibility. Many of the participants viewed the unfairness of siting a repository as a problem, and suggested that the sites be dispersed into <10 sites, or 47 sites—one for each prefecture. It would follow that the idea of dispersed repositories should be given a (perhaps qualitative) technical evaluation. When asked what sort of message might be left to future generations, many of the participants said a warning would be necessary, of the fact that high level radioactive waste is present, and of its storage place. Some offered to say that an apology to the

Table 1 Primary results of the group interviews

Opinions[a]	General implication
"It is better for the waste to be re-used"	Means to minimize the fundamental problems should be reconsidered
"It is better for the production of waste be reduced"	
"Can't the radiation be reduced?"	
"Can't the radioactivity to be shortened?"	
"Somewhere isolated from the human environment would be better for a repository"	The general idea of a geologic repository may not be far off, but other possibilities should also be evaluated in these terms
"Somewhere that has little to do with humans for years to come"	
"The waste should be stored so that the unfairness of siting is minimized"	Means to minimize "unfairness" should be reconsidered
"Monitoring of the waste should be continued for as long as possible"	Even for a repository with passive safety designs, maintaining active safety measures should be considered
"Effort should be made to be able to do something if anything happens"	
"Effort to pass on messages of the presence of the waste and of apology should be taken"	Means to pass on warnings and apologies should be reconsidered

[a]Revised and translated

future generations was due, yet others openly stated that they thought this unnecessary.

Overall, it can be said that once people started thinking about the high level radioactive waste, most did not opt for trying to have as little to do with it as possible, instead coming up with ideas about how the problematic could be reduced, what the generation could actually do in face of the waste.

At least 10 of the arguments held implications to the present policy in that they raised questions which had not been adequately investigated in Japan's context. These initiatives can be directly translated into 'homework' for the experts, or discussed further with citizens by giving more information about the general concerns for the ideas that the experts hold. For example, informing citizens about the technical difficulties of separating radioisotopes, or the technology level of transmutation may yield different feelings for the idea of decreasing radioactivity, or call for extra conditions on the desirability of waste re-use.

From the results of the interview, it can be said that a general background understanding of high level radioactive waste seemed to have been conveyed to most of the participants, although a more detailed explanation of the nature of radiation may have been necessary. There may also be room for improvement in the explanations to some of the technical questions raised during the interview, which could have affected the range of ideas the participant could voice, however, this point requires more trials to confirm. The setting of the issue itself can also be considered appropriate (neither too technical, nor too abstract) if we note that a certain number of opinions concerning values were obtained through this interview.

3 Citizen Input When Engineering Resilience

3.1 Defining "Resilience" in Relation with Engineering and "Values"

It seems important to clarify: What does it mean to "engineer resilience"? As described above, this study is based on the belief that engineering is the act of realizing values in society by integrating fields of expertise, including social, political or economic know-how. Where does resilience fit in this picture? Is it another tool, or is it a value? Is it something whose usefulness engineers can master and take for granted, or is it something which can only arrive into the engineering realm from outside? Yet another way of asking this question: Is resilience something which realizes (for example) "health", or is it something which competes with, requires a striking of balance with "health"?

As the last interrogative suggests, resilience is neither a societal value nor entirely a tool. Rather, it can be considered as an "added value" to the engineered societal value in question. If we tentatively phrase the definition of resilience as the (hopefully long-term and society-wide) *enduring effectiveness of an engineering act during and after adverse events* (keeping in mind common denominators offered in prior conceptualizations [7]), we could say, for example, that a health-resilient solution is likely better than a solution just meant to realize health. So health is the value which is being engineered, and "health-resilience" (the enduring conservation of health even during and after adverse events) is the "added value". In other words, an engineer who tries to achieve a healthy solution does not guarantee that the state of achieved health will be a long-term one for society at large, during and after disasters.

It is also noteworthy that aiming for "resilient" solutions instead of just healthy ones or uncostly ones is something which makes engineering more valuable, and is something which the engineering community should aspire for itself. Here, lessons learnt from the past century urge the engineers to recall that, the societal value of a technology (whether resilient or not) should be decided by its citizens. Resilience will not be required for a societally unrequired branch of expertise (no matter how those experts may believe that the technology will be more valuable once performing resiliently). Care is also required in conceiving of "resilient communities". The notion—just like any other non-resilient paradigm for modern societies—sees communities in light of engineering interests. These are interests which tend to repel exceptions or transitions of lifestyle, and retain control over society.

3.2 How This Research Approach Can Contribute to Engineering Resilience

As aforementioned, resilience can be considered as an "added value" to a solution realizing a societal value. Care is needed not to assume the value in question actually is valued, just because the solution is resilient. This means that significance remains in the attempt of this research to "discussing values" and clarify "what matters and why" for citizens upon deciding societal solutions. There may also be the need to give special attention to social values in disaster situations, since priorities may change. There may be multiple lifesaving triages, perhaps with heavier priorities on younger generations or the socially vulnerable. While it is unrealistic to attempt a thorough discussion of values during crises, it is possible to go over past cases and obtain citizen input into reconsideration of measures taken during emergencies. This is likely a key research agenda for Fukushima studies, and is necessary to investigate engineering challenges for "resilient communities", yet efforts heretofore seem fragmented and insufficient.

Another point to make is that introducing "resilience" into solutions (healthy solutions -> health-resilient solutions) requires more collaboration between "experts" and "non-experts" upon considering how they may be engineered. This is because more input of those who use the system (the majority of who are "non-experts") becomes more important when considering engineering in abnormal situations. For instance, looking for components in the techno-social system which may be susceptible to changes in working conditions is something which requires wisdom of "living in" the system as well as "planning and controlling it from outside". This feature of "lay people" being the "users" and "day-to-day supervisors" of the systems experts devise, is what makes their input more valuable upon considering what is too costly to plan at the normal engineering scale of time and probability. The method suggested in this study may contribute to such efforts in its approach to clarify what input we are asking from whom.

This is not to say that conventional expertise is neutralized. Resilience is essentially about the efficiency of the system, so we are acting as those "planning and controlling the system from outside" when we feel the need for resilience. When considering how to formulate resilient solutions, we need to integrate expertise, and consider in systems and balances. Here, the importance of a deep understanding of how much we know/don't know should not be diminished.

4 Conclusions

High level radioactive waste disposal policy has not worked out in Japan as it has been designed, and a study of the criticism it has received shows that those responsible for maintaining its policy have not been able to respond to such debate due to its lack of consideration about the value judgments incorporated in issues of technology use. This understanding of the problem has been applied to fieldwork in Tokai village, where group interviews were conducted with the cooperation of young citizens, to ask for ideas on "What would be a desirable means of 'disposal' of high level radioactive waste". The results centered on ideas to tackle the "fundamental" problems of the waste, such as its presence, amount, radiotoxicity and "uselessness" or "unwanted-ness", rather than just cope with the safety risks the waste imposes. There were also opinions on what sort of environment would be best for storing the waste, which shared points in common with the present policy but also added that the "unfairness" of concentrated siting should be reconsidered. Other opinions opted for persisting in R&D and maintenance, for reducing or reusing the waste, hauling it to outer space, etc. These opinions on the desirable means of dealing with the waste hold respective implications for those with the know-how to consider how they may be realized, or to evaluate their technical feasibilities into the future so that they can be reconsidered as alternative options.

A consideration of what it means to "engineer resilience" led to the idea that resilience can be considered as the "added value" of enduring efficiency during crises, added to the societal value realized in an engineered solution. The approach in this study of asking citizens for input on discussions of value also applies to "resilient" solutions, perhaps in the same way. One marked research agenda would be to obtain input of "what matters and why" in reconsideration of how societies have addressed passed disaster situations, especially to grasp what has developed in Fukushima. Another point made was that obtaining input from "non-experts" as the "users" and "day-to-day supervisors" of the systems experts devise, may be valuable upon considering resilient solutions.

Acknowledgements The fieldwork in this research was performed with Kohta Juraku (then an Assistant Professor at Tokyo Denki University) as part of the "Chiiki Shakai to Genshiryoku ni Kansuru Shakai Kagaku Kenkyuu Shien Jigyo (Funding Project for Social Scientific Research on Local Communities and Nuclear Power)" supported by Tokai village, Ibaraki prefecture.

References

1. See for example: JAPAN ATOMIC ENERGY COMMISSION, Hoshasei Haikibutsu Shobun Kondankai no Houkokusho (Reports of the Discussion Sessions on High-Level Radioactive Waste Disposal). Japan Atomic Energy Commission (1998)
2. See for example: Koremade Itadaita Goiken: Souhoukou Shinpojiumu -Dousuru Koureberu Hoshasei Haikibutsu? 2013 (Opinions and Questions received thus far: Two-way Symposium - How should we manage High-Level Radioactive Waste? 2013). Agency for Natural Resources and Energy, Ministry of Economy, Trade and Industry (2013)
3. R. Watanabe, A Study of the Debate Concerning High-Level Radioactive Waste Disposal. Graduate Thesis advised by Prof. S. Tanaka at the Dept. of Systems Innovation, Faculty of Engineering, University of Tokyo, not published (2014 < Japanese A.Y. 2013 >)
4. H.M. Collins, R.J. Evans, The third wave of science studies: studies of expertise and experience. Soc. Stud. Sci. **32**(2), 235–296 (2002)
5. W.C. Gunderson, B.G. Rabe, Voluntarism and its limits: Canada's search for radioactive waste-siting candidates. Can. Public Adm./Administration Publique du Canada. **42** (Summer/Eté), 193–214 (1999)
6. K. Juraku, To deal with the 'difficulties' of high-level radioactive waste management. The necessity of 'value judgment' discussions and knowledge sharing in the society. Kagaku **83**(10) (2013)
7. E. Hollnagel, D. Woods, Resilience engineering precepts, in *Resilience Engineering: Concepts and Precepts*, Epilogue, eds. by E. Hollnagel, D. Woods, N. Leveson (2006), pp. 348–358

Part V
Epilogue

Hybrid Disasters—Hybrid Knowledge

Charlotte Mazel-Cabasse

Abstract Drawing from in-depth anthropological research in the San Francisco Bay Area, looking at a community of scientists, experts, and other risk-conscious residents who are preparing for the next large earthquake, this article argues for an understanding of resilience as an overarching heuristic concept with the potential to articulate multiple forms of knowledge into a collaborative approach, associating scientists, experts, and residents. Building on the corpus of literature coming from Science and Technology Studies (STS), Geography and risk Disaster Studies, this article discusses the emergence of the concept of resilience and its articulation with the existing literature. Following this exploration, I will look at the implication such concept in the re-definition of knowledge and the categories of expertise as observed during my field research in the Bay Area of San Francisco. I find that resilience can be a useful concept only if the rigid definitions that have separated academic disciplines, as well as the concepts of "science" and "experience," are recomposed in favor of a more integrated approach taking into account the multiple, and emerging, dimensions of knowledge.

Keywords Disaster · Knowledge · Integration · Expertise

1 Introduction

Each disaster is unique. To prepare for the worst and increase the resilience of space "at risk," researchers and experts have been documenting the multiples dimensions of disaster, documenting the complex and interdependent systems that a disaster can break down. These studies have focused both on the technico-socio-economic preconditions that tend to worsen the effects of disasters; but also—as the contributions of this book have shown—the cultural, political, philosophical and even

C. Mazel-Cabasse (✉)
Berkeley Institute for Data Science, Berkeley, UC, USA
e-mail: charlottecabasse@berkeley.edu

© The Author(s) 2017
J. Ahn et al. (eds.), *Resilience: A New Paradigm of Nuclear Safety*,
DOI 10.1007/978-3-319-58768-4_30

metaphysical parameters that influence the conditions—and quality—of a particular response to a specific catastrophe.

In this chapter, I would like to discuss the potential and limits of resilience as an operative concept. Building on the idea that knowledge about disaster to come is necessarily incomplete, I will first give an overview of the emergence of the notion of resilience in the different fields of social science. Then, building on the field research conducted in the Bay Area of San Francisco between 2009 and 2013, I will show how experts and scientists of the Bay Area have learnt to use their empirical experience the earthquake risk and combine it with state of the art science production, creating de facto a corpus of knowledge that I will describe as "hybridized."

2   A Brief Overview

Risk and disaster are hard to conceptualize and apprehend. As Oliver-Smith [1] writes:

> Disasters are both socially constructed and experienced differently by different groups and individuals, generating multiple interpretations of an event process. A single disaster can fragment into different and conflicting sets of circumstances and interpretations according to the experience and identity of those affected. Disasters force researchers to confront the many and shifting faces of socially imagined realities. Disasters disclose in their unfolding the linkages and interpenetrations of natural forces or agents (pp. 25–26).

Recent major disasters, including a great number of earthquakes, have not only deeply transformed the actual space they impacted but also the world around them.[1] They forced us to face the human capacity to deal with, and to respond to, such events. As French geographer Michel Lussault [2] observed in the aftermath of the 2004 Indian Ocean tsunami, when engaged with destructive events, thinking becomes complicated by the influx of overwhelming information that seems incompatible with science's need for impartiality and objectivity. He writes:

> Soon, emotion facing what appeared as incomparable tragedies became global. [...] The extension at the scale of the globe of the rumors of the disaster came together with the diffusion of a spectacular dramaturgy associating narratives, pictures of professional and amateurs, more or less scientific description[s] of the tsunami and its consequences, science fiction discourses on the conceivable future replicas of such phenomena.[2] (My translation, [2])

---

[1]Among others, they include Hurricane Katrina in 2005, the Haiti Earthquake in 2008, and the Christchurch Earthquake.

[2]My translation. In the original: "Rapidement, l'émotion face a ce qui apparut comme une tragédie incomparable devint mondiale (…) La dilatation à l'échelle du globe et l'écho de la catastrophe s'accompagna de la diffusion d'une dramaturgie spectaculaire, associant des récits, des images de professionnels et d'amateurs, des descriptions plus ou moins scientifiques du tsunami et de ses conséquences, des discours d'anticipation sur les future répliques envisageables d'un tel phénomène" [2].

The attention toward examining the multiple impacts of disasters, and the challenges they make visible for large metropolitan areas, is not recent. One of the most famous historical examples is the 1755 Lisbon earthquake, which had a considerable influence on the emergence of scientific knowledge. The Great Lisbon Earthquake, estimated at M.8.5 to 9.0 on the Richter scale, is credited with transforming the social, philosophical, and metaphysical paradigms of the time. Some have gone so far as to state that this moment marked the beginning of the era of European Enlightenment [3–9]. Discussing the importance of the Great Lisbon Earthquake for continental philosophy, Gilles Deleuze argues that the event had an intellectual and metaphysical impact equivalent to the one of Holocaust in the twentieth century:

> It is very curious that in the eighteenth century, it is the Lisbon earthquake which assumes something like that, when across Europe people said: how is it still possible to maintain a certain optimism founded on God? You see, after Auschwitz raised the question: how it is possible to maintain a fading optimism about human reason. After the Lisbon earthquake, how is it possible to maintain the fading belief of rationality in divine origin?[3] (My translation, [10]).

In the twenty-first century, large size earthquakes such as the 2011 *Tōhoku earthquake* still provoke shifts in our understanding of the world, forcing us to question the modernist categories inherited from the Enlightenment.

Today, experts and lay people alike have also come to question the horizon of a world without disaster, as well as the place that disasters can take in our everyday lives [11–15]. Acknowledging that the balance between dangers and safety precautions is constantly renegotiated in contemporary societies [16–19], researchers have also emphasized the role of expertise in the definition of such concepts [20, 21].[4] Moving away from positivist definitions, these authors have engaged with larger questions posed by the complex relation between science, expertise and society [23–26]. Experts and scientists have—indeed—been often criticized for creating tensions within the democratic process [27] and not taking into account

---

[3]My translation. In the original, it reads: "Il est très curieux que au dix-huitième siècle, ce soit le tremblement de terre de Lisbonne qui assume quelque chose de cela, où toute l'Europe s'est dite: comment est-il encore possible de maintenir un certain optimisme fondé sur Dieu? Vous voyez, après Auschwitz retentit la question: comment est-il possible de maintenir le moindre optimisme sur ce qu'est la raison humaine? Après le tremblement de terre de Lisbonne, comment est-il possible de maintenir la moindre croyance en une rationalité d'origine divine?" [10].

[4]Wynne critiques are aimed mostly toward the reproduction of dichotomies, "which are key part of the problem of modernity: natural knowledge vs. 'social' knowledge, nature versus society, expert versus lay knowledge. It's also reflects—and reinforces—a more basic lack of recognition of the cultural/hermeneutic of scientific knowledge itself, as well as of social interactions and cognitive construction generally. […] I also thus problematize their uncritical conception of science and knowledge per se. It is important to distinguish here between their recognition of the (in recent years only) contested nature of scientific knowledge, and their uncritical reproduction of realist concept of scientific knowledge. This realist epistemology also, I argue, gives rise to an unduly one dimensional understanding of the underlying dynamics of the nature of 'risk' in the risk society" [22].

local knowledge [22]. But they have also been praised when they base their expertise on empirical knowledge [28] and when they coproduced knowledge along beyond disciplinary and scientific boundaries [29]. In fact, authors have noted, in many recent disasters, experts have been the beneficiaries of their own expertise [30–32].

The difficulty to deploy the entangled interactions and co-dependencies that composed the what is usually referred as "the" disaster, has been one of the many research programs tackled by researchers in Anthropology, Disaster Studies, and more recently, in Sciences and Technology Studies (STS).[5] Social scientists have greatly improved the granular understanding of the risk and disaster life-cycle (before, during, after) insisting on the importance of culturally situated knowledge and exposing tensions between political and scientific expertise, cultural production and regimen of governance [34–38].

Studied through the frame of socially constituted and culturally meaningful practices, contemporary social sciences have enabled researchers to create an index of adaptation—or often mal-adaptation—between human and non-human facing disasters. These studies have tackled questions coming from very different disciplinary fields, from the physical, engineering, social, and political sciences to the humanities, using multiple methodological approaches anchored in diverse epistemologies within academia [17, 39], (Guarnieri, Sato, Pecaud, this issue), and beyond when engaging with the more operational level of technical expertise [40, 41], this volume) and policy-making [42, 43].

Questions related to the threat of potential disasters, which come together under the concept of "risk," have become an extensive field of research, in which the seminal work of Beck [44] has been both celebrated and criticized [22, 45–48]. Following these trends, studies of risk that lacked empirical evidences—as it has been the case for Beck at the time of his first publication—were sharply criticized. As Latour [47] noted, *"like most sociologists, Beck suffers from anthropology blindness"* (p. 453). More specifically, Boudia and Jas [45] acknowledged that researchers reacted strongly to the publication of Beck's book, documenting what seems to have been blind spots in the book, namely, citizen science, public participation in scientific research (PPSR), and the role of concerned individuals in shaping the definition of risk. Indeed, in more than 50 years of fieldwork [49–53],[6] social scientists working on risk and disaster studies have shown that, when examined in detail, the conditions for the emergence of a risk and the characterization of a disaster, rather than being a homogeneous set of concepts, practices, and methods, have *"been continually fraught with internal tensions"* [55].

The Tōhoku Earthquake, Tsunami, and nuclear disaster of 2011 echoed this complexity [56, 57], (this volume) and has been broadly studied and commented

---

[5]STS can be defined as an "interdisciplinary field that examines the creation, development, and consequences of science and technology in their cultural, historical, and social contexts" [33].

[6]And this not mentioning the work done by United nation agencies in collaboration with social scientists see for instance Cabasse et al. [54].

upon, with interested parties asking questions about both the causes of the nuclear disaster and Japan's capacity to recover from such a large catastrophe [58–63]. Reflecting on the consequences of a disaster—death, destruction, the challenges of recovery [60, 64] and the long lasting-trauma [65–67]; resilience studies forces us to revisit our understanding of the events, focusing on the relationships between individual and collective experience and complex scientific, political and engineering systems. The possibilities of interactions, if clearly defined, could reconnect with previous effort to conceptualize risk and disasters multi-dimensionally.

3 Thinking with Resilience

For almost a decade now, researchers have noted that: "Like vulnerability, multiple definitions of resilience exist within the literature, with no broadly accepted single definition…While numerous research efforts have assessed various dimensions of community resilience, challenges remain in the development of consistent factors or standard metrics that can be used to evaluate the disaster resilience of communities" [68].

Building on the research of "vulnerability", which has tended to emphasize the subject as an agent, and his specific capacity or lack thereof, to adapt or adjust to a identified risk; resilience studies have favored a more systematic approach [69–71]. Using a specific case study located in the Sekhukhune District, Limpopo province of South Africa, Miller and al. have made explicit the divergences in the modes of engagement between the two concepts pointing the nature of this paradigmatic change:

> In the vulnerability study, the aim was to understand how different stakeholders view their vulnerability to support decision-making at the village and municipal scales in Sekhukhune District. […] In the resilience study, the aim was to establish an overall picture of system function, including qualitative system dynamics and vulnerability analysis, in the Sand river sub-catchment using resilience theory. […] This study made explicit the linkages between the social and ecological system on the one hand, and the time scales at which certain drivers proved more important than others. The vulnerability approach placed more emphasis on agency and on the identification of hooks for responding to adaptation and development challenges [72].

Miller et al. are insisting on the divergence of "analytical tools, scales, and indicators of interest" [72]. But these divergences remain unclear about the political implications and controversies that arise when considering both the unit and the scale of analysis (a system, an individual?) and the moral and political dimension of the distinction as pointed by Reghezza-Zitt et al.:

> The recent shift from vulnerability to resilience gives a glimpse of a radical change in the approach of international agencies to disaster management. On one side, a social vulnerability, mainly suffered by the poorest populations but that can be anticipated and reduced by various aids relying on solidarity and states' participation. On the other, a desired resilience that is only validated long after the crisis and that sanctions adaptation at an individual level [73].

Following this argument, orientation of these conceptual investigations (re-silience or vulnerability) should not be oblivious of political analysis which might help understand why and how, "environmental risk in the city is interpreted as an outcome of the political interests and struggles over power that shapes the urban environment and society" [74]. As Comfort et al. recall:

Some cities do better in the face of disaster than others. It is tempting to describe apparent success in terms of resilience and apparent failure in terms of a shopping list of explanatory variables. Resilience then becomes the synonym for survival and the prescribed antidote for administrative shortcomings. This is too simple (...) Far from a fix-it-and-forget-it approach, resilience is the outcome of a long-term process, enduring resilience is a bal-ancing act between risk and resources, between vulnerabilities and escalating or unman-ageable risk [75]: 272–273.

In 2009, the United Nations Office for Disaster Risk Reduction (UNISDR) have tried to unifies these array of definitions, proposing the following synthesis: "The ability of a system, community or society exposed to hazards to resist, absorb, accommodate to and recover from the effects of a hazard in a timely and efficient manner, including through the preservation and restoration of its essential basic structures and functions" [76].[7] Trying to account for the multiplicity of situations in which resilience could emerge, UNISDR definition was building years of cases studies and an extensive body of literature and modelization, both empirical and theoretical, that had seen a succession of conceptual apparatus trying to account for the complexity of the phenomenon to be both described and tackled. This heuristic dimension was confirmed in 2014 when the National Science Foundation (NSF) announced that 50 researchers had been awarded a total of 17 million dollars [77] to "investigate questions related to vulnerability, risk and resilience in the face of various hazards as well as the everyday degradation that infrastructures face."

Indeed, and perhaps because of this polysemic epistemology[8,9] the concept of resilience seems to have been successful in fostering discussions between scientists

---

[7]During first conference in 2005 in Kobe Japan was adopted The Hyogo Framework for Action 2005–2015: Building the Resilience of Nations and Communities to Disasters which contributed to the diffusion of the concepts of resilience define as "the ability of a system, community or society exposed to hazards to resist, absorb, accommodate to and recover from the effects of a hazard in a timely and efficient manner, including through the preservation and restoration of its essential basic structures and functions" [76].

[8]"There is no universal definition of the concept of resilience that can be applied to all domains. That said, the English term resilience, itself derived from the Latin verb resilire (to bounce), is made up of re (again) and salire (rise), which implies a retroactive effect [10]. While in the 1970s the term was associated with the ability to absorb and overcome the effects of significant, unex-pected and brutal disruption to ecological systems, hybrid definitions have since emerged in many disciplines including geography, psychology, sociology, organizational sciences, ergo-psychology, etc. Within this smorgasbord of definitions, two fundamental ideas prevail: community and the process." [43]: 3 this volume.

[9]Alexander [78] noted: "In theory, the term can be applied to any phenomenon that involves shocks to a system, whether it be physical or social, and whether the shock involve disasters or merely a hard knock in the literal or figurative sense" (p. 2713).

and practitioners from various disciplinary background. Since then, the concept has been largely adopted, "as much with scientists as with the administrators and international authorities in charge of preventing disasters" [73]. But, as pointed by Reghezza-Zitt, for social scientists working on risk and disasters, resilience is the latest iteration—and maybe not the least problematic—of a long list of overarching concepts.

4 Knowledge, Risk, and Building Resilience in the Bay Area

In the next section I will explore how experts, scientists and concerned residents of the Bay Area of San Francisco have collectively—and for decades—patiently built a awareness of risk that, they hope, will suffice to make the place they live and work in, resilient. Looking at the history of earthquake science, historian Deborah Coen [4] recalls that, in the nineteenth century, the "scientific description of an earthquake was built of stories—stories from as many people, in as many different situations, as possible" (p. 3). To define earthquakes, physicists and seismologists used their own perceptions of the event (e.g., "How did the earthquake feel?") and their senses of observation (e.g., "What did it produce?"). They also used as many indirect sources as they could locate, including accounts from magazines and newspapers, testimonies of other observers, and, when available, measurements from scientific equipment. Yet, in seismology, like in many others scientific domains, progress of research over the last century has favored the movement from subjective accounts to instrument-produced data, allowing for the development of predictive models and probabilistic conceptions of earthquake risk. In the process, earthquakes have become more abstract objects of science, defined mainly by complex mathematical operations and modeling. In such a context, it might be expected that scientists and experts working on earthquake risk grow more and more distant from residents' experiences [26, 79]. In a "world at risk," to borrow one of Ulrich Beck's book titles [80], earthquakes might finally have become like any other ungrounded, immaterial threat. Yet, for the Bay Area community of experts and scientists, dedicated to the understanding and mitigation of earthquake risk, knowledge about the latter remained hybrid and grounded in experience. In fact, this diversity of perspectives is precisely what makes this endeavor scientific.

The specialists of earthquakes who are also residents of the active seismic zone of the San Francisco Bay Area frame earthquake risk both as an object of science and an object of experience, pursuing a tradition of knowledge making now considered "hybrid." Just as it was for seismologists of the area throughout the past century, today seismologist, urban planers, disaster preparedness specialists consider earthquakes as phenomena that "cannot be comprehended exactly" [4] and need to be understood from different perspectives [81]. Therefore, for these experts, getting ready for "the Big One" is an active posture, encapsulating a large set of practices that articulate past, present, and future knowledge at different geographical scales: how memories of past disasters resonate with risks of future ones, how expert knowledge emerges from past and distant experiences, and, finally, how invisible threats transform space, subjectivity, knowledge, and politics in the SF Bay Area. This event horizon—the next Big One—provides an important perspective on a set of operations that happen each and every day in this area, and which, from residential choices to seismic-mapping activities, from memorialization to projection, define and help bring the potential consequences of a major earthquake into existence—into visibility.

This ethnography is built on my own research in the Bay Area of San Francisco, where I have identified the network of scientists and experts who connect science with their own experience of living in a seismic place and help making the Bay Area resilient [82]. As discussed above, when it come to evaluating a risk or thinking about a disaster, "knowing" is a complex operation. In the academic context, it is often taken for granted that scientific and expert knowledge surpasses lay knowledge, and that risk would be better prevented if only residents could think more as scientists do. Here, I want to show that, in the Bay Area, scientists and experts have also learned to think as residents do; re-defining their hierarchical categories, comparing and discussing everyday risks to re-situate the earthquake threat in a time frame and context of an individual's life span. As discussed by one respondent living in San Francisco, involved in risk prevention and the development of building codes,

> With earthquakes, they're so rare and extreme that to understand them, you have to think of them in the spectrum of everyday risks, monthly risks, and yearly risks. These all get compiled together and, most people, whether they articulate it or not, they're aware of that difference. You rarely find people that dumb that they don't understand risk in their daily life. [It] doesn't mean they always make the informed decisions, but they have an innate understanding of the rarity of things.[10]

Further, as a one-time non-expert resident, but now an expert in the field, recalled, the risk of an earthquake is not, and never was, "a given." Rather, it has been progressively instaured[11] by research work, specific practices, and attention to the matter. She stated, "I'm from Massachusetts and New York State. When I moved here, it was the 1970s; [The idea of a major earthquake] wasn't in anybody's

---

[10]Respondent D.21, personal communication, 2011.

[11]Toward a definition of instauration, Latour [83] writes: "Instauration and construction are clearly synonyms. But instauration has the distinct advantage of not dragging along all the metaphorical baggage of constructivism—which would in any case be an easy and almost automatic association given that an artwork is so obviously 'constructed' by the artist. To speak of 'instauration' is to prepare the mind to engage with the question of modality in quite the opposite way from constructivism. To say, for example, that a fact is 'constructed' is inevitably (and they paid me good money to know this) to designate the knowing subject as the origin of the vector, as in the image of God the potter. But the opposite move, of saying of a work of art that it results from an instauration, is to get oneself ready to see the potter as the one who welcomes, gathers, prepares, explores, and invents the form of the work, just as one discovers or 'invents' a treasure" (p. 10).

awareness."[12] This ongoing construction of earthquake risk was also not a one-way street; instead, it has been a slow elaboration of the capacities needed to understand both earth science and resident behavior, and many things in-between, such as the legacy of past and distant disasters, as well as cultural, philosophical, and metaphysical questions, often viewed in a reflective manner. In this context, experts and scientists have learned not only how to be rational subjects but also beneficiaries of their own expertise, specifically aware of the inherent tensions between forms of knowledge. In this process, scientists and experts have learned how to deal with, and even respect, residents' understandings and practices.

Moving away from the easily-taken-for-granted discourse regarding the lack of preparation and the irrationality of the residents[13] [84, 85], and also taking their distance with infructuous attempts to detach irrational thinking from idealistic, "pure" scientific knowledge, these experts have accepted that several "frames can be considered rational yet lead to radically different solutions" [86], p. 35). Taking this perspective spurs large possibilities, as this expert explained: "We need to define what is rational by what people do, rather than decide what's rational and say that they're not being rational. They *are* the definition of rational, and therefore we have to rethink what rational is" (emphasis added).[14] This new perspective also changes preconceived narratives about people's relations to risk,[15] and in a broader sense, experts and scientists understandings of individual and collective dynamics and social phenomena. As one respondent who has worked in the field of hazards mitigation for 30 years recalled, the process of defining priorities for earthquake preparedness is often full of surprises.[16] Taking the time to listen to the people they

---

[12]Respondent J.12, personal communication, 2011.

[13]As one, a structural engineer by training, summarized, "If it does not make sense for the people to retrofit their home, then it does not make sense by scaring them into doing it." This thought is echoed as a common Bay Area sentiment.

[14]Respondent D.21, personal communication, 2011.

[15]Here, I mean "risk" in the literal sense. See, for instance, Merchants of Doubt: How a Handful of Scientists Obscured the Truth on Issues from Tobacco Smoke to Global Warming [87].

[16]As one respondent stated: "We did a mail survey of people in the mid-1990s. It was intended to find out why people would choose do structural retrofit in their homes, and as part of that, we wanted to see the correlation with whether or not people have done the Red Cross kind of things, like food, water, and first aid. And it turned out—as a side-line, because we also asked their age and income—that the more educated you are, the less likely it was that you're going to have food, water, [and] first-aid training; and the less likely it was that you would have made the structural changes to retrofit your house, regardless of income! [Laughs] And we thought, 'Okay ... somehow, when people get a lot of education, they tend to have more blind faith that the utility companies are going to come through and they're going to have food and water. And [they think] they don't need to do this, because they know that their house is going to fall on the ground and therefore they're going to fix it. Whereas the other group, which was less well-educated, was convinced that it was going back to that basic survival training.' We were trying to hypothesize why this was going on, the basic survival training that they knew: that food and water were important on a day-to-day basis because they're having to deal with it weekly, as they did their budgeting. And therefore: 'I need to make sure that I have set aside a little extra so that I will have food and water in case of any emergency, not just a disaster'" (Respondent J.8, personal communication, 2011).

serve (the residents) and—more important maybe—accepting to be challenged, Bay Area experts have reframed their hypotheses and methods. Even more, building on their own experiences as residents, they redefines the meaning of "normal," which for those living outside of major earthquake areas is often seen as a world without hazards. Understanding that earthquake risk can overlap in situations that were previously thought to be unconnected—like science and personal experience—they framed the contours of a situated, acceptable, but still moving norm of what it means to "live with earthquakes," which never seems to reach a perfect and definitive conclusion.

Instauring the risk of an earthquake is a mental exercise (who can really say that he is prepared for a Tsunami or a large earthquake?) that allows experts to improve their knowledge about residential practices in a space of risk. It also allows for the residents' capacities to define—specific and personal—knowledge of the danger that they are dealing with. This never-ending work in progress is continually renegotiated between residents and experts, moving the cursor of risk acceptability. In this process, many things are taken into account and all of the micro as well as macro events re-defined the spectrum of the earthquake risk and the condition of resilience: when a disaster at the scale of the Tōhoku earthquake, tsunami, and nuclear disaster of 2011 hit, but also when a building construction is planned, a child is born, a new scientific discovery is unearthed, a new law is voted upon, or the time comes to choose a new house. Experts and scientists of the SF Bay Area maintain the memories of and knowledge about earthquakes, helping to ensure that the rest of us never forget this invisible presence.

The co-construction of the risk is not the implementation of zero safety risk—for really, how could that be?—but the renegotiation of what is an acceptable level of threat that people can afford and agree upon, knowing that what can be done might never be sufficient to cope with the extent of the damage and destruction. In many ways, the incapacity to think *of* the danger frames the limits of this tightrope-walk mental exercise. How, then, do experts and residents articulate the known and the unknown, and how does that articulation add another layer to the construction of earthquake risks? In this case, knowledge is supported by the capacity to imagine the unthinkable, and to expect and accept the consequences of a large-scale earthquake.

5 Conclusion

To make a place more resilient is to accept that the polysemy of the concept is both a strength and a weakness. A weakness because resilience cannot be declared, it has to be slowly co-constructed and instaured. A strength because it has the potential to allow each actor of a catastrophe to define the concept in their own words. To conclude I would argue that recognizing the hybridization of knowledge not only changes the definition of science and knowledge, but also allows for the subjectivity of experts, scientists, and residents and what should be taken into account, or listen to when trying to evaluate the condition under which a specific place could be considered resilient. And because they pose complex problems, knowledge about disaster have become more fully integrated.[17] As Gerson [88] states, this integration does require some adjustments in the ways scientists operate:

> One of the most important effects of integration processes is their encouragement of new specialties that separate from their parent specialties and become organizationally independent to some degree while continuing to increase their degree of epistemic integration [...]. This segmentation process [...] is one of the most important ways in which new specialties form (p. 5).

Yet, in the face of such statements, the experts' open secret is that many preventive actions cannot be accomplished preemptively. In such cases, scientists have to recognize that their scientific knowledge and their capacity for action to prevent, and respond to, major damage is limited, and that a potential future disaster can go way beyond—or simply be very different from—anything for which they had planned. In this process, thinking about resilience could be a way to think the unthinkable, to be attentive to the "other," and to provide a platform for discussion beyond traditional disciplinary boundaries as well as the long-held scientific/laity divides. and are the criteria that should facing and dealing with a disaster, redefining the figure of the concerned scientist into a knowing subject, grounded in her environment, circulating between the layers of knowledge that, prior, had been used to define the categories of the scientist, the expert, the lay person, or the amateur.

References

1. A. Oliver-Smith, S.M. Hoffman, The angry Earth, in Disaster in *Anthropological Perspective*, ed by A. Oliver-Smith, S.M. Hoffman (Routledge, 1999)
2. M. Lussault, *L'homme spatial, la construction sociale de l'espace humain* (Editions du Seuil, 2007)
3. F. Amador, The causes of 1755 Lisbon Earthquake on Kant, in *Actas VIII Congreso de La Sociedad Espanola de Historia de Las Ciencias Y de Los Tecnicos*, 485–495 (2004). http://dialnet.unirioja.es/servlet/dcfichero_articulo?codigo=1090088
4. D. Coen, *The Earthquake Observers—Disaster Science from Lisbon to Richter* (The University of Chicago Press, 2013)
5. R.R. Dynes, *The Lisbon Earthquake in 1755: Contested Meaning in the First Modern Disaster.* (No. #255) (1997)
6. R. Favier, A.-M. Granet-Abisset, in *Society and Natural Risk in France, 1500–2000: Changing Historical Perspectives*, eds. by C. Mauch, C. Pfister. Natural Disasters Cultural Response (Lexigton Book, 2009), pp. 103–136
7. J. Fressoz, *L'apocalypse joyeuse : Une histoire du risque technologique* (Seuil, 2012)

---

[17]Gerson [88] writes: "Integration thus means more than simple juxtaposition of efforts in the same location, and more than relationships that consist solely of market relations. Rather, it includes coordinated efforts to pose and solve new research problems that can redefine specialty boundaries. Because of its cooperative and coordinated nature, integration contrasts with the division of labor into independent lines of effort and the standardization of interfaces among line of work" (p. 516).

8. G. Quenet, *Les tremblements de terre aux XVIIe et XVIIIe siècles* (La naissance d'un risque, Seyssel, Champ Vallon, 2005)
9. F. Walter, *Catastrophes* (Une histoire culturelle XVI-XXIème siècle, Editions du Seuil, 2008)
10. G. Deleuze, Leibniz. Logique de l'évènement (1987). http://www.webdeleuze.com
11. J.-P. Dupuy, *Pour un catastrophisme éclairé. Quand l'impossible est certain* (Point Essa) (Seuil, 2002)
12. B. Latour, *An Inquiry Into Modes of Existence* (Harvard University Press, Cambridge, Mass., 2013)
13. F. Neyrat, *Bio-politique des catastrophes* (Dehors). Editions MF (2008)
14. M. Serres, *Temps des crises* (Manifestes). Le Pommier (2009)
15. I. Stengers, *Au temps des catastrophes ; résister a la barbarie qui vient* (les empêch). La Découverte (2009)
16. D. Lupton, *Risk* (Routledge, 1999)
17. D. Lupton, *Risk and Sociocultural Theory* (Cambridge University Press, New Directions and Perspectives, 1999)
18. J.B. Wiener, *The Real Pattern of Precaution* (2010)
19. J.B. Wiener, M.D. Rogers, Comparing precaution in the United States and Europe. J. Risk Res. **5**(4), 317–349 (2002). doi:10.1080/1366987021015368
20. M. Callon, P. Lascoumes, Y. Barthe, *Acting in an Uncertain World* (MIT Press, An Essay on Technological Democracy, 2009)
21. P. Haas, Epistemic communities and international policy coordination. Int. Org. **46**(1), 1–35 (1992)
22. B. Wynne, in *May the Sheep Safely Graze? A Reflexive View of the Expert–Lay Knowledge Divide*, eds. by S. Lash, B. Szersynsky, B. Wynne, *Risk Environment and Modernity. Toward a New Ecology* (Sage Publications, Published, 1996), pp. 27–44
23. F. Chateauraynaud, Les mobiles de l'expertise Entretien avec Francis Chateauraynaud. Experts **78**, 4 (2008)
24. H.M. Collins, R. Evans, The third wave of science studies: studies of expertise and experience. Soc. Stud. Sci. **32**(2), 235–296 (2002). doi:10.1177/0306312702032002003
25. S. Jasanoff, *The Political Science of Risk Perception, Reliability Engineering and System Safety* (1998)
26. T. Mitchell, *Rule of Experts: Egypt, Techno-Politics* (University of California Press, Modernity, 2002)
27. F. Fischer, *Citizens, Experts, and the Environment: The Politics of Local Knowledge* (Duke University Press Books, 2000)
28. R. Lidskog, Scientised citizens and democratised science. Re-assessing the expert-lay divide. J. Risk Res. **11**(1), 69–86 (2008)
29. S. Lane, N. Odoni, C. Landstrom, S. Whatmore, N. Ward, S. Bardley, Doing flood risk science differently: an experiment in radical scientific method. Trans. Inst. Br. Geogr. **36**, 15–36 (2010)
30. G. Atkinson, D. Wald, "Did You Feel It?" Intensity data: a suprisingly good measure of earthquake ground motion. Seismol. Res. Lett. **78**(3), 362–368 (2007). http://earthquake.usgs.gov/research/pager/prodandref/AtkinsonWaldDYFI.pdf
31. M. Bohy, Vers une sismologie citoyenne ? *Face Aux Risques* (2013, pp. 20–25)
32. D. Walde, V. Quitoriano, J. Dewey, USGS "Did you feel it?" community internet intensity maps: macroseismic data collection via the internet, in *First European Conference on Earthquake Engineering and Seismology* (2006). http://earthquake.usgs.gov/research/pager/prodandref/WaldEtAlECEESDYFI.pdf
33. E. Hackett, O. Amsterdamska, M. Lynch, J. Wajcman (eds.), *The Handbook of Science and Technology Studies* (MIT Press, 2008)
34. D. Henry, in *Anthropology and Disasters*, eds. by D. McEntire, W. Blanchard. Disciplines, Disasters and Emergency Management: the Convergence and Divergence of Concepts, Issues and Trends from the Research Literature (Federal Emergency Management Agency, 2005), p. 1. doi:10.2307/3032499

35. S.M. Hoffman, A. Oliver-Smith, in *Catastrophe and Culture. The Anthropology of Disasters*, eds. by S.M. Hoffman, A. Oliver-Smith (School of American Research Press, James Curry, 2002)

36. J. Langumier, *Survivre à l'innondation* (Pour une ethnologie de la catastrophe, ENS éditions, 2008)

37. V. November, Y. Leanza, *Risk, Disaster and Crisis Reduction* (Springer International Publishing, Cham, 2015). doi:10.1007/978-3-319-08542-5

38. S. Revet, *Anthropologie d'une catastrophe. Les coulées de boue de 1999 au Venezuela* (Presses Sorbonne Nouvelle, 2007)

39. J. Tulloch, D. Lupton, *Risk and Everyday Life* (Sage Publications, 2003)

40. J. Ahn, Exploring new paradigm of nuclear safety: from accident mitigation to resilient society facing extreme situations, in *International Workshop on Nuclear Safety: From Accident Mitigation to Resilient Society Facing Extreme Situations* (Berkeley, 2015)

41. K. Furuta, How the Fukushima Daiichi accident changed (or not) the nuclear safety fundamentals? in *International Workshop on Nuclear Safety: From Accident Mitigation to Resilient Society Facing Extreme Situations* (Berkeley, 2015)

42. H. Blazsin, F. Guarnieri, Practical safety, an ethical contribution to resilience To cite this version , in *The 6th REA Symposium : Managing Resilience, Learning to be Adaptable and Proactive in an Unpredictable World* (Lisbon, Portugal, 2015)

43. F. Guarnieri, S. Travadel, C. Martin, A. Portelli, A. Afrouss, *L'accident de Fukushima Daiichi, Le récit du directeur de la centrale*, vol. 1 (Presses des MINES, L'anéantissement, 2015)

44. U. Beck, in *Risk society, toward a new modernity*, ed. by M. Ritter. Nation, vol. 2 (Sage, 1992). doi:10.2307/2579937

45. S. Boudia, N. Jas, Risk and risk society in historical perspective. Hist. Technol. **4**, 317–331 (2007)

46. D. Bourg, P.-B. Joly, A. Kaufmann, in *Du risque à la Menace*, eds. by D. Bourg, P.-B. Joly, A. Kaufmann. (Ecologie e). (Presses Universitaires de France, 2013)

47. B. Latour, Whose cosmos? which cosmopolitics? a commentary on Ulrich Beck's peace proposal? Common Knowl. **10**(3), 450–462 (2004)

48. S. Lash, B. Szerszynski, B. Wynne (1996). Risk environment and modernity, in *Toward a New Ecology* (SAGE Publications Ltd, 1996)

49. I. Burton, R. Kates, G. Whites, *The Environment As Hazard, Second Edition* (The Guilford Press, 1993)

50. C. Fritz, The NORC studies of human behavior disasters. Journal of Social Issues 3(10), 26–41 (1954)

51. K. Hewitt, *Regions of Risk: A Geographical Introduction to Disasters* (Routledge, 1997)

52. M.K. Lindell, Disaster studies, 1–18 (2011). doi:10.1177/205684601111

53. G. White, *Human Adjustment to Floods—A geographical Approach to the Flood Problem in the United-States* (The University of Chicago, 1945). http://www.colorado.edu/hazards/gfw/images/Human_Adj_Floods.pdf

54. C. Cabasse, V. November, Y. Leanza, B. Barbey, K. De Conto, *Risk in Situ: la prévention de situations de risques et de crises: diffuser, mobiliser et saisir l'information sanitaire* (Lausanne, 2008)

55. S.J. Collier, A. Lakoff, Distributed preparedness: the spatial logic of domestic security in the United States. Environ. Plann. D: Soc. Space **26**(1), 7–28 (2008). doi:10.1068/d446t

56. D. Pécaud, Does the concept of loss orient risk prevention policy ? in *International Workshop on Nuclear Safety: From Accident Mitigation to Resilient Society Facing Extreme Situations* (2015)

57. K. Sato, Japan's nuclear imaginaries before and after Fukushima: visions of science, technology, and society, in *International Workshop on Nuclear Safety: From Accident Mitigation to Resilient Society Facing Extreme Situations* (Berkeley, 2015), pp 1–6. http://berkeleynuclearsafetyworkshop.weebly.com/uploads/2/4/7/9/24793500/sato_draft.pdf

58. P. Jobin, Dying for TEPCO? Fukushima's Nuclear Contract Workers 東京電力のため に死ぬ?福島の原発請負労働者. *The Asia-Pacific Journal : Japan Focus* **9**(18) (2011)
59. K. Juraku, "Made in Japan" Fukushima Nuclear Accident: A Critical Review for Accident Investigation Activities in Japan (2013), https://fukushimaforum.wordpress.com/workshops/sts-forum-on-the-2011-fukushima-east-japan-disaster/manuscripts/session-1/made-in-japan-fukushima-nuclear-accident-a-critical-review-for-accident-investigation-activities-in-japan/. Accessed 4 May 2015
60. N. Maki, How resilient is Japan? Response and recovery lessons from the 1995 Kobe and the 2011 Tohoku disasters (2013), https://fukushimaforum.wordpress.com/workshops/sts-forum-on-the-2011-fukushima-east-japan-disaster/manuscripts/session-1/how-resilient-is-japan-response-and-recovery-lessons-from-the-1995-kobe-and-the-2011-tohoku-disasters/. Accessed 5 Apr 2015
61. R. Shineha, Variety of Gaps: The Case of the 3.11 Japanese Triple Disasters (2013), https://fukushimaforum.wordpress.com/workshops/sts-forum-on-the-2011-fukushima-east-japan-disaster/manuscripts/. Accessed 5 Apr 2015
62. D. Slater, N. Keiko, L. Kindstrand, Social media, information and political activism in Japan's 3.11 crisis. Asia-Pac. J.: Jpn Focus **10**(24) (2012)
63. S. Traweek, Privileged science, citizen science, and radiation in Japan and the US: exposure, outsourcing, secrecy, and un/authorized knowing, in *An STS Forum on Fukushima* (2013)
64. C. Cabasse, *Le Mythe de Sophiatown et la Reconstruction Post-Apartheid de Johannesburg* (University de Reims Champagne-Ardenne, 2004)
65. C. Caruth, *Unclaimed Experience, Trauma, Narrative, and History* (The Johns Hopkins University Press, 1996)
66. G. Clavandier, *La mort collective* (CNRS Socio). CNRS Editions (2004)
67. V. Das, *Life and the Ordinary* (University of California Press, 2006)
68. S.L. Cutter, L. Barnes, M. Berry, C. Burton, E. Evans, E. Tate, J. Webb, A place-based model for understanding community resilience to natural disasters. Glob. Environ. Change **18**, 598–606 (2008). doi:10.1016/j.gloenvcha.2008.07.013
69. E. Hollnagel, J. Pariès, D. Woods, J. Wreathall, *Resilience Engineering in Practice—A Guidebook* (Ashgate St) (Ashgate, 2010)
70. M. Reghezza-Zitt, Utiliser la polysémie de la résilience pour comprendre les différentes approches du risque et leur possible articulation (2013)
71. L. Vale, T. Capagnella, *The Resilient City: How Modern Cities Recover from Disaster* (Oxford University Press, USA, 2005)
72. F. Miller, H. Osbahr, E. Boyd, F. Thomalla, S. Bharwani, G. Ziervogel, … D. Nelson, *Resilience andvulnerability: complementary or conflicting concepts?* (2010). Retrieved from http://www.ecologyandsociety.org/vol15/iss3/art11/
73. M. Reghezza-Zitt, S. Rufat, G. Djament-Tran, What Resilience Is Not: Uses and Abuses (2012). http://cybergeo.revues.org/25554
74. M. Pelling, Tracing the roots of urban risk and vulnerability, in *The Vulnerability of Cities: Natural Disasters and Social Resilience* (Routledge, p. 224)
75. L.K. Comfort, A. Boin, C.C. Demchak, *Designing Resilience: Preparing for Extreme Events* (University of Pittsburgh Press; 1 edition, 2010)
76. ISDR, *2009 UNISDR Terminology* (2009)
77. NSF. *Designing Infrastructure with Resilience from Disruptions and Disasters* (2014)
78. D.E. Alexander, Resilience and disaster risk reduction: an etymological journey. Nat. Hazards Earth Syst. Sci. Discuss. **13**, 2707–2716 (2013). doi:10.5194/nhess-13-2707-2013
79. C. Bessy, F. Chateauraynaud, *Experts et Faussaires*. (Métailié, 1995)
80. U. Beck, *World at risk. New York* (Vol. The Report). Polity Publisher (2008). http://scholar.google.com/scholar?hl=en&btnG=Search&q=intitle:World+at+risk:+The+report+of+the+Commission+on+the+Prevention+of+WMD+Proliferation+and+Terrorism#0
81. W. James, On Some Mental Effects of the Earthquake. Youth's Companion, Reprinted in James, H. Jr, 1911, Memories and Studies, ed. by Henry James, Jr., (Longmans, Green, & Co, 1906)

82. ABAG, Resilience initiative—building a disaster resilient bay area. http://quake.abag.ca.gov/projects/resilience_initiative/
83. B. Latour, Reflections on Etienne Souriau's Les différents modes d'existence, in *The Speculative Turn Continental Materialism and Realism*, eds. by G. Harman, L. Bryant, N. Srnicek, pp. 304–333 (re press, 2011). Retrieved from http://www.re-press.org/book-files/OA_Version_Speculative_Turn_9780980668346.pdf
84. C.-H. Geschwind, *California Earthquake, Science, Risk and the politic of Hazard Mitigation* (The Johns Hopkins University Press, 2001)
85. R.A. Stallings, *Promoting Risk: Constructing the Earthquake Threat (Social Problems and Social Issues)*. Aldine de Gruyter (1995)
86. D. Von Winterfieldt, N. Roselund, A. Kisuse, *Framing Earthquake retrofitting Decisions : The Case of Hillside Home in Los Angeles. Pacific Earthquake Engineering Research Center* (2000). http://peer.berkeley.edu/publications/peer_reports/reports_2000/0003.pdf
87. N. Oreskes, E. Conway, *Merchants of Doubt: How a Handful of Scientists Obscured the Truth on Issues from Tobacco Smoke to Global Warming* (Bloomsbury Press; Reprint edition, 2011)
88. E.M. Gerson, Integration of specialties: an institutional and organizational view. Stud. Hist. Philos. Sci. Part C: Stud. Hist. Philos. Biol. Biomed. Sci. **44**(4), 515–524 (2012). doi:10.1016/j.shpsc.2012.10.002
89. F. Guarnieri, S. Travadel, Engineering thinking in emergency situations: a new nuclear safety concept. Bull. At. Sci. **70**(6), 79–86 (2014)
90. V. November, Spatiality of risk. Environ. Plann. A **40**(7), 1523–1527 (2008). doi:10.1068/a4194
91. V. November, K. de Konto, Politiques des pandémies. De la détection des risques à l'action publique, in *Humains, non humains. Comment repeupler les sciences sociales* (2011)

Index

Printed in the United States
By Bookmasters